Studies in Systems, Decision and Control

Volume 340

The series "Studies in Systems, Decision and Control" (SSDC) covers both new developments and advances, as well as the state of the art, in the various areas of broadly perceived systems, decision making and control–quickly, up to date and with a high quality. The intent is to cover the theory, applications, and perspectives on the state of the art and future developments relevant to systems, decision making, control, complex processes and related areas, as embedded in the fields of engineering, computer science, physics, economics, social and life sciences, as well as the paradigms and methodologies behind them. The series contains monographs, textbooks, lecture notes and edited volumes in systems, decision making and control spanning the areas of Cyber-Physical Systems, Autonomous Systems, Sensor Networks, Control Systems, Energy Systems, Automotive Systems, Biological Systems, Vehicular Networking and Connected Vehicles, Aerospace Systems, Automation, Manufacturing, Smart Grids, Nonlinear Systems, Power Systems, Robotics, Social Systems, Economic Systems and other. Of particular value to both the contributors and the readership are the short publication timeframe and the world-wide distribution and exposure which enable both a wide and rapid dissemination of research output.

Indexed by SCOPUS, DBLP, WTI Frankfurt eG, zbMATH, SCImago.

All books published in the series are submitted for consideration in Web of Science.

More information about this series at http://www.springer.com/series/13304

Harendra Singh · Hemen Dutta ·
Marcelo M. Cavalcanti
Editors

Topics in Integral and Integro-Differential Equations

Theory and Applications

Springer

Editors
Harendra Singh
Department of Mathematics
Post Graduate College
Ghazipur, Uttar Pradesh, India

Hemen Dutta
Department of Mathematics
Gauhati University
Guwahati, Assam, India

Marcelo M. Cavalcanti
Departamento de Matemática
Universidade Estadual de Maringá
Maringá, Brazil

ISSN 2198-4182 ISSN 2198-4190 (electronic)
Studies in Systems, Decision and Control
ISBN 978-3-030-65508-2 ISBN 978-3-030-65509-9 (eBook)
https://doi.org/10.1007/978-3-030-65509-9

This Springer imprint is published by the registered company Springer Nature Switzerland AG
The registered company address is: Gewerbestrasse 11, 6330 Cham, Switzerland

Preface

The book is designed to include several topics in the areas of integral and integro-differential equations. Integral and Integro-differential equations are found to be useful in modeling many situations from science and engineering. The chapters include useful methods for solving various types of such equations as well as their significance and relevance in other scientific areas of study and research. There are nine chapters in this book and they are organized as follows.

Chapter "Wavelet-Galerkin Method for Second-Order Integro-differential Equations on Product Domains" deals with the wavelet-Galerkin method for numerical solution of second-order partial integro-differential equations on product domains with Dirichlet- or Neumann-type boundary conditions. It discussed the existence and uniqueness of the weak solution and constructed multi-dimensional wavelet bases to use in the Galerkin method for finding numerical solution of the integro-differential equations. Convergence of the method has been established as well as estimation of error. It also discussed the advantages of the method and performed numerical experiments to support the theoretical results.

Chapter "Analysis and Spectral Element Solution of Nonlinear Integral Equations of Hammerstein Type" aims to employ the spectral element method with Gauss-Lobatto-Legendre collocation points to approximate Hammerstein-type nonlinear integral equations. It used the Banach fixed point theorem to establish the existence and uniqueness of solutions in the L^2 norm as well as shown the convergence of the proposed method. It also performed numerical experiments involving one- and two-dimensional nonlinear equations to illustrate the effectiveness of the approach adopted.

Chapter "Approximate Methods for Solving Hypersingular Integral Equations" presents a review of analytical and numerical methods for solving linear hypersingular integral equations of the first and second kind on closed and open integration intervals with a particular attention on equations with second-order singularities. The proofs of the convergence of approximate methods are based on the general theory of approximate methods of analysis and on the continuous method for solving operator equations. It also discussed sufficient conditions for the solvability of computational schemes and estimations of the accuracy of the proposed methods.

Chapter "Solutions of Integral Equations by Reproducing Kernel Hilbert Space Method" aims to apply the reproducing kernel Hilbert space method to the integral equations and obtained the solutions in the form of a series in the reproducing kernel Hilbert space. It also demonstrated some numerical examples to ascertain the accuracy of the method.

Chapter "Restricted Global Convergence Domains for Integral Equations of the Fredholm-Hammerstein Type" focuses on obtaining restricted global convergence domains for Newton's method that usually are balls, when it is applied to solve Fredholm-Hammerstein-type nonlinear integral equations. It aims also to address the solution of nonlinear integral equations of the Fredholm-Hammerstein type having several solutions and obtain results concerning the location and separation of different solutions. It further included numerical examples to support the results obtained.

Chapter "Boundary Integral Equation Formulation for Fractional Order Theory of Thermo-Viscoelasticity" presents a formulation of the boundary integral equation method in connection with fractional order theory of thermo-viscoelasticity. It investigated for fundamental solutions of the corresponding differential equations in the Laplace transform domain and established also a reciprocity theorem. The implementation of the boundary element method has been discussed further for the solution of the equations and also included an example illustrating the formulation of boundary integral equation.

Chapter "Spectral Methods for Solving Integro-differential Equations and Bibiliometric Analysis" aims first to present bibliometric study of the integro-differential equations. It also discussed the application of such equations with arbitrary order in science. Then it presented certain numerical methods for finding solution of different kinds of integro-differential equations like the fractional integro-differential equations and the variable-order fractional partial integro-differential equations with estimation of errors for each of the method discussed. It further included numerical examples to demonstrate the efficiency of the techniques discussed.

Chapter "An Efficient Numerical Algorithm to Solve Volterra Integral Equation of Second Kind" deals with construction of exponential fitting of Volterra Runge-Kutta method for finding numerical solution of Volterra integral equation of the second kind. The work is based on the assumption that the linear operators corresponding to the internal and external stages annihilate the set of basis exponential functions with unknown frequencies where the optimum values of frequencies are computed by minimizing the local truncation error.

Chapter "An Integral Equation Method for Wave Scattering by a Pair of Horizontal Porous Plates" aims to develop a mathematical method comprising of solutions to a pair of coupled hypersingular integral equations for studying the interaction of water waves with a pair of horizontal porous plates submerged at different depths in case of finite as well as infinite depth water domain in the context of linear water wave theory. It examined the effect of a pair of horizontal permeable plates of different orientations situated at different depths on the propagation of surface waves. It studied the hydrodynamic performances of a dual porous plate system for the three cases, viz., when the plates are kept in a step-like manner, when there is a partial overlap between the plates and when the plates are fully overlapped. The coupled integral equations have been solved numerically and with the help of these solutions, the numerical estimates for the reflection coefficient, the transmission coefficient, vertical wave forces acting on the plates, and the energy-loss coefficient have also been computed. The correctness of the numerical results has been examined through an energy identity relation for porous plates and also by comparing the results with available results in the literature. It also discussed new results related to different orientations of the plates, different submergence depths, varying lengths, and different values of permeability of the plates.

The editors sincerely acknowledge the cooperation of contributors while dealing with their works for possible publication in this book. Reviewers deserve deep gratitude for offering their kind help in selecting and finalizing the contents of the book. The editors also thankfully acknowledge the support of editorial staff at Springer. The editors gratefully acknowledge the encouragement received from many colleagues and friends for this book project.

Ghazipur, India — Harendra Singh
Guwahati, India — Hemen Dutta
Maringá, Brazil — Marcelo M. Cavalcanti
October 2020

Contents

Wavelet-Galerkin Method for Second-Order Integro-differential Equations on Product Domains

Dana Černá and Václav Finěk

Abstract This chapter is concerned with the study of the wavelet-Galerkin method for the numerical solution of the second-order partial integro-differential equations on the product domains. Prescribed boundary conditions are of Dirichlet or Neumann type on each facet of the domain. The variational formulation is derived, and the existence and uniqueness of the weak solution are discussed. Multi-dimensional wavelet bases satisfying boundary conditions are constructed by the tensor product of wavelet bases on the interval using isotropic and anisotropic approaches. The constructed wavelet bases are used in the Galerkin method to find the numerical solution of the integro-differential equations. The convergence of the method is proven, and error estimates are derived. The advantage of the method consists in the uniform boundedness of the condition numbers of discretization matrices and in the fact that these matrices exhibit an exponential decay of their elements away from the main diagonal. Based on the decay estimates, we propose a compression strategy for an approximation of the discretization matrices by sparse or quasi-sparse matrices. Numerical experiments are presented to confirm the theoretical results and illustrate the efficiency and applicability of the method.

Keywords Wavelet · Spline · Riesz basis · Boundary conditions · Condition number · Integro-differential equation · Error estimate

Mathematics Subject Classification (2010) 47G20 · 65N12 · 65T60

D. Černá (✉) · V. Finěk
Department of Mathematics and Didactics of Mathematics, Technical University of Liberec, Studentská 2, 461 17 Liberec, Czech Republic
e-mail: dana.cerna@tul.cz

V. Finěk
e-mail: vaclav.finek@tul.cz

H. Singh et al. (eds.), *Topics in Integral and Integro-Differential Equations*,
Studies in Systems, Decision and Control 340,
https://doi.org/10.1007/978-3-030-65509-9_1

1 Introduction

This chapter is concerned with the wavelet-Galerkin method for the numerical solution of partial integro-differential equations. The differential operator is the second-order elliptic operator, and the integral operator is the linear operator of the Fredholm type.

We consider the product domain

$$\Omega = (a_1, b_1) \times (a_2, b_2) \times \cdots \times (a_d, b_d), \tag{1}$$

where $a_i, b_i \in \mathbb{R}$, $a_i < b_i$, $i = 1, \ldots d$, $d \in \mathbb{N}$. The aim is to find a solution u of the equation

$$-\varepsilon \Delta u + p(t) \cdot \nabla u + q(t) u + \mathcal{K} u = f(t) \quad \text{on } \Omega, \tag{2}$$

where $\varepsilon > 0$ is a constant, $p = (p_1, \ldots, p_d)$, ∇u denotes the gradient of u, Δ is the Laplace operator, and $\mathcal{K} : L^2(\Omega) \to L^2(\Omega)$ is an integral operator given by

$$(\mathcal{K} u)(t) = \int_\Omega K(t, x)\, u(x)\, dx, \tag{3}$$

with the kernel $K \in L^2(\Omega \times \Omega)$. The boundary of Ω is divided into two disjoint pieces $\Gamma^D \neq \emptyset$ and Γ^N such that each facet of Ω belongs either to Γ^D or Γ^N. Equation (2) is equipped with the boundary conditions

$$u = 0 \text{ on } \Gamma^D \quad \text{and} \quad \frac{\partial u}{\partial \mathbf{n}} = 0 \text{ on } \Gamma^N, \tag{4}$$

where $\mathbf{n}$ stands for the outward pointing unit normal vector. The efficient numerical solution of equations of this type is an important task because they represent many problems from physics, engineering, and financial mathematics, and they can also be obtained when using time scheme for semidiscretization of parabolic problems, see Sect. 5, or using linearization or quasilinearization of certain nonlinear problems [5, 52].

Since many physical phenomena, as well as problems from financial mathematics, are modeled by integro-differential equations, many authors have studied their numerical solutions, and nowadays, many methods are available for a wide range of integro-differential equations. Besides well-established classical methods such as the finite difference method, the finite element method, the Galerkin method with B-splines, and the colocation method, some advanced modern methods have also been developed [3, 13, 29, 32, 36, 37, 44, 47, 48, 50, 53]. However, the drawback of the mentioned classical methods for the numerical solution of the Eq. (2) is that the matrices arising from discretization are full due to the presence of the integral operator, which is nonlocal. Therefore, it can be difficult to solve the discrete problem efficiently or even to store the matrices. In contrast, the wavelet-Galerkin method leads to matrices that have many very small entries. Therefore, these matrices can be efficiently approximated by quasi-sparse or sparse matrices. Here, we call the matrices sparse if the number of nonzero entries in each row is bounded by a constant C that is independent of N and significantly smaller than N, where $N \times N$ is the size of the matrix. We call the matrices quasi-sparse, if the number of its nonzero entries is bounded by a constant $CN \log N$.

The fact that integral operators have sparse representations in wavelet bases was first studied in [2, 6], where it was shown that discretization of singular integral operators using orthogonal wavelets leads to sparse matrices. In [14, 42], the Galerkin method with orthogonal wavelets was applied on systems of integral equations of the second kind, and a compression strategy for the discretization matrix was proposed. The discretization of integral operators using biorthogonal wavelets was studied, e.g., in [22]. Fredholm integral equations of the second kind were also solved using spline wavelets, e.g., in [13, 40]. In the case of integral equations, the wavelet-Galerkin method is typically used with orthogonal or semiorthogonal wavelet bases, because these bases lead to diagonal or banded matrices.

Application of wavelets to the numerical solution of integro-differential equations seems to be much less explored than wavelet methods for various types of integral equations. Wavelet-based methods for PIDEs were used in [3, 13, 44], and for solving equations arising in quantitative finance in [32, 53]. In the case of integro-differential equations, the situation is somewhat different than in the case of integral equations. Here, the L^2-orthogonality or semiorthogonality of the bases is no longer a significant advantage because, due to the differential term, the resulting matrices are not banded for these bases. Moreover, to be able to apply a differential operator, the smoother basis functions are required. The quantitative properties of the wavelet-Galerkin method crucially depend on the wavelet basis used, namely on its polynomial exactness, the smoothness of basis functions, the number of vanishing wavelet moments, its condition number and the length of the support of wavelets. Therefore, the choice of an appropriate wavelet basis is a very important issue.

For the numerical solution of the integro-differential equation (2), several wavelet bases are available. The spline wavelet bases seem to be the most appropriate choice because they can be expressed in the explicit form, and since they are formed by piecewise polynomial basis functions, one is able to compute basis function values, derivatives, and integrals efficiently. The very well-known construction of biorthogonal wavelets on the real line from [17] was first adapted to the bounded interval in [21]. In this case, both the primal and dual wavelets are local, i.e., their support length is exponentially decreasing with the level of the wavelet. The condition numbers of these wavelet bases are relatively large, especially for the bases formed by splines of higher orders, and therefore several modifications of this construction were proposed, e.g., in [7, 8, 43]. Since the dual functions are not directly used in the computation, it can be beneficial to give up their locality and to deal with wavelet bases, where only the primal basis functions are local while the support of duals is the whole domain. Such wavelet bases were constructed on the real line in [31] and they were also constructed on the bounded interval and adapted to various types of boundary conditions, see [9–13, 15, 20, 30, 33–35, 45, 46]. The main advantage of these types of bases in comparison to bases with local duals is typically the shorter support of primal wavelets, the lower condition numbers of the bases and of the corresponding discretization matrices as well as the simplicity of the construction.

Starting from wavelets on the unit interval, appropriate wavelet bases on the product domain Ω can be constructed using the tensor product. There exist several approaches, an isotropic approach [35, 51], an anisotropic approach [27], and a sparse tensor product [26]. Constructions of wavelet bases on irregular grids and on more general domains, manifolds, and other objects are not so straightforward, especially if the Riesz basis property and vanishing wavelet moments are required, however, several methods have been proposed, e.g., [23–25, 39, 49].

This chapter is organized as follows. In Sect. 2 variational formulation and a weak solution are introduced, assumptions are made, and existence, uniqueness, and stability of the weak solution are studied. In Sect. 3, the concept of a wavelet basis is presented, and two approaches for the construction of multidimensional wavelet bases on a product domain based on the tensor product are reviewed. Furthermore, two concrete examples of spline wavelet bases are given. In Sect. 4, the wavelet-Galerkin method is introduced, the convergence rate of the method is discussed, and the condition number and decay estimates for entries of the discretization matrices are studied. The proposed method, combined with the Crank-Nicolson scheme, is used for option pricing under the Merton jump-diffusion model in Sect. 5. To illustrate the reliability and applicability of the method, we provide several numerical experiments in Sect. 6.

2 Existence and Uniqueness of the Weak Solution

First, we introduce notations used in this chapter. Let the domain $\Omega \in \mathbb{R}^d$ be the product domain defined by (1). The set of all m-times continuously differentiable functions on $\overline{\Omega}$ is denoted as $C^m\left(\overline{\Omega}\right)$.

We denote by $L^2(\Omega)$ the space of real-valued square-integrable functions on Ω equipped with the inner product and the norm

$$\langle f, g\rangle = \int_\Omega f(x)\, g(x)\, dx, \quad \|f\| = \sqrt{\langle f, f\rangle}, \tag{5}$$

respectively.

Let $H^1(\Omega)$ denote the Sobolev space, i.e., the space of all functions from $L^2(\Omega)$ for which their first-order weak derivatives also belong to $L^2(\Omega)$. This space is equipped with the inner product and the norm

$$\langle f, g\rangle_{H^1} = \sum_{i=1}^{d} \left\langle \frac{\partial f}{\partial x_i}, \frac{\partial g}{\partial x_i}\right\rangle + \langle f, g\rangle, \quad \|f\|_{H^1} = \sqrt{\langle f, f\rangle_{H^1}}, \tag{6}$$

respectively. We also use the seminorm in this space defined as

$$|f|_{H^1} = \sqrt{\sum_{i=1}^{d} \left\| \frac{\partial f}{\partial x_i} \right\|^2}. \quad (7)$$

Let H_0^1 be the closure in $H^1(\Omega)$ of the set of all functions f from $H^1(\Omega)$ such that supp $f \subset \Omega$. We also consider the space H of functions adapted to boundary conditions (4) defined as

$$H = \left\{ f \in H^1(\Omega),\ f = 0 \text{ on } \Gamma^D \right\} \quad (8)$$

and its dual H'.

Furthermore, the symbol $L^2(0, T; B)$ stands for the Bochner space of functions f such that $f(t) = f(\cdot, t) \in B$ for $t \in (0, T)$ and

$$\|f\|_{L^2(0,T;B)} = \left(\int_0^T \|f(t)\|_B^2 \, dt \right)^{1/2} < \infty \quad (9)$$

with $\|\cdot\|_B$ being the norm in Banach space B.

In the following, we formulate assumptions, introduce variational formulation, and study existence, uniqueness, and stability of the weak solution.

Let a bilinear form $a : H \times H \to \mathbb{R}$ be defined as

$$a(u, v) = \varepsilon \sum_{i=1}^{d} \left\langle \frac{\partial u}{\partial t_i}, \frac{\partial v}{\partial t_i} \right\rangle + \sum_{i=1}^{d} \left\langle p_i \frac{\partial u}{\partial t_i}, v \right\rangle + \langle qu, v \rangle + \langle \mathcal{K}u, v \rangle \quad (10)$$

for all $u, v \in H$. Then, the variational formulation of Eq. (2) reads as: Find $u \in H$ such that

$$a(u, v) = \langle f, v \rangle \quad \text{for all} \quad v \in H. \quad (11)$$

The solution u of (11) is called a *weak solution* of Eq. (2). We make the following assumptions:

(A1) The functions $p_i, i = 1, \ldots, d$, and q satisfy $p_i, q \in C\left(\overline{\Omega}\right)$.
(A2) The kernel K is smooth enough, i.e., $K \in C^m\left(\overline{\Omega \times \Omega}\right)$ for some $m \in \mathbb{N}$.
(A3) The function f belongs to the space $L^2(\Omega)$.
(A4) The bilinear form a is coercive, which means that there exists a constant $\alpha > 0$ such that

$$a(u, u) \geq \alpha \|u\|_{H^1}^2 \quad \text{for all} \quad u \in H. \quad (12)$$

The property (12) is also called H-ellipticity. The smoothness of the kernel K is important for the proof of decay estimates of the entries of discretization matrices, which enables approximation of the discretization matrices by quasi-sparse or sparse matrices. The coercivity is a crucial property for the uniform boundedness of the condition numbers of the discretization matrices as well as for the proof of existence, uniqueness, and stability of the weak solution.

Theorem 1 *If the assumptions (A1)–(A2) are satisfied, then the bilinear form a defined by (10) is continuous, i.e., there exists a constant β such that*

$$|a(u,v)| \leq \beta \|u\|_H \|v\|_H \quad \forall u, v \in H. \tag{13}$$

Proof Due to (A1) and (A2) there exist constants P, Q, and $\tilde{K}$ such that

$$P = \max_{i=1,\ldots,d} \max_{x \in \overline{\Omega}} |p_i(x)|, \quad Q = \max_{x \in \overline{\Omega}} |q(x)|, \quad \tilde{K} = \max_{x,t \in \overline{\Omega}} |K(x,t)|. \tag{14}$$

The Cauchy-Schwartz inequality implies that

$$\int_\Omega |u(x)|\, dx \leq \sqrt{|\Omega|}\, \|u\|, \quad |\Omega| = \int_\Omega 1 dx. \tag{15}$$

Furthermore, we have

$$\left| \sum_{i=1}^{d} \left\langle p_i \frac{\partial u}{\partial t_i}, v \right\rangle \right| \leq P \sum_{i=1}^{d} \left\langle \left| \frac{\partial u}{\partial t_i} \right|, |v| \right\rangle \leq P \sum_{i=1}^{d} \left\| \frac{\partial u}{\partial t_i} \right\| \|v\| \leq \sqrt{2} P \|v\| \, |u|_{H^1}, \tag{16}$$

where in the last step we used the inequality

$$\left(\sum_{i=1}^{d} a_i \right)^2 \leq 2 \sum_{i=1}^{d} a_i^2. \tag{17}$$

Combining (10), (15), and (16), we obtain

$$\begin{aligned} |a(u,v)| &\leq \varepsilon\, |u|_{H^1} |v|_{H^1} + \sqrt{2} P\, |u|_{H^1} \|v\| + Q \|u\| \|v\| + \tilde{K} |\Omega| \|u\| \|v\| \\ &\leq \beta \|u\|_{H^1} \|v\|_{H^1}, \end{aligned} \tag{18}$$

with $\beta = \varepsilon + \sqrt{2} P + Q + \tilde{K} |\Omega|$. □

Under the assumption that the bilinear form a is continuous and coercive, the existence, uniqueness, and stability of the solution of (11) follow from the well-known Lax-Milgram theorem, for the proof see [16, 38].

Theorem 2 (Lax–Milgram) *Let H be a Hilbert space, let the bilinear form $a : H \times H \to \mathbb{R}$ be continuous and coercive with the coercivity constant α as in (12), and let $f \in H'$. Then, the solution u of the equation*

$$a(u,v) = \langle f, v \rangle \quad \forall v \in H$$

exists and is unique, and the stability estimate

$$\|u\|_H \leq \frac{1}{\alpha} \|f\|_{H'}$$

holds.

Theorem 3 *If the assumptions (A1)–(A4) are satisfied, then there exists a unique solution $u \in H$ of Eq. (11).*

Proof The assertion of this theorem is a direct consequence of the assumption (A4) and Theorems 1 and 2. □

First, we study coercivity for the part of the bilinear form a that corresponds to the differential term.

Theorem 4 *If the bilinear form $a_D : H \times H \to \mathbb{R}$ defined by*

$$a_D(u, v) = \varepsilon \sum_{i=1}^{d} \left\langle \frac{\partial u}{\partial t_i}, \frac{\partial v}{\partial t_i} \right\rangle + \sum_{i=1}^{d} \left\langle p_i \frac{\partial u}{\partial t_i}, v \right\rangle + \langle qu, v \rangle \quad \forall u, v \in H \tag{19}$$

satisfies

$$q - \frac{1}{2} \sum_{i=1}^{d} \frac{\partial p_i}{\partial t_i} \geq \omega > 0, \quad p \cdot \mathbf{n} \geq 0 \text{ on } \Gamma^N, \tag{20}$$

then a_D is coercive.

Proof By Gauss theorem, for $u \in H$ it holds that

$$\left\langle \sum_{i=1}^{d} p_i \frac{\partial u}{\partial t_i}, u \right\rangle = \int_{\Gamma^N} \frac{u^2}{2} p \cdot \mathbf{n} \, dS - \left\langle \sum_{i=1}^{d} \frac{1}{2} \frac{\partial p_i}{\partial t_i}, u^2 \right\rangle. \tag{21}$$

Therefore, the term $a_D(u, u)$, $u \in H$, can be expressed as

$$a_D(u, u) = \varepsilon \sum_{i=1}^{d} \left\langle \frac{\partial u}{\partial t_i}, \frac{\partial u}{\partial t_i} \right\rangle + \int_{\Gamma^N} \frac{u^2}{2} p \cdot \mathbf{n} \, dS + \left\langle q - \sum_{i=1}^{d} \frac{1}{2} \frac{\partial p_i}{\partial t_i}, u^2 \right\rangle. \tag{22}$$

Using (20) and (22), the coercivity of a_D is obtained since

$$a_D(u, u) \geq \varepsilon |u|_{H^1}^2 + \omega \|u\|^2 \geq \min(\varepsilon, \omega) \|u\|_{H^1}^2. \tag{23}$$

□

The following lemma gives a sufficient condition on the kernel K under which the bilinear form a corresponding to the integro-differential operator is coercive, under the assumption that the bilinear form a_D corresponding to the differential operator is already coercive.

Lemma 1 *Let us assume that the bilinear form* $a_D : H \times H \to \mathbb{R}$ *defined by (19) is coercive with the coercivity constant* α_D*, i.e.,*

$$a_D(u, u) \geq \alpha_D \|u\|_{H^1}^2 \quad \text{for all} \quad u \in H. \tag{24}$$

Furthermore, let $\tilde{K}$ *be given by (14) and let* $\alpha_D - \tilde{K}|\Omega| > 0$*. Then, the bilinear form a is also coercive.*

Proof Using (15) we obtain for $u \in H$

$$\begin{aligned}\langle \mathcal{K}u, u\rangle &= \int_\Omega \int_\Omega K(x,t)\, u(x)\, u(t)\, dx dt \geq -\tilde{K} \int_\Omega \int_\Omega |u(x)\, u(t)|\, dx dt \\ &\geq -\tilde{K} \left(\int_\Omega |u(x)|\, dx\right)^2 \geq -\tilde{K}\, |\Omega|\, \|u\|^2 .\end{aligned} \tag{25}$$

Hence,

$$\begin{aligned} a(u,u) &\geq \alpha_D \|u\|_{H^1}^2 - \tilde{K}\,|\Omega|\, \|u\|^2 \geq \alpha_D \|u\|_{H^1}^2 - \tilde{K}\,|\Omega|\, \|u\|_{H^1}^2 \\ &\geq \left(\alpha_D - \tilde{K}\,|\Omega|\right) \|u\|_{H^1}^2 . \end{aligned} \tag{26}$$

□

3 Wavelet Bases on Product Domains

This section introduces the concept of a wavelet basis on the product domain Ω and constructions of spline wavelet bases. For more details about wavelet bases refer to [14, 32, 51].

We use the following notations. Consider the space H given by (8). Let $\mathcal{J}$ be at most countable index set such that each index $\lambda \in \mathcal{J}$ takes the form $\lambda = (j, k)$, where $|\lambda| = j \in \mathbb{Z}$ denotes a level. We define

$$\|\mathbf{v}\| = \sqrt{\sum_{\lambda \in \mathcal{J}} v_\lambda^2}, \quad \text{for} \quad \mathbf{v} = \{v_\lambda\}_{\lambda \in \mathcal{J}},\ v_\lambda \in \mathbb{R}, \tag{27}$$

and

$$l^2(\mathcal{J}) = \{\mathbf{v} : \|\mathbf{v}\| < \infty\}. \tag{28}$$

3.1 Concept of a Wavelet Basis

We formulate the conditions on the family of functions $\Psi = \{\psi_\lambda, \lambda \in \mathcal{J}\}$ to be a wavelet basis. However, the concept of a wavelet basis is not unified in the mathematical literature, and some papers present the conditions in a different form.

First, a wavelet basis is a Riesz basis for the space H.

Definition 1 A family $\Psi = \{\psi_\lambda, \lambda \in \mathcal{J}\}$ is called a *Riesz basis* of H if the span of Ψ is dense in H, and there exist constants $c, C \in (0, \infty)$ such that

$$c\,\|\mathbf{b}\| \le \left\| \sum_{\lambda \in \mathcal{J}} b_\lambda \psi_\lambda \right\|_H \le C\,\|\mathbf{b}\| \quad \forall\, \mathbf{b} = \{b_\lambda\}_{\lambda \in \mathcal{J}} \in l^2(\mathcal{J}). \tag{29}$$

We refer to the constants

$$c_\Psi = \sup\{c : c \text{ satisfies (29)}\} \quad \text{and} \quad C_\Psi = \inf\{C : C \text{ satisfies (29)}\} \tag{30}$$

as lower and upper Riesz bounds (with respect to the H-norm), respectively, and to the number cond $\Psi = C_\Psi / c_\Psi$ as the *condition number* of Ψ. In some papers, the squares of the norms are used in (29) and the Riesz bounds are defined as c_Ψ^2 and C_Ψ^2. Riesz basis property (29) plays an important role in the wavelet-Galerkin method, because it is a crucial assumption for a uniform boundedness of condition numbers of the discretization matrices, for more details see Sect. 4.

Furthermore, the functions ψ_λ have to be *local*. This means that wavelet systems on the interval and isotropic wavelet systems, see Sect. 3.2, satisfy

$$\text{diam}\, \mathit{supp}\, \psi_\lambda \le C 2^{-|\lambda|}, \quad \lambda \in \mathcal{J}, \tag{31}$$

where the constant C does not depend on λ, and at a given level j the supports of only finitely many wavelets overlap at any point $x \in \Omega$. In some applications, it is beneficial to deal with anisotropic wavelet systems, for their description see Sect. 3.3. In this case, a basis function is constructed as a tensor product of functions $\psi_{\lambda_1}, \ldots, \psi_{\lambda_d}$ from one-dimensional wavelet systems, and locality is understood in the sense that

$$\text{diam supp}\, \psi_\lambda \le C 2^{-[\lambda]}, \quad [\lambda] = \min_{i=1,\ldots,d} |\lambda_i|. \tag{32}$$

This means that the parameter $[\lambda]$ corresponding to the basis function

$$\psi_\lambda = \psi_{\lambda_1} \otimes \ldots \otimes \psi_{\lambda_d} \tag{33}$$

is the minimum over all the levels $|\lambda_i|$ of one-variable functions ψ_{λ_i}.

A wavelet basis Ψ has typically the hierarchical structure, i.e., Ψ is of the form

$$\Psi = \Phi_{j_0} \cup \bigcup_{j=j_0}^{\infty} \Psi_j, \tag{34}$$

j_0 being the coarsest level. The functions from the set Φ_{j_0} are called *scaling functions* and the functions from the set Ψ_j, $j \geq j_0$, are called *wavelets* on the level j. Wavelets in the inner part of the interval called *inner wavelets* are typically translations and dilations of one function ψ or several functions $\psi^1, \ldots, \psi^p$ also called *wavelets* (or *mother wavelets, wavelet generators*), i.e.,

$$\psi_{j,k}(x) = 2^{j/2} \psi^l \left(2^j x - m\right), \tag{35}$$

for some $l \in \{1, \ldots, p\}$ and some $k, m \in \mathbb{Z}$, k dependent on m and l. Similarly, the wavelets near the boundary are derived from functions called *boundary wavelets.*

Furthermore, we require a *polynomial exactness* of multiresolution spaces. More precisely, the subset Ψ^k of Ψ defined as

$$\Psi^k = \Phi_{j_0} \cup \bigcup_{j=j_0}^{k} \Psi_j, \; j_0 \leq k, \tag{36}$$

is called a *multiscale basis*. We say that the spaces $X_k = \text{span}\, \Psi^k$ have polynomial exactness of order $M \geq 1$, if span $\Psi^k \subset H \cap \Pi_M$, where Π_M is the set of all polynomials of degree at most $M - 1$ defined on Ω. Polynomial exactness determines the convergence rate of the wavelet-Galerkin method.

Finally, we assume that wavelets have $L \geq 1$ *vanishing moments*. This means that all wavelets $\psi_\lambda \in \Psi_j$, $j_0 \leq j$, satisfy

$$\int_\Omega p(x)\, \psi_\lambda(x)\, dx = 0, \tag{37}$$

for any polynomial p of degree less than L. Vanishing wavelet moments are essential for decay estimates of the entries of discretization matrices corresponding to the integral term, i.e., for sparse representations of integro-differential operators.

To construct a wavelet basis on the product domain Ω, one usually starts with one-dimensional wavelet bases on bounded intervals and then applies the tensor product. There are several approaches based on the tensor product, for example an isotropic approach [35, 51], an anisotropic approach [27] or a sparse tensor product [26]. In the following subsections, we introduce an isotropic and an anisotropic approach. Both constructions preserve important properties of one-dimensional wavelets such as the Riesz basis property, the polynomial exactness, the hierarchical structure, and vanishing wavelet moments.

3.2 Isotropic Wavelet Bases

We consider a product domain Ω defined by (1) and denote its facets by the following way

$$\Gamma_{a_i} = \left\{x = (x_1, \ldots, x_d) \in \overline{\Omega} : x_i = a_i\right\},$$
$$\Gamma_{b_i} = \left\{x = (x_1, \ldots, x_d) \in \overline{\Omega} : x_i = b_i\right\}, \tag{38}$$

where $i = 1, \ldots, d$. Let us define the one-dimensional spaces incorporating boundary conditions as

$$H_{m,n}(a_i, b_i) = \left\{f \in H^1(a_i, b_i) : f(a_i) = 0 \text{ if } m = 1,\ f(b_i) = 0 \text{ if } n = 1\right\}, \tag{39}$$

where $m, n \in \{0, 1\}$.

The construction starts with a wavelet basis on the interval $(0, 1)$. Let $\Psi^{0,0}$ denotes a wavelet basis on this interval, where no boundary conditions are prescribed. Similarly, let $\Psi^{1,0}$ denotes a wavelet basis adapted to homogeneous boundary condition in the point 0, $\Psi^{0,1}$ denotes a wavelet basis adapted to homogeneous boundary condition in the point 1, and $\Psi^{1,1}$ denotes a wavelet basis adapted to homogeneous boundary condition in both the points 0 and 1. Let these bases have the structure

$$\Psi^{m,n} = \left\{\phi^{m,n}_{j_0,k},\ k \in \mathcal{I}^{m,n}_{j_0}\right\} \cup \left\{\psi^{m,n}_{j,k},\ k \in \mathcal{J}_j,\ j \geq j_0\right\}, \tag{40}$$

where $m, n \in \{0, 1\}$. Let us assume that the set $\Psi^{m,n}$ is a Riesz basis in $L^2(0, 1)$ and when normalized in the H^1-norm then it is a Riesz basis of the space $H_{m,n}(0, 1)$.

First, we use a simple linear transformation to obtain a wavelet basis for the space $H_{m,n}(a_i, b_i)$. Let us define

$$\phi^i_{j,k}(x) = \phi^{m,n}_{j,k}\left(\frac{x - a_i}{b_i - a_i}\right), \quad \psi^i_{j,k}(x) = \psi^{m,n}_{j,k}\left(\frac{x - a_i}{b_i - a_i}\right), \quad x \in (a_i, b_i), \tag{41}$$

and $\mathcal{I}^i_{j_0} = \mathcal{I}^{m,n}_{j_0}$, where

$$[m, n] = \begin{cases} [0, 0] & \text{if } \Gamma_{a_i}, \Gamma_{b_i} \subset \Gamma^N, \\ [0, 1] & \text{if } \Gamma_{a_i}, \subset \Gamma^N,\ \Gamma_{b_i} \subset \Gamma^D, \\ [1, 0] & \text{if } \Gamma_{a_i} \subset \Gamma^D,\ \Gamma_{b_i} \subset \Gamma^N, \\ [1, 1] & \text{if } \Gamma_{a_i}, \Gamma_{b_i} \subset \Gamma^D. \end{cases} \tag{42}$$

Then,

$$\Psi^{(i)} = \left\{\phi^i_{j_0,k}, k \in \mathcal{I}_{j_0}\right\} \cup \left\{\psi^i_{j,k}, k \in \mathcal{J}_j, j \geq j_0\right\} \tag{43}$$

forms a Riesz basis in $H_{m,n}(a_i, b_i)$.

We define multivariate scaling functions by

$$\phi_{j,k}(x) = \prod_{l=1}^{d} \phi^{l}_{j,k_l}(x_l), \quad x = (x_1, \ldots, x_d) \in \Omega, \tag{44}$$

with $k = (k_1, \ldots, k_d)$ now being a multi-index, $k \in \mathcal{I}_j^{\Omega} = \mathcal{I}_j^1 \times \cdots \times \mathcal{I}_j^d$. We introduce the abbreviation

$$\mathcal{J}_{j,e} = \begin{cases} \mathcal{J}_j, & e = 1, \\ \mathcal{I}_j, & e = 0, \end{cases} \tag{45}$$

i.e., the parameter e allows distinguishing between scaling functions and wavelets. Furthermore, we denote

$$E = \{e = (e_1, \ldots, e_d), e_i \in \{0, 1\}, e \neq (0, 0, \ldots, 0)\}, \tag{46}$$

and

$$\mathcal{J}_{j,e}^{\Omega} = \mathcal{J}_{j,e_1} \times \cdots \times \mathcal{J}_{j,e_d}, \quad \mathcal{J}_j^{\Omega} = \bigcup_{e \in E} \mathcal{J}_{j,e}. \tag{47}$$

For any $e = (e_1, \ldots, e_d) \in E$, $j \geq j_0$, and $k = (k_1, \ldots, k_d) \in \mathcal{J}_j^{\Omega}$, we define the multivariate wavelet

$$\psi_{\lambda}(x) = \prod_{l=1}^{d} \psi^{l}_{j,e_l,k_l}(x_l), \quad x = (x_1, \ldots, x_d) \in \Omega, \quad \lambda = (j, e, k), \tag{48}$$

where

$$\psi^{l}_{j,e_l,k_l} = \begin{cases} \phi^{l}_{j,k_l}, & e_l = 1, \\ \psi^{l}_{j,k_l}, & e_l = 0. \end{cases} \tag{49}$$

The wavelet basis on the hyperrectangle Ω is then given by

$$\Psi = \left\{\psi_{j,e,k}, \; e \in E, \; k \in \mathcal{J}_j^{\Omega}, j \geq j_0\right\} \cup \left\{\phi_{j_0,k}, \; k \in \mathcal{I}_j^{\Omega}\right\}. \tag{50}$$

We denote the multiscale basis containing wavelets up to level J as

$$\Psi^J = \left\{\psi_{j,e,k}, \; e \in E, \; k \in \mathcal{J}_j^{\Omega}, j_0 \leq j \leq J\right\} \cup \left\{\phi_{j_0,k}, \; k \in \mathcal{I}_j^{\Omega}\right\}. \tag{51}$$

If we start with the Riesz bases $\Psi^{m,n}$, $m, n \in \{0, 1\}$, in the space $L^2(0, 1)$ such that $\Psi^{m,n}$ when normalized with respect to the H^1-norm is the Riesz basis of $H^{m,n}(0, 1)$, then the resulting basis Ψ is a Riesz basis in the space $L^2(\Omega)$. Moreover, Ψ when normalized with respect to the H^1-norm is the Riesz basis of H. This approach preserves the regularity of basis functions, the full degree of polynomial exactness, vanishing wavelet moments, as well as locality of bases functions. For more details see e.g., [35, 51].

3.3 Anisotropic Wavelet Basis

Let $\Psi^{(i)}$ be a wavelet basis on the interval (a_i, b_i) defined by (41) and (43). For notational simplicity, we denote $\mathcal{J}_{j_0-1} = \mathcal{I}_{j_0}$ and

$$\psi^i_{j_0-1,k} = \phi^i_{j_0,k}, \quad k \in \mathcal{J}_{j_0-1}, \quad \mathcal{J} = \left\{(j,k), \ j \geq j_0 - 1, \ k \in \mathcal{J}_j\right\}. \tag{52}$$

Then $\Psi^{(i)}$ can also be expressed as

$$\Psi^{(i)} = \left\{\psi^i_{j,k}, \ j \geq j_0 - 1, \ k \in \mathcal{J}_j\right\} = \left\{\psi^i_\lambda, \ \lambda \in \mathcal{J}\right\}. \tag{53}$$

Recall that for the index $\lambda = (j,k)$ we denote $|\lambda| = j$. We use $u \otimes v$ to denote the tensor product of functions u and v, i.e. $(u \otimes v)(x_1, x_2) = u(x_1) v(x_2)$. For $d \geq 1$ we generalize the definition of the index set $\mathcal{J}$:

$$\mathcal{J} = \left\{\lambda = (\lambda_1, \ldots, \lambda_d) : \lambda_i = (j_i, k_i), \ j_i \geq j_0 - 1, k_i \in \mathcal{J}_{j_i}\right\}. \tag{54}$$

We define multivariate basis functions as

$$\psi_\lambda = \bigotimes_{i=1}^{d} \psi^i_{\lambda_i}, \quad \lambda = (\lambda_1, \ldots, \lambda_d) \in \mathcal{J}. \tag{55}$$

Then the parameter

$$|\lambda| = \max_{i=1,\ldots,d} |\lambda_i| \tag{56}$$

represents a level of ψ_λ. We also define the parameter

$$[\lambda] = \min_{i=1,\ldots,d} |\lambda_i|, \tag{57}$$

which characterizes the length of the support of ψ_λ in the following sense. Due to locality of the one-dimensional basis functions, i.e.,

$$\text{diam supp}\, \psi^i_{\lambda_i} \leq C_i 2^{-|\lambda_i|}, \tag{58}$$

we have

$$\text{diam supp}\, \psi_\lambda \leq \sqrt{\sum_{i=1}^{d} C_i^2 2^{-2|\lambda_i|}} \leq C\sqrt{d} 2^{-[\lambda]}, \quad C = \max_{i=1,\ldots,d} C_i. \tag{59}$$

In this case, basis functions are not local in the sense that $\text{diam supp}\, \psi_\lambda \leq C 2^{-|\lambda|}$, but only in the sense that (59) holds. We define the set $\Psi = \{\psi_\lambda, \lambda \in \mathcal{J}\}$, and the set

$$\Psi^k = \{\psi_\lambda : \lambda = (\lambda_1, \ldots, \lambda_d),\ |\lambda_i| \leq k\}. \tag{60}$$

If we start with a univariate Riesz basis in the space $L^2(0, 1)$, then the set Ψ is a Riesz basis of the space $L^2(\Omega)$, see [28]. This approach also preserves the properties of the univariate basis, such as polynomial exactness, smoothness of basis functions, and vanishing moments, but as already mentioned the resulting functions are local only in the sense of (59). For more details see [27, 28].

3.4 Construction of Spline Wavelet Bases

In the following, two examples of wavelet bases are presented. The first wavelet basis is a quadratic spline wavelet basis designed in [13]. The wavelets have three vanishing moments and the shortest possible support among all quadratic spline wavelets with the same number of vanishing wavelet moments generated from quadratic B-spline scaling functions. The second wavelet basis is a cubic spline wavelet basis from [11]. The wavelets have four vanishing moments and also the shortest possible support among all wavelets of the same type. Both wavelet bases are adapted to homogeneous Dirichlet boundary conditions.

Example 1 Quadratic spline wavelets with short support
The construction starts with the scaling basis, which is formed by quadratic B-splines as in [7, 15, 43]. The generator of inner scaling functions is a quadratic B-spline defined on knots [0, 1, 2, 3], which can be written explicitly as

$$\phi(x) = \begin{cases} \frac{x^2}{2}, & x \in [0, 1], \\ -x^2 + 3x - \frac{3}{2}, & x \in [1, 2], \\ \frac{x^2}{2} - 3x + \frac{9}{2}, & x \in [2, 3], \\ 0, & \text{otherwise}. \end{cases} \tag{61}$$

The generator ϕ_b of boundary scaling functions is a quadratic B-spline defined on knots [0, 0, 1, 2], i.e.,

$$\phi_b(x) = \begin{cases} -\frac{3x^2}{2} + 2x, & x \in [0, 1], \\ \frac{x^2}{2} - 2x + 2, & x \in [1, 2], \\ 0, & \text{otherwise}. \end{cases} \tag{62}$$

Scaling functions on the level $j \geq 2$ are then constructed using translations and dilations of the generators ϕ and ϕ_b by the following way. For $x \in [0, 1]$, we set

$$\begin{aligned} &\phi_{j,k}(x) = 2^{j/2}\phi(2^j x - k + 2),\ k = 2, \ldots, 2^j - 1, \\ &\phi_{j,1}(x) = 2^{j/2}\phi_b(2^j x), \quad \phi_{j,2^j}(x) = 2^{j/2}\phi_b(2^j(1 - x)). \end{aligned} \tag{63}$$

A wavelet generator ψ and a boundary wavelet generator ψ_b are defined as linear combinations of scaling functions as

$$\begin{aligned}\psi(x) &= -\frac{1}{4}\phi(2x) + \frac{3}{4}\phi(2x-1) - \frac{3}{4}\phi(2x-2) + \frac{1}{4}\phi(2x-3), \\ \psi_b(x) &= -\frac{\phi_b(2x)}{4} + \frac{47\phi(2x)}{120} - \frac{13\phi(2x-1)}{40} + \frac{\phi(2x-2)}{10}. \end{aligned} \tag{64}$$

The graphs of the scaling and wavelet generators are displayed in Fig. 1.

These wavelets have three vanishing moments, $\operatorname{supp}\psi = [0, 3]$ and $\operatorname{supp}\psi_b = [0, 2.5]$. It was proved in [13] that the support of ψ is the shortest possible in the sense that the quadratic spline wavelets which are defined as linear combinations of the functions $\phi(2x - k)$, $k \in \mathbb{Z}$, and have three vanishing moments cannot have shorter supports. Similarly, ψ_b has also the shortest possible support among all boundary wavelets satisfying homogeneous Dirichlet boundary conditions and having three vanishing moments.

For $j \geq 2$ and $x \in [0, 1]$, wavelets are defined by

$$\begin{aligned}&\psi_{j,k}(x) = 2^{j/2}\psi(2^j x - k + 2),\ k = 2, \ldots, 2^j - 1, \\ &\psi_{j,1}(x) = 2^{j/2}\psi_b(2^j x), \quad \psi_{j,2^j}(x) = 2^{j/2}\psi_b(2^j(1-x)). \end{aligned} \tag{65}$$

Then, the scaling and wavelet bases on the level $j \geq 2$ are the sets

$$\Phi_j = \left\{\phi_{j,k}, k \in \mathcal{I}_j\right\}, \quad \Psi_j = \left\{\psi_{j,k}, k \in \mathcal{I}_j\right\}, \tag{66}$$

where the index set $\mathcal{I}_j$ is given by

$$\mathcal{I}_j = \left\{k \in \mathbb{Z} : 1 \leq k \leq 2^j\right\}. \tag{67}$$

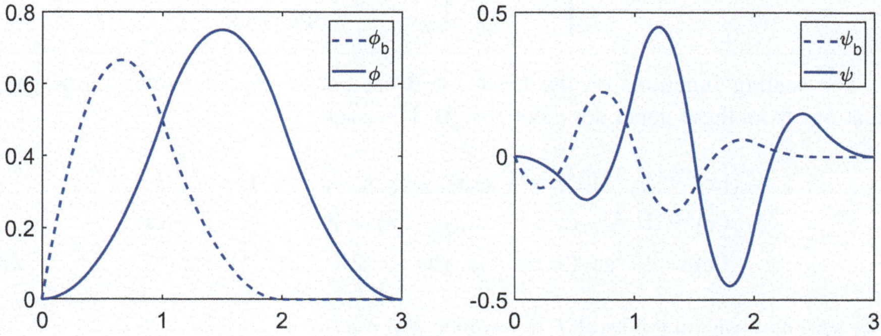

Fig. 1 The scaling functions ϕ_b and ϕ (left) and the wavelets ψ_b and ψ (right)

It was proved in [13] that the set

$$\Psi = \Phi_2 \cup \bigcup_{j=2}^{\infty} \Psi_j \tag{68}$$

is a wavelet basis of the space $L^2(0,1)$ and Ψ normalized with respect to the H^1 norm is a wavelet basis of the space $H_0^1(0,1)$.

Example 2 Cubic spline wavelets with short support

In this example, the scaling basis consists of cubic B-splines similarly as in [7, 15, 43]. The generator of inner scaling functions is a function ϕ that is a cubic B-spline on knots [0, 1, 2, 3, 4]. This B-spline has an explicit expression

$$\phi(x) = \begin{cases} \frac{x^3}{6}, & x \in [0,1], \\ -\frac{x^3}{2} + 2x^2 - 2x + \frac{2}{3}, & x \in [1,2], \\ \frac{x^3}{2} - 4x^2 + 10x - \frac{22}{3}, & x \in [2,3], \\ \frac{(4-x)^3}{6}, & x \in [3,4], \\ 0, & \text{otherwise}. \end{cases} \tag{69}$$

There are two boundary generators; ϕ_{b1} is a cubic B-spline defined on knots [0, 0, 0, 1, 2] and ϕ_{b2} is a cubic B-spline defined on knots [0, 0, 1, 2, 3]. Their explicit forms are

$$\phi_{b1}(x) = \begin{cases} \frac{7x^3}{4} - \frac{9x^2}{2} + 3x, & x \in [0,1], \\ \frac{(2-x)^3}{4}, & x \in [1,2], \\ 0, & \text{otherwise}, \end{cases} \tag{70}$$

and

$$\phi_{b2}(x) = \begin{cases} -\frac{11x^3}{12} + \frac{3x^2}{2}, & x \in [0,1], \\ \frac{7x^3}{12} - 3x^2 + \frac{9x}{2} - \frac{3}{2}, & x \in [1,2], \\ \frac{(3-x)^3}{6}, & x \in [2,3], \\ 0, & \text{otherwise}. \end{cases} \tag{71}$$

The scaling functions on the level $j \geq 3$ are constructed via translations and dilations from these generators. For $x \in [0,1]$ we set

$$\begin{aligned} \phi_{j,k}(x) &= 2^{j/2}\phi(2^j x - k + 3), \; k = 3, ..., 2^j - 1, \\ \phi_{j,1}(x) &= 2^{j/2}\phi_{b1}(2^j x), \quad \phi_{j,2^j+1}(x) = 2^{j/2}\phi_{b1}(2^j(1-x)), \\ \phi_{j,2}(x) &= 2^{j/2}\phi_{b2}(2^j x), \quad \phi_{j,2^j}(x) = 2^{j/2}\phi_{b2}(2^j(1-x)). \end{aligned} \tag{72}$$

The scaling basis on the level $j \geq 3$ is then given by

$$\Phi_j = \left\{\phi_{j,k}, k = 1, \ldots, 2^j + 1\right\}. \tag{73}$$

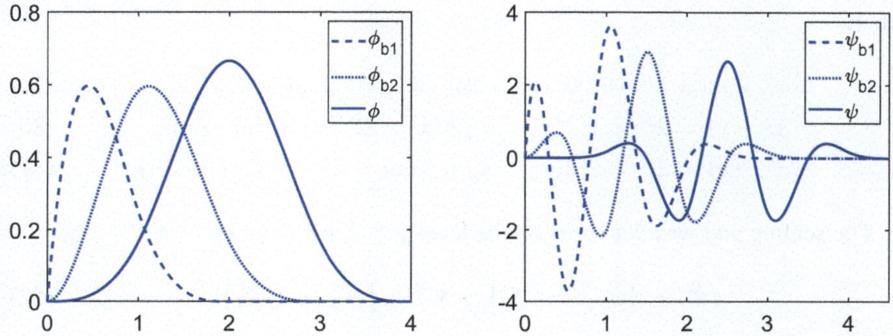

Fig. 2 The scaling functions ϕ_{b1}, ϕ_{b2}, and ϕ (left) and the wavelets ψ_{b1}, ψ_{b2}, and ψ (right)

In this construction, there is one inner wavelet generator ψ and two boundary generator ψ_{b1} and ψ_{b2}, which are defined as linear combinations of scaling functions on the first level. The wavelet ψ is defined as

$$\psi(x) = \phi(2x-1) - 4\phi(2x-2) + 6\phi(2x-3) - 4\phi(2x-4) + \phi(2x-5). \tag{74}$$

The wavelet ψ has four vanishing moments and its support is the interval [0.5, 4.5]. It was proved in [11] that the wavelet ψ has the shortest possible support in the sense that any function g with four vanishing moments defined as a linear combination of cubic B-splines, i.e.,

$$g \in \overline{\text{span}}\,\{\phi\,(2\cdot -k)\,,\; k \in \mathbb{Z}\}\,, \tag{75}$$

has the length of the support at least four.

The boundary wavelets ψ_{b1} and ψ_{b2} are defined by

$$\psi_{b1}(x) = 6\phi_{b1}(2x) - \frac{57}{5}\phi_{b2}(2x) + \frac{919}{100}\phi(2x) - \frac{116}{25}\phi(2x-1) + \phi(2x-2) \tag{76}$$

and

$$\psi_{b2}(x) = \frac{7}{3}\phi_{b2}(2x) - \frac{319}{60}\phi(2x) + \frac{101}{15}\phi(2x-1) - \frac{25}{6}\phi(2x-2) + \phi(2x-3). \tag{77}$$

The graphs of scaling and wavelet generators are displayed in Fig. 2.

Then, both boundary wavelets have four vanishing moments, the support of ψ_{b1} is the interval [0, 3], and the support of ψ_{b2} is the interval [0, 3.5]. The boundary wavelets are adapted to homogeneous Dirichlet boundary condition at the point 0 and they also have the shortest possible support.

The wavelet basis on the level $j \geq 3$ is defined as

$$\Psi_j = \left\{\psi_{j,k}, k = 1, \ldots, 2^j\right\}, \tag{78}$$

where

$$\psi_{j,k}(x) = 2^{j/2}\psi(2^j x - k + 2), \quad k = 3, ..., 2^j - 2, \tag{79}$$

$$\psi_{j,1}(x) = 2^{j/2}\psi_{b1}(2^j x), \quad \psi_{j,2^j}(x) = 2^{j/2}\psi_{b1}(2^j(1-x)), \tag{80}$$

$$\psi_{j,2}(x) = 2^{j/2}\psi_{b2}(2^j x), \quad \psi_{j,2^j-1}(x) = 2^{j/2}\psi_{b2}(2^j(1-x)). \tag{81}$$

The scaling and wavelet bases on the level $j \geq 3$ are given by

$$\Phi_j = \left\{\phi_{j,k}, k \in \mathcal{I}_j\right\}, \quad \Psi_j = \left\{\psi_{j,k}, k \in \mathcal{J}_j\right\}, \tag{82}$$

where the index sets are given by

$$\mathcal{I}_j = \left\{k \in \mathbb{Z} : 1 \leq k \leq 2^j + 1\right\}, \quad \mathcal{J}_j = \left\{k \in \mathbb{Z} : 1 \leq k \leq 2^j\right\}. \tag{83}$$

It was proved in [11] that the family

$$\Psi = \Phi_3 \cup \bigcup_{j=3}^{\infty} \Psi_j \tag{84}$$

is a wavelet basis of the space $L^2(0,1)$ and Ψ normalized with respect to the H^1-norm is a wavelet basis of the space $H_0^1(0,1)$.

4 Wavelet-Galerkin Method

We now introduce the wavelet-Galerkin method for solving Eq. (11). Let Ψ be a wavelet basis of the space $L^2(\Omega)$ such that Ψ when normalized with respect to the H^1-norm is a wavelet basis in the space H. Let $\Psi^k \subset \Psi$ be a multiscale basis that contains the scaling functions on the coarsest level and the wavelets up to the level k. Then

$$X_k = \text{span}\,\Psi^k \tag{85}$$

are the finite-dimensional subspaces of H that are nested, i.e., $X_k \subset X_{k+1}$, and

$$H = \overline{\bigcup_{k=j_0}^{\infty} X_k}. \tag{86}$$

The Galerkin formulation of (11) reads: Find $u_k \in X_k$ such that

$$a(u_k, v) = \langle f, v \rangle \quad \text{for all} \quad v \in X_k. \tag{87}$$

Theorem 5 (Existence, uniqueness, and stability of the numerical solution)
If the assumptions (A1)–(A4) are satisfied and α *is the coercivity constant from (12), then there exists a unique solution* u_k *of Eq. (87) and*

$$\|u_k\|_{H^1} \le \frac{1}{\alpha} \|f\|_{H'} . \tag{88}$$

Proof The existence and uniqueness of the solution u_k and the stability estimate (88) are consequences of the continuity and coercivity of the bilinear form a and Theorem 2.

Now, we focus on the study of the convergence rate of the wavelet-Galerkin method, which is based on the well-known Céa's theorem.

Theorem 6 (Céa's) *Let the spaces* X_k *be given by (85),* u *be the solution of (11), and* u_k *be the solution of (87). If the bilinear form* $a : H \times H \to \mathbb{R}$ *is continuous and coercive with constants* α *and* β *as in (12) and (13), then*

$$\|u - u_k\|_H \le \frac{\beta}{\alpha} \inf_{v \in X_k} \|u - v\|_H .$$

Hence, Céa's theorem shows that the convergence rate of the Galerkin method depends on the approximation power of the spaces X_k. The term

$$E_k(u) = \inf_{v \in X_k} \|u - v\|_H \tag{89}$$

is known as the *error of the best approximation* in X_k. The study of this error is a subject of approximation theory. Nowadays, approximation order is known for several kinds of spaces X_k. As already mentioned in the introduction, it is advantageous to use wavelet bases formed by splines. Therefore, we present here the approximation properties of spline spaces.

Theorem 7 (Approximation properties of spline spaces)
Let $\Pi_r(a, b)$ *be the space of all polynomials on* (a, b) *of degree less than* r*. Let* V_k *be the spline space for the unit interval given by*

$$V_k = \left\{ v \in C^m(0, 1) : v|_{\left(\frac{l}{2^k}, \frac{l+1}{2^k}\right)} \in \Pi_r\left(\frac{l}{2^k}, \frac{l+1}{2^k}\right), l = 0, \ldots, 2^k - 1 \right\}, \tag{90}$$

where $0 < m < r$*, and* X_k *be the spline space for the product domain* $(0, 1)^d$ *given by* $X_k = V_k \otimes \cdots \otimes V_k$*. Then,*

$$\inf_{v \in X_k} \|u - v\|_{H^s} \le C 2^{-(r-s)k} |u|_{H^r} ,$$

for any $u \in H^r(\Omega)$ *provided that* $0 \le s < r$ *and* X_k *is contained in* $H^s(\Omega)$*. Here, we view* $H^s(\Omega)$ *for* $s = 0$ *as the space* $L^2(\Omega)$*.*

Similar results hold for spaces of piecewise polynomial functions incorporating boundary conditions [18]. Hence, the convergence rate for the wavelet-Galerkin method is $r = 3$ if quadratic spline wavelets are used and $r = 4$ if cubic spline wavelets are used.

Note that due to Theorems 2 and 6, the convergence rate depends on the chosen discretization spaces X_k and not directly on the chosen bases of these spaces. Since spline wavelet bases typically generate the same spaces as the corresponding B-splines, it can be expected that the error will be similar for the wavelet-Galerkin method and for the standard Galerkin method with B-splines. However, the wavelet-Galerkin method has several advantages, and it seems to be superior to classical methods when the integral term is nonzero because the discretization matrices can be approximated by sparse matrices, which is not the case for most other methods, see [6, 11–14]. The second advantage is that a simple diagonal preconditioner is optimal in the sense that diagonally normalized discretization matrices have uniformly bounded condition numbers. For some iterative methods such as the conjugate gradient method, it was proved that this guarantees that the number of iterations needed to resolve the discrete problem with the desired accuracy is also uniformly bounded.

Now, we transform the Eq. (87) into a system of linear algebraic equations using a wavelet basis Ψ^k. We write the function u_k as

$$u_k = \sum_{\psi_\lambda \in \Psi^k} c_\lambda^k \psi_\lambda. \tag{91}$$

Let $\mathbf{G}^k$ and $\mathbf{K}^k$ be matrices with the entries

$$\mathbf{G}_{\mu,\lambda}^k = \varepsilon \langle \nabla\psi_\lambda, \nabla\psi_\mu \rangle + \langle p \cdot \nabla\psi_\lambda, \psi_\mu \rangle + \langle q\psi_\lambda, \psi_\mu \rangle, \quad \mathbf{K}_{\mu,\lambda}^k = \langle \mathcal{K}\psi_\lambda, \psi_\mu \rangle, \tag{92}$$

for $\psi_\lambda, \psi_\mu \in \Psi^k$. Let $\mathbf{f}^k$ be a vector with entries

$$f_\mu^k = \langle f, \psi_\mu \rangle, \quad \psi_\mu \in \Psi^k, \tag{93}$$

and $\mathbf{c}^k$ be the column vector of coefficients c_λ^k. We obtain the system

$$\mathbf{A}^k \mathbf{c}^k = \mathbf{f}^k, \quad \text{where} \quad \mathbf{A}^k = \mathbf{G}^k + \mathbf{K}^k. \tag{94}$$

We apply the standard Jacobi diagonal preconditioning on the system (94). Let $\mathbf{D}^k$ be a diagonal matrix with diagonal elements

$$\mathbf{D}_{\lambda,\lambda}^k = \sqrt{\mathbf{A}_{\lambda,\lambda}^k} = \sqrt{a(\psi_\lambda, \psi_\lambda)}. \tag{95}$$

Then, we obtain the preconditioned system

$$\tilde{\mathbf{A}}^k \tilde{\mathbf{c}}^k = \tilde{\mathbf{f}}^k \tag{96}$$

with

$$\tilde{\mathbf{A}}^k = \left(\mathbf{D}^k\right)^{-1} \mathbf{A}^k \left(\mathbf{D}^k\right)^{-1}, \quad \tilde{\mathbf{f}}^k = \left(\mathbf{D}^k\right)^{-1} \mathbf{f}^k, \quad \tilde{\mathbf{c}}^k = \mathbf{D}^k \mathbf{c}^k. \tag{97}$$

We solve the system (96) by some iterative method, e.g., by the method of generalized residuals (GMRES) or, in the case the system matrix is symmetric and positive definite, by the conjugate gradient method (CG).

It was proved in [19] that diagonally preconditioned matrices arising from discretization of elliptic PDEs have uniformly bounded condition numbers. Due to coercivity of the bilinear form a and the Riesz basis property of the basis, a similar result also holds for the matrices $\tilde{\mathbf{A}}^k$ corresponding to integro-differential equation (2).

Theorem 8 *Let the assumptions (A1)–(A4) be satisfied. Then there exist constants C_1, C_2, and C_3 independent of k such that for all $k \geq j_0$ it holds that*

$$\left\|\tilde{\mathbf{A}}^k\right\| \leq C_1, \quad \left\|\left(\tilde{\mathbf{A}}^k\right)^{-1}\right\| \leq C_2, \quad \operatorname{cond} \tilde{\mathbf{A}}^k \leq C_3, \tag{98}$$

where $\|\cdot\|$ denotes the spectral norm.

Proof The proof follows the lines of the proof of Theorem 14 from [13].

Now we study the structure of the matrices $\mathbf{G}^k$ and $\mathbf{K}^k$. Let the size of these matrices be $N_k \times N_k$. Due to the locality property of wavelet bases, the matrix $\mathbf{G}^k$ has $\mathcal{O}(N_k \ln N_k)$ nonzero entries. For wavelet bases on the interval and anisotropic wavelet bases, it has a so-called finger-band pattern. For $\varepsilon = 1$, $p = 0$, $q = 1$, and wavelet bases from [12, 13], the patterns are displayed in Fig. 3. In [12], wavelets have been constructed such that they are L^2 and H^1 orthogonal if their levels differ by more than two. Due to this property, the resulting matrix is sparse and not only quasi-sparse, and it has a simplified structure, as can be seen in Fig. 3.

For most standard methods, the matrix $\mathbf{K}^k$ is full. The advantage of wavelet methods is that for some classes of operators and some types of wavelet bases, many entries of the matrix $\mathbf{K}^k$ are small and can be thresholded and the matrix $\mathbf{K}^k$

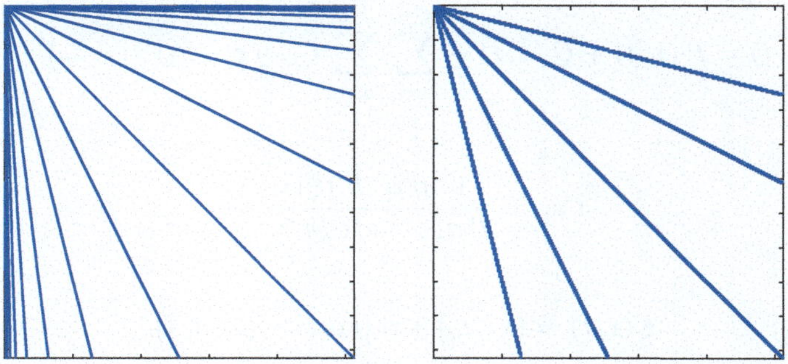

Fig. 3 The sparsity pattern of the matrix $\mathbf{G}^k$ for wavelet bases from [13] (left) and [12] (right)

can be approximated with a matrix that is quasi-sparse or sparse, see e.g., [6, 13, 14, 22, 42]. The decay estimates of the entries of the matrix $\mathbf{K}^k$ are both for the isotropic wavelet systems and anisotropic wavelet systems and for a smooth kernel K presented in the following theorem.

Theorem 9 *Let Ψ be a wavelet basis of the space $L^2(\Omega)$, $\psi_\lambda, \psi_\mu \in \Psi$ be wavelets with L vanishing moments and let $K \in C^{2L}\left(\overline{\Omega} \times \overline{\Omega}\right)$. Let us denote the elements of the matrix $\mathbf{K}^k$ as*

$$K_{\mu,\lambda} = \int_\Omega \int_\Omega K(x,t)\, \psi_\lambda(x)\, \psi_\mu(t)\, dx\, dt. \tag{99}$$

(a) If Ψ is an isotropic wavelet system constructed by the method from Sect. 3.2, then

$$\left|K_{\mu,\lambda}\right| \le C 2^{-(|\lambda|+|\mu|)(L+d/2)} \tag{100}$$

for some constant C is independent of λ and μ.

(b) If Ψ is an anisotropic wavelet system constructed by the method from Sect. 3.3, then

$$\left|K_{\mu,\lambda}\right| \le C 2^{-([\lambda]+[\mu])(L+d/2)} \tag{101}$$

for some constant C is independent of λ and μ.

Proof Let x_λ and t_λ be the centres of the supports of ψ_λ and ψ_μ, respectively. For

$$l = (l_1, \ldots, l_d) \in \mathbb{N}_0^d, \quad \mathbb{N}_0 = \mathbb{N} \cup \{0\}, \quad x = (x_1, \ldots, x_d) \in \mathbb{R}^d, \tag{102}$$

we denote

$$|l| = l_1 + \ldots + l_d, \quad l! = l_1! \ldots l_d!, \quad x^l = x_1^{l_1} \ldots x_d^{l_d}. \tag{103}$$

We employ the smoothness of the kernel K and the Taylor Theorem. There exists a function P such that for t fixed the function $P(x,t)$ is a polynomial with respect to x of degree at most $L-1$, and there exists a function Q such that for x fixed the function $Q(x,t)$ is a polynomial with respect to t of degree at most $L-1$ such that

$$K(x,t) = P(x,t) + Q(x,t) + \sum_{|l|=L} \sum_{|m|=L} C_{l,m} (x - x_\lambda)^l \left(t - t_\mu\right)^m \tag{104}$$

with

$$C_{l,m} = \frac{1}{l!m!} \frac{\partial^l \partial^m K(\xi(x,t))}{\partial x^l \partial t^m} \tag{105}$$

and

$$\xi(x,t) = \left(x_\lambda, t_\mu\right) + \alpha\left((x,t) - \left(x_\lambda, t_\mu\right)\right) \tag{106}$$

for some $\alpha \in [0, 1]$. Since Ψ has L vanishing moments, it holds that

$$\int_{\Omega}\int_{\Omega} \left(P\left(x, t\right) + Q\left(x, t\right)\right) \psi_{\lambda}\left(x\right) \psi_{\mu}\left(t\right) dx\, dt = 0. \tag{107}$$

Recall that the isotropic approach consists in tensorizing basis functions on the same level. Hence, if Ψ is isotropic wavelet basis, then

$$\left|(x_i - (x_{\lambda})_i)^{l_i}\right| \leq C2^{-l_i|\lambda|} \tag{108}$$

and thus

$$|x - x_{\lambda}|^{l} \leq C\prod_{i=1}^{d} 2^{-l_i|\lambda|} \leq C2^{-|l||\lambda|}. \tag{109}$$

In this proof, C denotes a generic constant, which is not necessarily the same in all the formulas.

In the case that Ψ is an anisotropic wavelet basis, the support length depends on the parameter $[\lambda]$ defined by (57) and we obtain

$$|x - x_{\lambda}|^{l} \leq C\prod_{i=1}^{d} 2^{-|\lambda_i|l_i} \leq C2^{-|l|[\lambda]}. \tag{110}$$

Furthermore, we have

$$\int_{\Omega} |\psi_{\lambda}\left(x\right)| \, dx \leq \begin{cases} C2^{-|\lambda|d/2} & \text{for isotropic systems,} \\ C2^{-[\lambda]d/2} & \text{for anisotropic systems.} \end{cases} \tag{111}$$

Combining (104) and (107) we have

$$\left|K_{\mu,\lambda}\right| \leq \sum_{|l|=L}\sum_{|m|=L} C_{l,m} \int_{\Omega}\int_{\Omega} |x - x_{\lambda}|^{l} \left|t - t_{\mu}\right|^{m} \left|\psi_{\lambda}\left(x\right) \psi_{\mu}\left(t\right)\right| dx\, dt. \tag{112}$$

Using this relation together with inequalities (109), (110), and (111) the estimate (100) for isotropic systems and estimate (101) for anisotropic systems are proven. □

Remark 1 The main focus of this paper is the numerical solution of PIDEs with smooth kernels. However, from the proof of Theorem 9, it follows that the decay estimates remain valid also for non-smooth kernels but only locally. More precisely, if the assumptions on the wavelet basis in Theorem 9 are satisfied, and $K \in C^{2L}\left(\overline{\Omega_1} \times \overline{\Omega_1}\right)$, where $\Omega_1 \subset \Omega$, then the estimates (100) and (101) are valid for all indices λ and μ for which $\operatorname{supp} \psi_{\lambda,\mu} \subset \Omega_1$.

Due to Theorem 9 and the local support of wavelets, many entries of the matrix $\tilde{\mathbf{A}}^k$ are small and can be thresholded. Thus, the matrix $\tilde{\mathbf{A}}^k$ can be represented by a compressed matrix, which is sparse or quasi-sparse. More precisely, let T be a chosen threshold and let $\hat{\mathbf{A}}^k$ be defined as

$$\hat{\mathbf{A}}^k_{m,l} = \begin{cases} \tilde{\mathbf{A}}^k_{m,l}, & \text{if } \left|\tilde{\mathbf{A}}^k_{m,l}\right| > T, \\ 0, & \text{if } \left|\tilde{\mathbf{A}}^k_{m,l}\right| \leq T. \end{cases} \tag{113}$$

Then, the norms of the errors caused by compression can be expressed as

$$\left\|\tilde{\mathbf{A}}^k - \hat{\mathbf{A}}^k\right\|_1 = \max_m \sum_l \left|\tilde{\mathbf{A}}^k_{m,l} - \hat{\mathbf{A}}^k_{m,l}\right| \leq T N_k, \tag{114}$$

and

$$\left\|\tilde{\mathbf{A}}^k - \hat{\mathbf{A}}^k\right\|_\infty = \max_l \sum_m \left|\tilde{\mathbf{A}}^k_{m,l} - \hat{\mathbf{A}}^k_{m,l}\right| \leq T N_k. \tag{115}$$

Due to (114) and (115), the spectral norm of the error satisfies

$$\left\|\tilde{\mathbf{A}}^k - \hat{\mathbf{A}}^k\right\| \leq \sqrt{\left\|\tilde{\mathbf{A}}^k - \hat{\mathbf{A}}^k\right\|_1 \left\|\tilde{\mathbf{A}}^k - \hat{\mathbf{A}}^k\right\|_\infty} \leq T N_k. \tag{116}$$

Thus, choosing a threshold T small enough guarantees small error of compression.

Now, we focus on the effect of the compression on the solution of the discrete problem. In the following, we admit that the threshold T_k used for the compression of the matrix $\tilde{\mathbf{A}}^k$ depends on the level k.

Theorem 10 *Let the assumptions (A1)–(A4) be satisfied, let $\tilde{\mathbf{c}}^k$ be the solution of (96), and let $\hat{\mathbf{c}}^k$ be the solution of the system*

$$\hat{\mathbf{A}}^k \hat{\mathbf{c}}^k = \tilde{\mathbf{f}}^k. \tag{117}$$

Furthermore, let us assume that the threshold T_k satisfies

$$T_k N_k \left\|\left(\tilde{\mathbf{A}}^k\right)^{-1}\right\| < \gamma < 1 \tag{118}$$

for some constant γ independent of k. Then

$$\frac{\left\|\tilde{\mathbf{c}}^k - \hat{\mathbf{c}}^k\right\|}{\left\|\tilde{\mathbf{c}}^k\right\|} \leq C T_k N_k, \tag{119}$$

for some constant C is independent of k.

Proof We employ the theorem about perturbed systems, see e.g. Theorem 7.12 in [4]. It states that, if

$$\gamma_k = \left\| \left(\tilde{\mathbf{A}}^k \right)^{-1} \left(\tilde{\mathbf{A}}^k - \hat{\mathbf{A}}^k \right) \right\| < 1, \tag{120}$$

then

$$\frac{\left\| \tilde{\mathbf{c}}^k - \hat{\mathbf{c}}^k \right\|}{\left\| \tilde{\mathbf{c}}^k \right\|} \le \frac{\operatorname{cond} \tilde{\mathbf{A}}^k}{1 - \gamma_k} \frac{\left\| \tilde{\mathbf{A}}^k - \hat{\mathbf{A}}^k \right\|}{\left\| \tilde{\mathbf{A}}^k \right\|}. \tag{121}$$

Hence, we have

$$\frac{\left\| \tilde{\mathbf{c}}^k - \hat{\mathbf{c}}^k \right\|}{\left\| \tilde{\mathbf{c}}^k \right\|} \le \frac{T_k N_k \operatorname{cond} \tilde{\mathbf{A}}^k}{(1 - \gamma) \left\| \tilde{\mathbf{A}}^k \right\|}. \tag{122}$$

Due to Theorem 8, we have

$$\operatorname{cond} \tilde{\mathbf{A}}^k < C_1, \quad \frac{1}{\left\| \tilde{\mathbf{A}}^k \right\|} \le \left\| \left(\tilde{\mathbf{A}}^k \right)^{-1} \right\| < C_2. \tag{123}$$

This proves the assertion of the theorem. □

The similar result holds for the approximate solution of the continuous problem.

Theorem 11 *Let the assumptions (A1)–(A4) be satisfied, let u^k be the solution of (87), and $\hat{u}^k$ be the approximate solution that is obtained when the matrix is compressed with the threshold T_k. If*

$$T_k N_k \left\| \left(\tilde{\mathbf{A}}^k \right)^{-1} \right\| < \gamma < 1 \tag{124}$$

for some constant γ independent of k, then

$$\left\| u_k - \hat{u}_k \right\|_{H^1} \le C T_k N_k, \tag{125}$$

for some constant C is independent of k.

Proof First, we express the functions u_k and $\hat{u}_k$ using elements of the vectors $\tilde{\mathbf{c}}^k$ and $\hat{\mathbf{c}}^k$. We have

$$u_k = \sum_{\lambda \in \Psi^k} c_\lambda^k \psi_\lambda = \sum_{\lambda \in \Psi^k} \frac{\tilde{c}_\lambda^k}{\mathbf{D}_{\lambda,\lambda}^k} \psi_\lambda \tag{126}$$

and similarly

$$\hat{u}_k = \sum_{\lambda \in \Psi^k} \frac{\hat{c}_\lambda^k}{\mathbf{D}_{\lambda,\lambda}^k} \psi_\lambda. \tag{127}$$

Therefore, we can write the error as

$$u_k - \hat{u}_k = \sum_{\lambda \in \Psi^k} \left(\tilde{c}_\lambda^k - \hat{c}_\lambda^k\right) \frac{\psi_\lambda}{\mathbf{D}_{\lambda,\lambda}^k}. \tag{128}$$

We rewrite the coefficients in this expansion by the following way. We define

$$d_\lambda^k = \frac{\left(\tilde{c}_\lambda^k - \hat{c}_\lambda^k\right) \|\psi_\lambda\|_{H^1}}{\mathbf{D}_{\lambda,\lambda}^k} \tag{129}$$

and we obtain

$$u_k - \hat{u}_k = \sum_{\lambda \in \Psi^k} d_\lambda^k \frac{\psi_\lambda}{\|\psi_\lambda\|_{H^1}}. \tag{130}$$

Due to the assumption that the functions ψ_λ normalized with respect to the H^1 norm form Riesz basis in $H^1(\Omega)$, we obtain that

$$\left\|u_k - \hat{u}_k\right\|_{H^1} \leq C_\Psi \left\|\mathbf{d}^k\right\|, \tag{131}$$

where $\mathbf{d}^k$ is the vector with elements d_λ^k and C_Ψ is the upper H^1 Riesz bound of the normalized basis Ψ. Let α be the coercivity constant defined by (12), then

$$\mathbf{D}_{\lambda,\lambda}^k = \sqrt{a(\psi_\lambda, \psi_\lambda)} \geq \sqrt{\alpha} \|\psi_\lambda\|_{H^1}. \tag{132}$$

Hence, we have

$$\frac{\|\psi_\lambda\|_{H^1}}{\mathbf{D}_{\lambda,\lambda}^k} \leq \frac{1}{\sqrt{\alpha}}, \tag{133}$$

which implies that

$$\left\|u_k - \hat{u}_k\right\|_{H^1} \leq \frac{C_\Psi}{\sqrt{\alpha}} \left\|\tilde{\mathbf{c}}^k - \hat{\mathbf{c}}^k\right\|. \tag{134}$$

Due to Theorem 8, the norms of the inverses of $\tilde{\mathbf{A}}^k$ are uniformly bounded and thus

$$\left\|\tilde{\mathbf{c}}^k\right\| \leq \left\|\left(\tilde{\mathbf{A}}^k\right)^{-1}\right\| \left\|\tilde{\mathbf{f}}_k\right\| < C, \tag{135}$$

where C is a constant independent of k.

By (134), (135), and the Theorem 10, the assertion of the theorem is proved. □

Finally, we show that the condition numbers of the compressed matrices are also uniformly bounded.

Theorem 12 *Let the assumptions (A1)–(A4) be satisfied. If the threshold T_k is chosen such that*

$$T_k N_k \left\|\left(\tilde{\mathbf{A}}^k\right)^{-1}\right\| < \gamma < 1 \tag{136}$$

for some constant γ independent of k, then there exists a constant C also independent of k such that cond $\hat{\mathbf{A}}^k \leq C$.

Proof Let **A** and **B** be two square matrices of the same size, let **A** be invertible, and let **I** be the identity matrix of the same size as **A** and **B**. Then,

$$\|\mathbf{A}+\mathbf{B}\| \leq \left\|\mathbf{A}\left(\mathbf{I}+\mathbf{A}^{-1}\mathbf{B}\right)\right\| \leq \|\mathbf{A}\|\left(1+\left\|\mathbf{A}^{-1}\right\|\|\mathbf{B}\|\right). \tag{137}$$

From the identity

$$(\mathbf{A}+\mathbf{B})(\mathbf{A}+\mathbf{B})^{-1} = \mathbf{I}, \tag{138}$$

we obtain

$$\mathbf{A}\left(\mathbf{I}+\mathbf{A}^{-1}\mathbf{B}\right)(\mathbf{A}+\mathbf{B})^{-1} = \mathbf{I} \tag{139}$$

and

$$\left(\mathbf{I}+\mathbf{A}^{-1}\mathbf{B}\right)(\mathbf{A}+\mathbf{B})^{-1} = \mathbf{A}^{-1}. \tag{140}$$

This implies that

$$(\mathbf{A}+\mathbf{B})^{-1} = \mathbf{A}^{-1} - \mathbf{A}^{-1}\mathbf{B}(\mathbf{A}+\mathbf{B})^{-1}. \tag{141}$$

From this, we have

$$\left\|(\mathbf{A}+\mathbf{B})^{-1}\right\| \leq \left\|\mathbf{A}^{-1}\right\| + \left\|\mathbf{A}^{-1}\right\|\|\mathbf{B}\|\left\|(\mathbf{A}+\mathbf{B})^{-1}\right\|, \tag{142}$$

which implies that

$$\left\|(\mathbf{A}+\mathbf{B})^{-1}\right\|\left(1-\left\|\mathbf{A}^{-1}\right\|\|\mathbf{B}\|\right) \leq \left\|\mathbf{A}^{-1}\right\|. \tag{143}$$

We assume that $\left\|\mathbf{A}^{-1}\right\|\|\mathbf{B}\| < 1$ and obtain

$$\left\|(\mathbf{A}+\mathbf{B})^{-1}\right\| \leq \frac{\left\|\mathbf{A}^{-1}\right\|}{1-\left\|\mathbf{A}^{-1}\right\|\|\mathbf{B}\|}. \tag{144}$$

From (137) and (144), it follows that

$$\text{cond}\,(\mathbf{A}+\mathbf{B}) = \|\mathbf{A}+\mathbf{B}\|\left\|(\mathbf{A}+\mathbf{B})^{-1}\right\| \leq \text{cond}\,\mathbf{A}\frac{1+\left\|\mathbf{A}^{-1}\right\|\|\mathbf{B}\|}{1-\left\|\mathbf{A}^{-1}\right\|\|\mathbf{B}\|}. \tag{145}$$

Setting $\mathbf{A} = \tilde{\mathbf{A}}^k$ and $\mathbf{B} = \hat{\mathbf{A}}^k - \tilde{\mathbf{A}}^k$ into (145), we obtain

$$\text{cond}\left(\hat{\mathbf{A}}^k\right) \leq \text{cond}\left(\tilde{\mathbf{A}}^k\right)\frac{1+\gamma}{1-\gamma} \leq C. \tag{146}$$

□

Hence, the estimates (100) and (101) together with the known structure of the matrices $\mathbf{G}^k$ can be used to compute only significant entries of the matrix $\mathbf{A}^k$, and the compression does not affect significantly the condition numbers of the matrices and the resulting approximate solution if the threshold is sufficiently small.

5 Merton Jump-Diffusion Option Pricing Model

In this section, the wavelet-Galerkin method is employed in the option pricing under the Merton jump-diffusion model.

Options constitute financial derivatives that give its holder the right to buy or sell, depending on the type of an option, the underlying asset at a specific price, and are therefore important financial instruments for hedging. The first model for the computation of the fair market price of the option was the famous Black–Scholes model. This model assumes that asset prices depend continuously on time and follow a log-normal distribution and that the volatility of the asset price is constant. Since these assumptions are not always consistent with real market behaviour, various models were designed to better capture several features of asset price dynamics such as stochastic volatility or the jumps in asset prices. In this section, we focus on a sophisticated jump-diffusion model suggested by Merton [41]. This model assumes that the spot price S of the underlying asset follows at time $\tilde{t}$ the jump-diffusion process

$$\frac{dS\left(\tilde{t}\right)}{S\left(\tilde{t}-\right)} = \mu\, d\tilde{t} + \sigma\, dW\left(\tilde{t}\right) + d\left(\sum_{i=1}^{N(\tilde{t})} \left(V_i - 1\right)\right), \tag{147}$$

where $S\left(\tilde{t}-\right)$ denotes the left-hand limit of S at $\tilde{t}$, μ stands for the drift rate, $W\left(\tilde{t}\right)$ is the Wiener process, and σ denotes the volatility of the diffusion component of the process. Furthermore, $N\left(\tilde{t}\right)$ is a Poisson process with intensity λ that is independent of $W\left(\tilde{t}\right)$, and $\{V_i\}$ is a sequence of independent identically distributed nonnegative random variables such that $Y_i = \log V_i$ has a distribution with the probability density function g. In the Merton model, it is assumed that random variables Y_i are normally distributed with the mean μ_J and the standard deviation σ_J and

$$g(x) = \frac{1}{\sqrt{2\pi}\sigma_J} e^{-\frac{(x-\mu_J)^2}{2\sigma_J^2}}. \tag{148}$$

Let the variable $t = T - \tilde{t}$ represent time to maturity and r be a risk-free interest rate. Using no arbitrage principle and Itô calculus, one can derive that the market price $U(S,t)$ of the option at time to maturity t is the solution of the equation [32, 41]:

$$\frac{\partial U}{\partial t} - \mathcal{D}(U) - \mathcal{I}(U) = 0, \;\; S > 0, \;\; t \in (0, T), \tag{149}$$

where the operators $\mathcal{D}$ and $\mathcal{I}$ are given by

$$\mathcal{D}(U) = \frac{\sigma^2 S^2}{2} \frac{\partial^2 U}{\partial S^2} + (r - \lambda\kappa) S \frac{\partial U}{\partial S} - (r + \lambda) U \qquad (150)$$

and

$$\mathcal{I}(U) = \lambda \int_{-\infty}^{\infty} U\left(Se^x, t\right) g(x)\, dx. \qquad (151)$$

The parameter κ represents the expected value $E(V_i - 1) = e^{\mu_J + \sigma_J^2/2} - 1$.

The initial and boundary conditions are prescribed in dependence on the type of the option. Here, only a European put option is considered because the value of a European call option can be simply computed using the well-known put-call parity, see, e.g., [1]. The initial conditions for a European vanilla put option represent values of the option at maturity and thus they are of the form

$$U(S, 0) = \max(K - S, 0), \qquad (152)$$

where K is the strike price. The boundary conditions are set as follows

$$U(0, t) = Ke^{-rt}, \quad U(S, t) \approx 0 \text{ for } S \to \infty. \qquad (153)$$

To solve the Eq. (149) numerically, the unbounded domain $(0, \infty)$ is approximated by a bounded domain $\Omega = \left(S^{min}, S^{max}\right)$. The value $0 < S^{min}$ should be small enough and the value of S^{max} large enough so that the error caused by this approximation is sufficiently small, usually S^{max} is at least $4K$. The boundary conditions have to be adapted to the bounded domain Ω. Using $U(S, t) \approx Ke^{-rt} - S$ for S close to zero, the new boundary conditions are formulated as follows

$$U\left(S^{min}, t\right) = Ke^{-rt} - S^{min}, \quad U\left(S^{max}, t\right) = 0. \qquad (154)$$

It is convenient to transform Eq. (149) to logarithmic prices $x = \log S$. Let denote

$$X = \left(x^{min}, x^{max}\right), \quad x^{min} = \log S^{min}, \quad x^{max} = \log S^{max}, \qquad (155)$$

and

$$\hat{U}(x, t) = U\left(e^x, t\right), \quad x \in X. \qquad (156)$$

The advantage is that the transformed differential operator

$$\hat{\mathcal{D}}\left(\hat{U}\right) = \frac{\sigma^2}{2} \frac{\partial^2 \hat{U}}{\partial x^2} + \left(r - \lambda\kappa - \frac{\sigma^2}{2}\right) \frac{\partial \hat{U}}{\partial x} - (r + \lambda) \hat{U} \qquad (157)$$

have constant coefficients. The integral operator is transformed to

$$\hat{\mathcal{I}}\left(\hat{U}\right) = \lambda \int_{-\infty}^{\infty} \hat{U}\left(x + y, t\right) g\left(y\right) dy = \lambda \int_{-\infty}^{\infty} \hat{U}\left(y, t\right) g\left(y - x\right) dy. \tag{158}$$

The equation corresponding to logarithmic prices is given by

$$\frac{\partial \hat{U}}{\partial t} - \hat{\mathcal{D}}\left(\hat{U}\right) - \hat{\mathcal{I}}\left(\hat{U}\right) = 0, \quad x \in X, \ t \in (0, T). \tag{159}$$

Due to $U(S, t) \approx Ke^{-rt} - S$ for S close to zero, we have

$$\hat{U}\left(x, t\right) \approx Ke^{-rt} - e^{x} \tag{160}$$

for $x \to -\infty$, and thus the boundary conditions for the Eq. (159) are of the form

$$\hat{U}\left(x^{min}, t\right) = Ke^{-rt} - e^{x^{min}} \quad \text{and} \quad \hat{U}\left(x^{max}, t\right) = 0. \tag{161}$$

The integral term $\hat{\mathcal{I}}$ has to be further modified because the solution $\hat{U}$ is now defined only on X and not on the whole real line. Since values of a put option and values of the probability density function are characterized by fast decay when x goes to infinity, one obtains

$$\int_{x^{max}}^{\infty} \hat{U}\left(y, t\right) g\left(y - x\right) dy \approx 0 \text{ for } x^{max} \to \infty. \tag{162}$$

Due to (160) and (162), the integral term $\hat{\mathcal{I}}$ can be approximated by

$$\hat{\mathcal{I}}\left(\hat{U}\right) \approx \mathcal{I}_1 + \mathcal{I}_2\left(\hat{U}\right), \tag{163}$$

where

$$\mathcal{I}_1 = \int_{-\infty}^{x^{min}} \left(Ke^{-rt} - e^{x}\right) g\left(y - x\right) dy, \quad \mathcal{I}_2\left(\hat{U}\right) = \int_{x^{min}}^{x^{max}} \hat{U}\left(y, t\right) g\left(y - x\right) dy. \tag{164}$$

The non-homogeneous Dirichlet boundary condition are prescribed at the point x^{min}. Therefore, we apply appropriate transform to obtain PIDE with homogeneous Dirichlet boundary conditions. Let $\tilde{U} = \hat{U} - W$, where $\hat{U}$ is the solution of the Eq. (159) satisfying the initial and boundary conditions defined above and W is a function satisfying boundary conditions that is smooth enough. A possible choice of the function W is

$$W(x,t) = \left(Ke^{-rt} - e^{x^{min}}\right) \frac{x^{max} - x}{x^{max} - x^{min}} \tag{165}$$

for $x \in \bar{X}$ and $t \in [0, T]$. Then $\tilde{U}$ is the solution of the equation

$$\frac{\partial \tilde{U}}{\partial t} - \hat{\mathcal{D}}\left(\tilde{U}\right) - \mathcal{I}_2\left(\tilde{U}\right) = f(W), \quad f(W) = -\frac{\partial W}{\partial t} + \hat{\mathcal{D}}(W) + \mathcal{I}_2(W) + \mathcal{I}_1 \tag{166}$$

satisfying the initial condition

$$\tilde{U}(x, 0) = U\left(e^x, 0\right) - W(x, 0), \quad x \in X, \tag{167}$$

and homogeneous Dirichlet boundary conditions

$$\tilde{U}\left(x^{min}, t\right) = 0, \quad \tilde{U}\left(x^{max}, t\right) = 0, \quad t \in [0, T]. \tag{168}$$

To get a variational formulation, let define a bilinear form b as

$$b(u, v) = \left\langle \hat{\mathcal{D}}(u), v\right\rangle + \langle \mathcal{I}_2(u), v\rangle, \tag{169}$$

for all $u, v \in H_0^1(X)$.

Then the variational formulation of Eq. (166) reads as:

Find $\tilde{U} \in L^2\left(0, T; H_0^1(X)\right)$ such that $\frac{\partial \tilde{U}}{\partial t} \in L^2\left(0, T; H^{-1}(X)\right)$ and $\tilde{U}$ satisfies (167) and

$$\left\langle \frac{\partial \tilde{U}}{\partial t}, v\right\rangle - b(u, v) = \langle f(W), v\rangle, \quad \forall v \in H_0^1(X), \tag{170}$$

almost everywhere in $(0, T)$.

It can be shown that the bilinear form b is continuous and satisfies a Gårding inequality, which implies the existence of a unique solution to this problem, see [32].

For temporal discretization, the Crank–Nicolson method is employed. Let

$$M \in \mathbb{N}, \ \tau = \frac{T}{M}, \ t_l = l\tau, \ l = 0, \ldots, M, \tag{171}$$

and let us denote

$$\tilde{U}_l(x) = \tilde{U}(x, t_l), \quad f_l(x) = f(W(x, t_l)). \tag{172}$$

The Crank–Nicolson method leads to the scheme

$$\frac{\left\langle \tilde{U}_{l+1}, v\right\rangle}{\tau} - \frac{b\left(\tilde{U}_{l+1}, v\right)}{2} = \frac{\left\langle \tilde{U}_l, v\right\rangle}{\tau} + \frac{b\left(\tilde{U}_l, v\right)}{2} + \left\langle \frac{f_l + f_{l+1}}{2}, v\right\rangle \tag{173}$$

for $l = 0, \ldots, M-1$.

In this scheme, the function $\tilde{U}_0$ is known from (167) and for l fixed the function $\tilde{U}_l$ is always known as the solution from the previus step. Therefore, one has to find the function $\tilde{U}_{l+1}$ as the solution of partial integro-differential equation (173), which is of the type (2). Hence, we use the wavelet-Galerkin method proposed in Sect. 4 for the numerical solution of (173).

6 Numerical Examples

In this section, several numerical experiments are provided. The following notations are used. The finest level of the used wavelet basis is denoted as k. The exact solution of the integro-differential equation is denoted as u and the numerical solution obtained by the proposed method with compressed matrices is denoted as $\hat{u}_k$. The error of the method is then defined as $e_k = \hat{u}_k - u$. In the numerical experiments, we study L^2 and H^1 norms of the error and also the L^∞ norm defined as

$$\|e_k\|_\infty = \max_{x \in \overline{\Omega}} |e_k(x)| . \tag{174}$$

The number of nonzero entries of the system matrix is denoted as *nz*, its condition number as *cond*, and $N_k \times N_k$ denotes the size of the matrix. The number of iterations for the conjugate gradient method is denoted by CG and for the generalized minimal residual method by GMRES. Furthermore, we define the compression ratio as CR $= nz/N_k^2$ and present CPU time in seconds including the computation of the right-hand side and the matrix, numerical solution of the system, and evaluation of the solution in points $k/2^J$, where J denotes the finest level of the wavelet basis used.

Example 3 We apply the proposed method on the integro-differential equation

$$-u''(t) + u(t) + \int_0^1 \sin \frac{(x+t)\pi}{2} u(x)\, dx = f(t), \quad t \in (0,1), \tag{175}$$

with the boundary conditions $u(0) = u(1) = 0$. The right-hand side f is set such that the solution is $u(t) = \sin(50\pi t)$. The used wavelet bases are quadratic and cubic spline wavelet bases with short support from [13] and [11], respectively, described also in Sect. 3.4.

In Fig. 4 we display structures of preconditioned matrices

$$\tilde{\mathbf{G}}^k = \left(\mathbf{D}^k\right)^{-1} \mathbf{G}^k \left(\mathbf{D}^k\right)^{-1}, \quad \tilde{\mathbf{K}}^k = \left(\mathbf{D}^k\right)^{-1} \mathbf{K}^k \left(\mathbf{D}^k\right)^{-1}. \tag{176}$$

to illustrate a difference in structures of discretization matrices corresponding to the differential and integral term. These matrices correspond to quadratic spline wavelets, they are compressed using the threshold $T = 10^{-10}$ and their size is 256×256. The

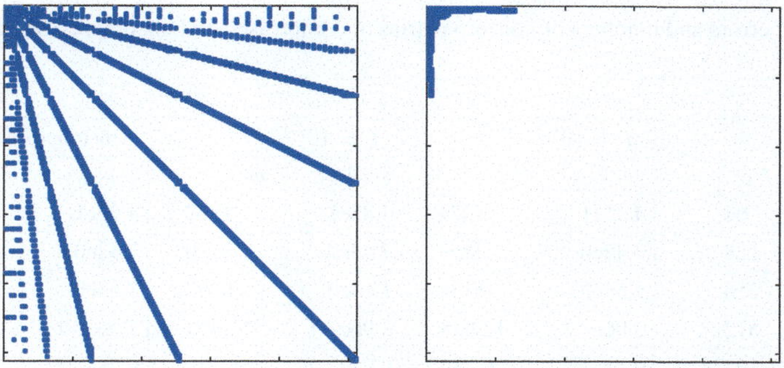

Fig. 4 The sparsity patterns of the compressed matrices of the size 256 × 256 corresponding to the differential term (left) and the integral term (right) for Eq. (175)

matrix $\tilde{\mathbf{G}}^k$ contains 6128 nonzero element after compression while the matrix $\tilde{\mathbf{K}}^k$ contains only 480 nonzero elements.

The resulting system of linear algebraic equations was solved by the conjugate gradient method. The stopping criterion was that the relative residual is smaller than 10^{-10}. In Table 1, the resulting H^1 errors and the numbers of nonzero entries of the compressed discretization matrices are presented for the choices of thresholds $T = 10^{-7}$ and $T = 10^{-10}$ and for comparison purposes also for the uncompressed matrices. It can be seen that for these choices of threshold the resulting errors were not significantly affected by the compression, but the compressed matrices have significantly smaller numbers of nonzero entries than the uncompressed ones. In Table 2, the further results are presented for the algorithm with the threshold $T = 10^{-10}$.

Example 4 We consider the equation

$$-\epsilon \Delta u(t_1, t_2) + u(t_1, t_2) + \iint\limits_{\Omega} K(x_1, x_2, t_1, t_2)\, u(x_1, x_2)\, dx_1 dx_2 = f(t_1, t_2), \tag{177}$$

where $(t_1, t_2) \in \Omega = (0, 1)^2$, $\epsilon = 10^{-2}$, and

$$K(x_1, x_2, t_1, t_2) = (\cos(x_1 + t_1) + \cos(x_1 - t_1))\, x_2 t_2. \tag{178}$$

Furthermore, the equation is equipped with the homogeneous Dirichlet boundary conditions and the right-hand side corresponds to the solution

$$u(t_1, t_2) = \left(1 - e^{20t_1 - 20} - e^{20t_2 - 20} + e^{20t_1 + 20t_2 - 40}\right) t_1 t_2. \tag{179}$$

We employ the quadratic and cubic spline wavelet bases from Sect. 3.4 again. Starting from these bases, we use both an isotropic and an anisotropic approach for constructions of two-dimensional wavelet bases. The threshold used for the matrix

Table 1 H^1 errors and numbers of nonzero entries of compressed and uncompressed matrices for Example 3

Quadratic

		$T = 10^{-7}$		$T = 10^{-10}$		Uncompressed	
k	N_k	$\|e_k\|_{H^1}$	nz	$\|e_k\|_{H^1}$	nz	$\|e_k\|_{H^1}$	nz
6	64	4.35e1	1274	4.35e1	1522	4.35e1	4096
7	128	7.42e0	2852	7.42e0	3100	7.42e0	16,384
8	256	1.64e0	6134	1.64e0	6382	1.63e0	65,536
9	512	3.96e−1	12,848	3.96e−1	13,096	3.96e−1	261,518
10	1024	9.78e−2	26,450	9.78e−2	26,698	9.78e−2	1,048,576
11	2048	2.45e−2	53,852	2.45e−2	54,100	2.45e−2	4,194,304
12	4096	6.22e−3	108,878	6.22e−3	109,126	6.22e−3	16,777,216

Cubic

		$T = 10^{-7}$		$T = 10^{-10}$		Uncompressed	
k	N_k	$\|e_k\|_{H^1}$	nz	$\|e_k\|_{H^1}$	nz	$\|e_k\|_{H^1}$	nz
6	65	2.77e1	1961	2.77e1	2001	2.77e1	4225
7	129	1.67e0	4585	1.67e0	4633	1.66e0	16,641
8	257	1.62e−1	10,105	1.62e−1	10,209	1.62e−1	66,049
9	513	1.89e−2	21,437	1.89e−2	21,737	1.89e−2	263,169
10	1025	2.32e−3	44,393	2.32e−3	45,233	2.32e−3	1,050,625
11	2049	2.89e−4	90,641	2.87e−4	92,721	2.89e−4	4,198,401

Table 2 The errors in the L^∞, L^2, and H^1 norm, the condition numbers of matrices, compression ratios, numbers of CG iterations, and CPU time in seconds for Example 3

Quadratic

N_k	$\|e_k\|_\infty$	$\|e_k\|$	$\|e_k\|_{H^1}$	cond	CR	CG	Time
64	3.68e−1	1.92e−1	4.35e1	10.8	0.372	27	0.19
128	1.97e−2	1.10e−2	7.42e0	10.9	0.189	29	0.40
256	1.97e−3	1.04e−3	1.64e0	10.9	0.097	30	0.85
512	2.35e−4	1.21e−4	3.96e−1	10.9	0.050	31	1.66
1024	2.90e−5	1.48e−5	9.82e−2	10.9	0.026	32	3.52
2048	3.62e−6	1.84e−6	2.45e−2	11.0	0.013	33	7.78
4096	4.93e−7	2.32e−7	6.22e−3	11.0	0.007	33	15.46

Cubic

N_k	$\|e_k\|_\infty$	$\|e_k\|$	$\|e_k\|_{H^1}$	cond	CR	CG	Time
65	2.32e−1	1.18e−1	2.77e1	4.44	0.473	20	0.20
129	4.97e−3	2.51e−3	1.67e0	4.53	0.278	21	0.40
257	2.13e−4	1.07e−4	1.62e−1	4.58	0.155	21	0.86
513	1.26e−5	5.94e−6	1.89e−2	4.60	0.083	21	1.81
1025	7.74e−7	3.60e−7	2.32e−3	4.62	0.043	21	4.39
2049	4.86e−8	2.23e−8	2.89e−4	4.63	0.022	21	8.82

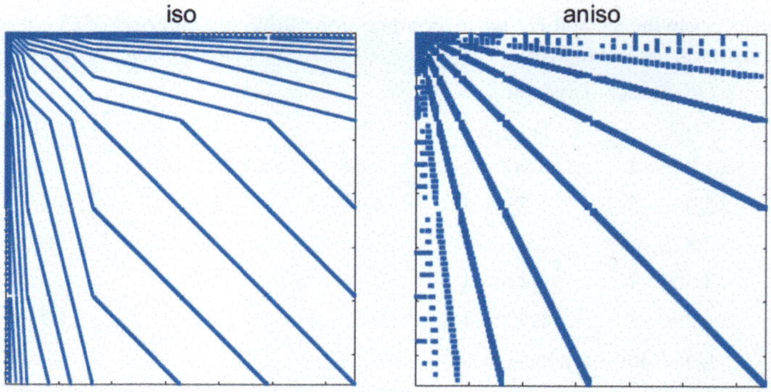

Fig. 5 The sparsity patterns of the compressed matrices of the size 65536 × 65536 for an isotropic (left) and an anisotropic (right) quadratic spline wavelet bases for equation (177)

compression is $T = 10^{-8}$. Since the system matrix is symmetric and positive definite, we use the conjugate gradient method for the numerical solution of corresponding linear equations. The iterations stop if the relative residual is less than 10^{-9}. The structures of discretization matrices for both isotropic and anisotropic quadratic spline wavelet bases are displayed in Fig. 5 and the numerical results are listed in Table 3. Since wavelet bases of both types, isotropic and anisotropic, generate the same multiresolution analysis, the errors are similar. Here, the numbers of iterations and the compression ratios were smaller for the isotropic wavelet bases than for the anisotropic wavelet bases.

Example 5 Based on Remark 1, we also apply the proposed method on the integro-differential equation with weakly singular kernel of the form

$$-u''(t) + 20u(t) + \int_0^1 \frac{u(x)}{5\sqrt{|x-t|}} dx = f(t), \quad t \in (0,1), \tag{180}$$

with the boundary conditions $u(0) = u(1) = 0$. The right-hand side f is set such that the solution is a high-order oscillatory polynomial

$$u(t) = (6t(6t-1)(6t-2)(6t-3)(6t-4)(6t-5)(6t-6))^2. \tag{181}$$

Moreover, the derivatives of u are relatively large and thus also the H^1-norm of u is large, namely $\|u\|_1 = 7.73e4$. This fact and the singular kernel make Eq. (180) relatively hard to resolve with the sufficient accuracy. The cubic spline wavelet basis with short support from [11] was used, the threshold was set to $T = 10^{-10}$, and the system was solved by the conjugate gradient method with the relative residual 10^{-10}. The results are presented in Table 4.

Table 3 The errors in the L^2 and H^1 norm, compression ratios, and numbers of CG iterations, and CPU time in seconds for Example 4

	Quadratic—isotropic				
N_k	$\|e_k\|$	$\|e_k\|_{H^1}$	CR	CG	Time
1024	1.94e−4	3.90e−2	1.6e−1	78	0.26
4096	2.33e−5	9.75e−3	4.9e−2	85	0.38
16,384	2.87e−6	2.47e−3	1.4e−2	90	0.64
65,536	3.56e−7	6.49e−4	3.7e−3	94	1.50
262,144	4.29e−8	1.90e−4	9.5e−4	94	3.21
	Quadratic—anisotropic				
N_k	$\|e_k\|$	$\|e_k\|_{H^1}$	CR	CG	Time
1024	1.94e−4	3.90e−2	2.7e−1	104	0.28
4096	2.33e−5	9.75e−3	9.5e−2	124	0.45
16,384	2.87e−6	2.47e−3	3.0e−2	136	0.78
65,536	3.56e−7	6.49e−4	8.7e−3	147	1.86
262,144	4.29e−8	1.90e−4	2.4e−3	154	3.74
	Cubic—isotropic				
N_k	$\|e_k\|$	$\|e_k\|_{H^1}$	CR	CG	Time
289	1.99e−4	2.29e−2	7.1e−1	63	0.19
1089	1.41e−5	3.16e−3	3.2e−1	67	0.24
4225	1.01e−6	4.31e−4	1.1e−1	69	0.37
16,641	6.85e−8	5.70e−5	3.2e−2	70	0.63
66,049	4.46e−9	7.37e−6	8.7e−3	70	1.46
	Cubic—anisotropic				
N_k	$\|e_k\|$	$\|e_k\|_{H^1}$	CR	CG	Time
289	1.99e−4	2.29e−2	8.3e−1	73	0.21
1089	1.41e−5	3.16e−3	4.3e−1	96	0.27
4225	1.10e−6	4.31e−4	2.0e−1	104	0.43
16,641	6.85e−8	5.70e−5	6.6e−2	109	0.78
66,049	4.45e−9	7.34e−6	2.0e−2	110	1.86

Table 4 The errors in the L^∞, L^2, and H^1 norm, the condition numbers of matrices, compression ratios, numbers of CG iterations, and CPU time in seconds for Example 5

Cubic							
N_k	$\|e_k\|_\infty$	$\|e_k\|$	$\|e_k\|_{H^1}$	cond	CR	CG	Time
17	9.04e2	2.89e2	2.66e4	3.45	0.917	13	0.09
33	9.05e1	1.64e1	3.80e3	4.13	0.724	15	0.18
65	6.70e0	1.07e0	4.88e2	4.42	0.472	18	0.48
129	5.85e−1	7.73e−2	6.50e1	4.52	0.277	19	1.54
257	4.56e−2	7.91e−3	8.51e0	4.57	0.153	19	5.61
513	4.60e−3	3.26e−3	1.09e0	4.60	0.082	19	16.23

Table 5 Pointwise errors and errors in L^∞ and L^2 norm, condition numbers and compression ratios of the matrices, and numbers of GMRES iterations

Quadratic

N_k	M_k	$\lvert e_k (100)\rvert$	$\Vert e_k\Vert_{L^\infty}$	$\Vert e_k\Vert_{L^2}$	cond	CR	GMRES
64	16	1.18e−4	3.34e−3	7.99e−3	25.6	4.8e−1	6(4)
128	46	2.15e−5	3.61e−4	8.79e−4	29.6	2.1e−1	6(9)
256	128	9.42e−6	4.52e−5	1.07e−4	32.9	9.9e−2	7(8)
512	363	1.41e−6	6.12e−6	2.02e−5	35.8	4.5e−2	8(2)

Cubic

N_k	M_k	$\lvert e_k (100)\rvert$	$\Vert e_k\Vert_{L^\infty}$	$\Vert e_k\Vert_{L^2}$	cond	CR	GMRES
33	4	1.96e−3	1.54e−2	4.10e−2	36.2	8.2e−1	5(2)
65	16	4.90e−4	7.06e−4	1.66e−3	40.9	5.1e−1	6(7)
129	64	3.91e−5	3.94e−5	9.64e−5	44.1	2.8e−1	8(7)
257	256	1.26e−7	2.71e−6	8.91e−6	46.3	1.5e−1	7(2)

Example 6 Merton jump-diffusion option pricing model
In this example, the method proposed in Sect. 5 is used for the computation of the market value of a European vanilla put option under the Merton model [41]. The parameters of the option are set similarly as in [12, 29, 36, 37], so that one can compare the numerical results for the wavelet method and the methods proposed therein. The parameters have the following values: option maturity is $T = 0.25$ year, the interest rate is $r = 0.05$, the volatility is $\sigma = 0.15$, intensity of the Poisson process is $\lambda = 0.1$, and the strike price is $K = 100$. The probability density function is given by (148) with the parameters $\mu_J = -0.9$ and $\sigma_J = 0.45$. The aim is to compute the option prices for the values of the asset $S \in [0, 2K]$. The maximal value $S_{max} = 4K$ and the minimal value $S_{min} = 30$ are set. The threshold for the matrix compression is $T = 10^{-9}$.

In Table 5, the results for both quadratic and cubic spline wavelet bases are listed. Here, M_k denotes the number of time steps and $|e_k (100)|$ stands for the pointwise errors for the asset price $S = 100$. Furthermore, the L^2 and H^1 errors, the condition numbers and the compression ratios of the diagonally preconditioned discretization matrices are listed. The resulting system was solved by GMRES method with the following settings: restart after 10 iterations and the iterations stop if the relative residual is smaller than 10^{-12}. The resulting number of outer iterations and number of iterations in the last cycle are presented in the last column of Table 5. In comparison with the results for the same numerical example presented in [29, 36, 37], it seems that the proposed wavelet based discretization scheme requires significantly smaller number of parameters representing the solution with required accuracy and thus smaller discretization matrices are involved in the computation.

7 Conclusions

The wavelet-Galerkin method for the numerical solution of the second-order partial integro-differential equation on product domains was studied. The appropriate assumptions under which the solution exists and is unique were formulated. The concept of the wavelet basis and constructions of multi-dimensional wavelet bases from a wavelet basis on the unit interval such that the resulting bases satisfy the prescribed boundary conditions were introduced. The constructions of quadratic and cubic spline wavelet bases with short support were reviewed. Furthermore, the wavelet-Galerkin method was analyzed, i.e., existence, uniqueness, and stability of the approximate solution were proved, and the rate of convergence for spline wavelet bases was discussed. The main focus was on the study of the decay estimates of matrix entries resulting in compressed quasi-sparse matrices and on the study of the condition numbers of compressed as well as uncompressed matrices. The sparsity of the discretization matrices is essential for the efficient application of iterative methods for large systems. It is also important for the efficient computation of the discretization matrix, especially its part corresponding to the integral term, because it enables computing only a relatively small number of significant entries which are computed as integrals over $2d$-dimensional domain, while the elements of the matrix corresponding to the differential term are computed as integrals over d-dimensional domain. Furthermore, the discretization scheme, which is based on the wavelet-Galerkin method combined with the time discretization by the Crank-Nicolson scheme, was proposed for the Merton jump-diffusion option pricing model. Several numerical experiments confirmed the theoretical results and illustrated the applicability and efficiency of the wavelet-Galerkin method.

Acknowledgements This work was supported by grant No. PURE-2020-4003 funded by the Technical University of Liberec.

References

1. Achdou, Y., Pironneau, O.: Computational Methods for Option Pricing. Society for Industrial and Applied Mathematics, Philadelphia (2005)
2. Alpert, B., Beylkin, G., Coifman, R., Rokhlin, V.: Wavelet-like bases for the fast solution of second-kind integral equations. SIAM J. Sci. Comput. **14**, 159–184 (1993)
3. Avudainayagam, A., Vani, C.: Wavelet-Galerkin method for integro-differential equations. Appl. Numer. Math. **32**(3), 247–254 (2000)
4. Beilina, L., Karchevskii, E., Karchevskii, M.: Numerical Linear Algebra: Theory and Applications. Springer International Publishing (2017)
5. Bellman, R.: Methods of Nonlinear Analysis, vol. II. Academic Press, New York (1973)
6. Beylkin, G., Coifman, R., Rokhlin, V.: Fast wavelet transforms and numerical algorithms I. Commun. Pure Appl. Math. **44**, 141–183 (1991)
7. Černá, D., Finěk, V.: Construction of optimally conditioned cubic spline wavelets on the interval. Adv. Comput. Math. **34**, 219–252 (2011)
8. Černá, D., Finěk, V.: Cubic spline wavelets with complementary boundary conditions. Appl. Math. Comput. **219**, 1853–1865 (2012)

9. Černá, D., Finěk, V.: Quadratic spline wavelets with short support for fourth-order problems. Result. Math. **66**, 525–540 (2014)
10. Černá, D., Finěk, V.: Quadratic spline wavelets with short support satisfying homogeneous boundary conditions. Electron. Trans. Numer. Anal. **48**, 15–39 (2018)
11. Černá, D.: Cubic spline wavelets with four vanishing moments on the interval and their applications to option pricing under Kou model. Int. J. Wavelets Multiresolut. Inf. Process. **17**, Article No. 1850061 (2019)
12. Černá, D.: Quadratic spline wavelets for sparse discretization of jump-diffusion models. Symmetry **11**, Article No. 999 (2019)
13. Černá, D., Finěk, V.: Galerkin method with new quadratic spline wavelets for integral and integro-differential equations. J. Comput. Appl. Math. **363**, 426–443 (2020)
14. Chen, Z., Micchelli, C.A., Xu, Y.: Multiscale Methods for Fredholm Integral Equations. Cambridge Monographs on Applied and Computational Mathematics, Cambridge (2015)
15. Chui, C.K., Quak, E.: Wavelets on a bounded interval. In: Braess, D., Schumaker, L.L. (eds.) Numerical Methods of Approximation Theory, pp. 53–75. Birkhäuser (1992)
16. Ciarlet, P.: The Finite Element Method for Elliptic Problems. Society for Industrial and Applied Mathematics (2002)
17. Cohen, A., Daubechies, I., Feauveau, J.-C.: Biorthogonal bases of compactly supported wavelets. Commun. Pure and Appl. Math. **45**, 485–560 (1992)
18. Cohen, A.: Numerical Analysis of Wavelet Methods. Studies in Mathematics and Its Applications. Elsevier (2003)
19. Dahmen, W., Kunoth, A.: Multilevel preconditioning. Numer. Math. **63**, 315–344 (1992)
20. Dahmen, W., Han, B., Jia, R.Q., Kunoth, A.: Biorthogonal multiwavelets on the interval: cubic Hermite splines. Constr. Approx. **16**, 221–259 (2000)
21. Dahmen, W., Kunoth, A., Urban, K.: Biorthogonal spline wavelets on the interval—stability and moment conditions. Appl. Comp. Harm. Anal. **6**, 132–196 (1999)
22. Dahmen, W., Harbrecht, H., Schneider, R.: Compression techniques for boundary integral equations—asymptotically optimal complexity estimates. SIAM J. Numer. Anal. **43**, 2251–2271 (2006)
23. Dahmen, W., Schneider, R.: Wavelets with complementary boundary conditions—function spaces on the cube. Results Math. **34**, 255–293 (1998)
24. Dahmen, W., Stevenson, R.: Element-by-element construction of wavelets satisfying stability and moment conditions. SIAM J. Numer. Anal. **37**, 319–325 (1999)
25. Dem'yanovich, Y.K.: Wavelets on a manifold. Dokl. Math. **79**, 21–24 (2009)
26. Dijkema, T.J.: Adaptive Tensor Product Wavelet Methods for Solving PDEs. Universiteit Utrecht (2009)
27. Dijkema, T.J., Stevenson, R.: A sparse Laplacian in tensor product wavelet coordinates. Numer. Math. **115**, 433–449 (2010)
28. Griebel, M., Oswald, P.: Tensor product type subspace splittings and multilevel iterative methods for anisotropic problems. Adv. Comput. Math. **4**, 171–206 (1995)
29. d'Halluin, Y., Forsyth, P.A., Vetzal, K.R.: Robust numerical methods for contingent claims under jump diffusion processes. IMA J. Numer. Anal. **25**, 87–112 (2005)
30. Han, B., Michelle, M.: Derivative-orthogonal Riesz wavelets in Sobolev spaces with applications to differential equations. Appl. Computat. Harm. Anal. **47**, 759–794 (2019)
31. Han, B., Shen, Z.: Wavelets with short support. SIAM J. Math. Anal. **38**, 530–556 (2006)
32. Hilber, N., Reichmann, O., Schwab, C., Winter, C.: Computational Methods for Quantitative Finance. Springer, Berlin (2013)
33. Jia, R.Q., Liu, S.T.: Wavelet bases of Hermite cubic splines on the interval. Adv. Comput. Math **25**, 23–39 (2006)
34. Jia, R.Q.: Spline wavelets on the interval with homogeneous boundary conditions. Adv. Comput. Math **30**, 177–200 (2009)
35. Jia, R.Q., Zhao, W.: Riesz bases of wavelets and applications to numerical solutions of elliptic equations. Math. Comput. **80**, 1525–1556 (2011)

36. Kadalbajoo, M.K., Tripathi, L.P., Kumar, A.: Second order accurate IMEX methods for option pricing under Merton and Kou jump-diffusion models. J. Sci. Comput. **65**, 979–1024 (2015)
37. Kwon, Y., Lee, Y.: A second-order finite difference method for option pricing under jump-diffusion models. SIAM J. Numer. Anal. **49**, 2598–2617 (2011)
38. Lax, P.D., Milgram, A.N.: Parabolic equations. Ann. Math. Stud. **33**, 167–190 (1954)
39. Makarov, A.A.: On two algorithms of wavelet decomposition for spaces of linear splines. J. Math. Sci. **232**, 926–937 (2018)
40. Maleknejad, K., Yousefi, M.: Numerical solution of the integral equation of the second kind by using wavelet bases of Hermite cubic splines. Appl. Math. Comput. **183**, 134–141 (2006)
41. Merton, R.C.: Option pricing when underlying stock returns are discontinuous. J. Financ. Econ. **3**, 125–144
42. Micchelli, C.A., Xu, Y., Zhao, Y.: Wavelet Galerkin methods for second-kind integral equations. J. Comput. Appl. Math. **86**, 251–270 (1997)
43. Primbs, M.: New stable biorthogonal spline-wavelets on the interval. Result. Math. **57**, 121–162 (2010)
44. Rostami, Y., Maleknejad, K.: Numerical solution of partial integro-differential equations by using projection method. Mediterr. J. Math. **14**, Article No. 113 (2017)
45. Schneider, A.: Biorthogonal cubic Hermite spline multiwavelets on the interval with complementary boundary conditions. Result. Math. **53**, 407–416 (2009)
46. Shumilov, B.M.: Semi-orthogonal spline-wavelets with derivatives and the algorithm with splitting. Numer. Anal. Appl. **10**, 90–100 (2017)
47. Singh, S., Patel, V.K., Kumar, V., Tohidi, S.E: Numerical solution of nonlinear weakly singular partial integro-differential equation via operational matrices. Appl. Math. Comput. **298**, 310–321 (2017)
48. Singh, S., Patel, V.K., Singh, V.K.: Convergence rate of wavelet collocation method for partial integro-differential equations arising from viscoelasticity. Numer. Meth. Part. Differ. Equ. **34**, 1781–1798 (2018)
49. Stevenson, R.: Divergence-free wavelets on the hypercube: general boundary conditions. Constr. Approx. **44**, 233–267 (2016)
50. Tahernezhad, T., Jalilian, R.: Exponential spline for the numerical solutions of linear Fredholm integro-differential equations. Adv. Differ. Equ. **2020**, Article No. 141 (2020)
51. Urban, K.: Wavelet Methods for Elliptic Partial Differential Equations. Oxford University Press, Oxford (2009)
52. Vatsala, A.S., Wang, L.: The generalized quasilinearization method for parabolic integro-differential equations. Q. Appl. Math. **59**, 459–470 (2001)
53. Winter, C.: Wavelet Galerkin Schemes for Option Pricing in Multidimensional Lévy Models. Ph.D. thesis, ETH Zurich (2009)

Analysis and Spectral Element Solution of Nonlinear Integral Equations of Hammerstein Type

Juarez S. Azevedo, Saulo P. Oliveira, Suzete M. Afonso, and Mariana P. G. da Silva

Abstract We employ the spectral element method with Gauss-Lobatto-Legendre collocation points to approximate nonlinear integral equations of Hammerstein type. Using the Banach Fixed Point Theorem, we establish sufficient conditions for the existence and uniqueness of solutions in the L^2 norm, as well as the convergence of the proposed method, under a different aspect of the existing works in the literature, indicating that the numerical error decays exponentially provided that the kernel function be smooth enough. The iterative Picard process was used to approximate the nonlinear problem. Numerical experiments involving one- and two-dimensional nonlinear equations illustrate the effectiveness of this approach.

Keywords Nonlinear Hammerstein integral equations · Spectral element method · Picard iteration

Mathematics Subject Classification (2010) 45D05 · 45G10 · 46E35 · 65R20

J. S. Azevedo
ICTI-UFBA Centro, Camaçari, BA 42809-000, Brazil
e-mail: jdazevedo@ufba.br

S. P. Oliveira (✉)
DMAT-UFPR and INCT-GP, Curitiba, PR 81531-980, Brazil
e-mail: saulopo@ufpr.br

S. M. Afonso
Universidade Estadual Paulista (UNESP) IGCE, Rio Claro, SP 13506-900, Brazil
e-mail: s.afonso@unesp.br

M. P. G. da Silva
CETEC-UFRB Centro, Cruz das Almas, BA 44380-000, Brazil
e-mail: mpinheiro@ufrb.edu.br

H. Singh et al. (eds.), *Topics in Integral and Integro-Differential Equations*,
Studies in Systems, Decision and Control 340,
https://doi.org/10.1007/978-3-030-65509-9_2

1 Introduction

The goal of this paper is the study of solutions to the nonlinear Hammerstein integral equation

$$u(x) = g(x) + \int_a^b \kappa(x, y) f(y, u(y))\, dy, \quad x \in [a, b], \tag{1}$$

where g, κ, and f are known functions, while u is the solution to be determined. Hammerstein integral equations play a significant role in several applications in physics, engineering, and chemistry [6, 9, 23], and thus have attracted much interest in the last decades, specially regarding existence and uniqueness of solutions and numerical approximation by discrete projection methods.

Analytical studies for the Hammerstein equation can be found in Dolph [12], for a particular case when the kernel is degenerate and symmetrical. More general cases were treated in Figueiredo and Gupta [11] using variational methods. Later on, Karoui and Jawahdou [16] have proved the existence of solutions based mainly on fixed point theorems and successive approximation techniques in L^p, however the authors demand the continuity of the kernel κ. This requirement is waived in our assumptions.

From the numerical point of view, several methods for approximating solutions of Hammerstein equations have been studied in the literature. A Nyström method was proposed by Lardy [20]. A collocation-type method was presented by Kumar and Sloan [19] and its superconvergence properties were obtained by Kumar [18]. High-order convergence may also be achieved through iterated Galerkin methods [15, 24].

Motivated by these analytical and numerical studies, we first seek a theoretical analysis for Hammerstein integral equations with a different strand than the existing works in the literature. Namely, we prove the existence and uniqueness of solutions on the function space $L^2([a, b])$ using the Banach Fixed Point Theorem, which is the natural framework for proving the convergence of schemes based on the Picard iteration.

Indeed, we apply the spectral element method (SEM) with the Gauss-Lobatto-Legendre (GLL) quadrature combined with the Picard iteration to numerically approximate the solutions of (1). To the best of our knowledge, spectral elements have not been used to solve Hammerstein integral equation in the literature. SEM is a high-order Galerkin method that combines the accuracy of spectral methods with the flexibility of finite elements [27] that has been recently adapted to integral equations [25, 30]. We show that the Galerkin solution converges in the L^2-norm with the order $\mathcal{O}(h^{N+1})$, where N is the polynomial degree employed in the approximation and h is the step size. Moreover, this order can be as high as $\mathcal{O}(h^{2N})$ when the error is measured at the GLL quadrature points.

In the next section, we highlight some preliminary concepts and theoretical results about the existence and uniqueness of solutions on the space $L^2([a, b])$. In the sequel, we outline the spectral element method to solve nonlinear integral equation (1) and analyze its convergence, following arguments presented in [26, 28, 29]. Numerical examples are presented to demonstrate the effectiveness of the proposed method, which includes an application to chemical reactor modeling [13, 23].

2 Preliminary Assumptions

Let us introduce some notations and concepts that will be used throughout this work. We recall the Hilbert space $L^2([a, b])$ defined with the usual inner product and its induced norm

$$(u, v) = \int_a^b u(x)v(x)\,dx, \quad \|v\|_0 = (v, v)^{1/2}, \tag{2}$$

along with the spaces

$$H^m([a, b]) = \left\{ v \in L^2([a, b]) : \frac{\partial^k v}{\partial x^k} \in L^2([a, b]),\ 0 \leq k \leq m \right\} \tag{3}$$

equipped with the norm $\|\cdot\|_m$ defined by

$$\|v\|_m^2 = \sum_{k=0}^{m} |v|_k^2, \qquad |v|_k = \left\| \frac{\partial^k v}{\partial x^k} \right\|_0. \tag{4}$$

In order to rewrite the nonlinear Hammerstein integral equation (1) in the form of a functional operator, we define the operator $H = KF$, where K is an integral linear operator expressed by

$$(Ku)(x) = \int_a^b \kappa(x, y)u(y)\,dy, \quad \text{with } \kappa : [a, b] \times [a, b] \longrightarrow \mathbb{R}, \tag{5}$$

and F is the superposition operator $(Fu)(x) = f(x, u(x))$ for each $x \in [a, b]$, i.e.,

$$(Hu)(x) = (KFu)(x) = \int_a^b \kappa(x, y) f(y, u(y))\,dy, \quad \text{for } x \in [a, b]. \tag{6}$$

We establish conditions to guarantee the existence and uniqueness of solutions to the equation

$$u = Au, \tag{7}$$

where the operator A is denoted by

$$Au = g + Hu. \tag{8}$$

Consequently, we obtain the existence and uniqueness of solutions to equation (1), since it takes form (7).

In the next section, the result of existence is based on determining a fixed point for the operator A. For this end, we present the following conditions on the functions g, f and κ:

Assumption 2.1

(A1) $g \in L^2([a, b])$.
(A2) $\kappa \in L^2([a, b] \times [a, b])$, i.e.,

$$\|\kappa\|_0 := \left(\int_a^b \int_a^b |k(x, y)|^2 \, dx \, dy \right)^{1/2} < \infty. \tag{9}$$

(A3) $f : [a, b] \times I\!R \longrightarrow I\!R$ satisfies the Carathéodory conditions and there exist a positive constant $\tau_1 > 0$ and a nonnegative function $\theta_1 \in L^2([a, b])$ such that

$$|f(x, u)| \leq \theta_1(x) + \tau_1 |u|, \quad u \in I\!R, \ x \in [a, b]. \tag{10}$$

(A4) f is Lipschitz with respect to the second variable with Lipschitz constant $\tau_2 > 0$, i.e.,

$$|f(x, u) - f(x, v)| \leq \tau_2 |u - v|, \tag{11}$$

for all $x \in [a, b]$ and $u, v \in \mathbb{R}$.
(A5) The kernel κ and the constant τ_2 satisfy

$$\tau_2 \|\kappa\|_0 < 1. \tag{12}$$

3 Existence and Uniqueness

We begin our study of existence and uniqueness of solutions on $L^2([a, b])$ for the nonlinear Hammerstein integral equation (1) with the following preliminary result:

Lemma 3.1 *If the functions g, f, and κ satisfy conditions (A1)–(A3), then A applies $L^2([a, b])$ on itself, i.e.,*

$$\forall u \in L^2([a, b]) \implies Au \in L^2([a, b]). \tag{13}$$

Proof It suffices to show that the operator H applies $L^2([a, b])$ on itself, since $g \in L^2([a, b])$ by (A1). Given $u \in L^2([a, b])$, using Hölder's inequality and assumptions (A2) and (A3), we have

$$\begin{aligned} \|Hu\|_0^2 &= \left(\int_a^b |(KFu)(x)|^2 \, dx \right) \\ &= \int_a^b \left| \int_a^b \kappa(x, y) f(y, u(y)) \, dy \right|^2 dx \end{aligned}$$

$$
\begin{aligned}
&\leq \int_a^b \left[\left(\int_a^b |\kappa(x,y)|^2 \, dy \right)^{1/2} \left(\int_a^b |f(y,u(y))|^2 \, dy \right)^{1/2} \right]^2 dx \\
&= \int_a^b \left(\int_a^b |\kappa(x,y)|^2 \, dy \right) dx \cdot \int_a^b |f(y,u(y))|^2 \, dy \\
&\leq \|\kappa\|_0^2 \int_a^b |\theta_1(y) + \tau_1 \, |u(y)| \, |^2 \, dy \\
&= \|\kappa\|_0^2 \left(\int_a^b |\theta_1(y)|^2 \, dy + 2\tau_1 \int_a^b |\theta_1(y)||u(y)| \, dy + \tau_1^2 \int_a^b |u(y)|^2 \, dy \right) \\
&\leq \|\kappa\|_0^2 \left(\int_a^b |\theta_1(y)|^2 \, dy + 2\tau_1 \|\theta_1\|_0 \|u\|_0 + \tau_1^2 \int_a^b |u(y)|^2 \, dy \right) \\
&= \|\kappa\|_0^2 \left(\|\theta_1\|_0 + \tau_1 \, \|u\|_0 \right)^2 .
\end{aligned}
$$

Therefore, $\|Hu\|_0 < \infty$ and $Hu \in L^2([a,b])$. □

The next result guarantees the existence and uniqueness of solution to nonlinear integral equation (1) in $L^2([a,b])$. Its proof is based on Banach Fixed Point Theorem. For more details on this fixed point theorem, we refer the reader to [21].

Theorem 3.1 *Consider the integral equation* (1)*, where g, f, and k satisfy conditions (A1)–(A5). Then, there exists a unique function $u \in L^2([a,b])$ such that $u = Au$.*

Proof It follows from Lemma 3.1 that A maps $L^2([a,b])$ to $L^2([a,b])$. By Banach Fixed Point Theorem, it is enough to prove that A is a contraction. Indeed, given $u, v \in L^2([a,b])$, using the Holder inequality and condition (A4), we obtain

$$
\begin{aligned}
&||Au - Av||_0^2 = \|Hu - Hv\|_0^2 \\
&= \left(\int_a^b |(KFu)(x) - (KFv)(x)|^2 \, dx \right) \\
&= \int_a^b \left| \int_a^b \kappa(x,y)[f(y,u(y)) - f(y,v(y))] \, dy \right|^2 dx \\
&\leq \int_a^b \left[\left(\int_a^b |\kappa(x,y)|^2 \, dy \right)^{1/2} \left(\int_a^b |f(y,u(y)) - f(y,v(y))|^2 \, dy \right)^{1/2} \right]^2 dx \\
&= \int_a^b \left(\int_a^b |\kappa(x,y)|^2 \, dy \right) dx \cdot \int_a^b |f(y,u(y)) - f(y,v(y))|^2 \, dy \\
&\leq \|\kappa\|_0^2 \left(\tau_2^2 \int_a^b |u(y) - v(y)|^2 \, dy \right) \\
&= \tau_2^2 \|\kappa\|_0^2 \|u - v\|_0^2,
\end{aligned}
$$

whence it follows that

$$\|Au - Av\|_0 \leq \tau_2 \|\kappa\|_0 \|u - v\|_0. \tag{14}$$

Condition (A5) implies that A is a contraction, which completes the proof. □

4 Spectral Element Approximation

Let us present the numerical implementation of the spectral element method for (1). We seek the Galerkin approximation $u_h \in V_h$ such that

$$\langle u_h, v_h \rangle = \langle g, v_h \rangle + \langle KFu_h, v_h \rangle \quad \forall\, v_h \in V_h, \tag{15}$$

where the finite-dimensional space $V_h = \text{span}\{\phi_1, \phi_2, \ldots, \phi_n\}$ is defined by the Lagrange piecewise polynomial functions ϕ_I $(1 \leq I \leq n)$ of degree N satisfying the relation $\phi_I(x_J) = \delta_{I,J}$.

The spectral element nodes x_J are defined as follows: we partition the interval $[a, b]$ into N_e elements with uniform spacing $h := (b - a)/N_e$ having $N + 1$ nodes (endpoint nodes are shared with adjacent elements) leading to a total of $n = N_e N + 1$ nodes. For any element index $1 \leq e \leq N_e$ and local index $0 \leq j \leq N$, the global nodal index is $IEN(j, e) = j + (e - 1)N + 1$. The global nodes are $x_J = F^e(\xi_j)$, where $J = IEN(j, e)$, F^e is the linear transformation from $[-1, 1]$ to the e-th element, and ξ_j is the j-th GLL collocation point.

The Gauss-Lobatto-Legendre collocation points ξ_j are the solutions to

$$(1 - \zeta^2)L'_N(\zeta) = 0, \tag{16}$$

where $L_N(\cdot)$ is the N-th degree Legendre polynomial. These points also provide the following compound quadrature rule:

$$\int_a^b v(x)\, dx \approx \sum_{e=1}^{N_e} \sum_{j=0}^{N} \hat{w}_j v(F^e(\xi_j)) J^e = \sum_{J=1}^{n} w_J v(x_J), \tag{17}$$

which is exact for piecewise polynomials of degree $2N - 1$, see [7]. The local weights are $\hat{w}_j = 2/(N(N + 1)L_N(\xi_j)^2)$, while J^e is the determinant of the Jacobian matrix of the transformation F^e. The global weights w_J are

$$w_J = \sum_{(j,e)\in IEN_J} \hat{w}_j J^e, \quad IEN_J = \{(j, e) : J = IEN(j, e)\}. \tag{18}$$

From quadrature (17), we define the discrete bilinear form

$$\langle u, v\rangle_{\mathrm{w}} = \sum_{J=1}^{n} w_J u(x_J) v(x_J). \tag{19}$$

Let us also define

$$\|v\|_{\mathrm{w}} = \sqrt{\langle v, v\rangle_{\mathrm{w}}}. \tag{20}$$

The SEM approximation corresponds to the discrete Galerkin equations obtained by applying the GLL quadrature (17) to (15):

$$\langle u_h, v_h\rangle_{\mathrm{w}} = \langle g, v_h\rangle_{\mathrm{w}} + \langle K_h F u_h, v_h\rangle_{\mathrm{w}} \quad \forall\, v_h \in V_h, \tag{21}$$

where K_h is the GLL quadrature approximation to (5). If we look for a Galerkin solution u_h of (21) in the form

$$u_h(x) = \sum_{J=1}^{n} u_J \phi_J(x), \tag{22}$$

and choose the test functions $v_h = \phi_I$, $1 \le I \le n$, we find the nonlinear system

$$\sum_{J=1}^{n} u_J \langle \phi_J, \phi_I\rangle_{\mathrm{w}} = \langle g, \phi_I\rangle + \left\langle K_h F\left(\sum_{J=1}^{n} u_J \phi_J\right), \phi_I\right\rangle_{\mathrm{w}} \quad (1 \le I \le n). \tag{23}$$

Recalling that $\phi_I(x_J) = \delta_{I,J}$, nonlinear equations (23) reduce to the collocation-type method

$$u_I = g(x_I) + \sum_{J=1}^{n} w_J \kappa(x_I, x_J) f(x_J, u_J), \quad I = 1, \dots, n. \tag{24}$$

Note that the expression on the right-hand side depends on the unknowns $\{u_J\}$, but not on the basis functions $\{\phi_J\}$ $(1 \le J \le n)$.

The Picard iteration [4, 28] and the Newton-Kantorovich method [14] are the most used iterative processes for nonlinear algebraic systems such as (24). Here we make use of the Picard iteration. Let $\boldsymbol{u} = [u_1, \dots, u_n]^T$ be the vector of coefficients of the Galerkin solution (22). We can rewrite (24) in vector form as

$$\boldsymbol{u} = \boldsymbol{A}(\boldsymbol{u}), \quad \boldsymbol{A}(\boldsymbol{u}) = \boldsymbol{g} + \boldsymbol{K}\boldsymbol{F}(\boldsymbol{u}), \tag{25}$$

with

$$\begin{cases} F_J(\mathbf{y}) = f(x_J, y_J), \\ g_I = g(x_I). \\ K_{I,J} = w_J \kappa(x_I, x_J), \qquad 1 \le I, J \le n. \end{cases} \tag{26}$$

The Picard iteration for (25) is simply

$$\boldsymbol{u}^{(k+1)} = \boldsymbol{A}(\boldsymbol{u}^{(k)}), \quad k = 0, 1, 2, \ldots, \tag{27}$$

where $\boldsymbol{u}^{(k)} = [u_1^{(k)}, \ldots, u_n^{(k)}]^T$.

Remark 4.1 The Galerkin method for the Hammerstein equation (1) can be numerically expensive to implement [2]. However, the reduction of equations (23) to the collocation equations (24) through the use of GLL quadrature decreases dramatically its computational cost.

Remark 4.2 The extension of the SEM described in the previous section towards two-dimensional problems is straightforward. In the particular case of a square domain, the basis functions are constructed from the one-dimensional tensor product, i.e.,

$$\phi_{IJ} = \phi_I \phi_J, \quad I, J = 1, \ldots, n. \tag{28}$$

The weights w_J of GLL quadrature are also a tensor-product extension of the one-dimensional quadrature (see [25] for details) defined by

$$w_{IJ} = w_I w_J, \quad I, J = 1, \ldots, n. \tag{29}$$

Thus the algebraic system given in (24) reads

$$u_{I'} = g(\boldsymbol{x}_{I'}) + \sum_{J'=1}^{n^2} w_{J'} \kappa(\boldsymbol{x}_{I'}, \boldsymbol{x}_{J'}) f(\boldsymbol{x}_{J'}, u_{J'}), \tag{30}$$

such that $u_{I'} = u_{IJ}$, and $\boldsymbol{x}_{I'} = (x_I, x_J)$, $I, J = 1, \ldots, n$. A similar construction can be obtained for three-dimensional problems. More general domains are handled with quadrilateral/hexahedral meshes, where the discrete orthogonality relation $\langle \phi'_J, \phi'_I \rangle_{\mathrm{w}} = 0$ $(J' \neq I')$ remains valid (see [17]).

5 Convergence Analysis

Under the same assumptions of Theorem 3.1 and additional regularity assumptions, we will show that the approximate solution obtained by the spectral element method (27) converges to the solution of (1).

In order to relate the coefficient vector $\mathbf{u}^{(k)} = [u_1^{(k)}, \ldots, u_n^{(k)}]^T$ from (27) and the exact solution u, which are not in the same vector space, we consider the function $u_h^{(k)} \in V_h$ defined by expression

$$u_h^{(k)}(x) = \sum_{J=1}^{n} u_J^{(k)} \phi_J(x). \tag{31}$$

Let us denote the global error as

$$e_{h,k} = \|u - u_h^{(k)}\|_0; \tag{32}$$

we can estimate it with the aid of the triangle inequality

$$e_{h,k} \le \|u - u_h\|_0 + \|u_h - u_h^{(k)}\|_0, \tag{33}$$

where u_h is the exact solution of (24).

In inequality (33), the first term on the right-hand side represents the error of the spectral element method, while the second term takes into account the error of the Picard iteration. We must analyze each term of inequality (33) separately.

5.1 Convergence of the Spectral Element Method

Let us review some useful lemmas that are important to the convergence analysis. Let us set C as a generic positive constant that does not depend on the element length h.

Lemma 5.1 ([8, Theorem 3.2.1]) *The interpolation operator*

$$I_h v(x) = \sum_{J=1}^{n} v(x_J)\phi_J(x) \tag{34}$$

satisfies the following estimate for any $v \in H^s([a,b])$ *with* $s \le N+1$*:*

$$\|v - I_h v\|_0 \le C h^s |v|_s. \tag{35}$$

Lemma 5.2 *If* $u, v \in H^k([a,b])$ *with* $k \le 2N$*, then*

$$|\langle u, v\rangle - \langle u, v\rangle_{\mathrm{w}}| \le C h^k |u|_k |v|_k, \tag{36}$$

In particular, we have

$$\left| \|u\|_0^2 - \|u\|_{\mathrm{w}}^2 \right| \le C h^k |u|_k^2, \tag{37}$$

Moreover, the two-dimensional counterpart of (37) *also holds if* $u \in H^k([a,b]^2)$.

Proof Recalling that the GLL quadrature is exact for piecewise polynomials of degree $2N-1$, the above error estimates can be verified at each element through Bramble-Hilbert Lemma, as in [1, Theorem 3.5]. □

Lemma 5.3 ([1, Lemma 3.2]) *The norm* $\|\cdot\|_0$ *is equivalent to* $\|\cdot\|_{\mathrm{w}}$ *in space* V_h, *i.e., there exist* $C_1, C_2 > 0$ *such that*

$$C_1\|v_h\|_0^2 \leq \|v_h\|_{\mathrm{w}}^2 \leq C_2\|v_h\|_0^2 \quad \forall v_h \in V_h. \tag{38}$$

We are now going to prove a theorem that provides us an error estimate for the first term of expression (33):

Theorem 5.1 *Let* u *be the exact solution of equation* (1) *and let* u_h *be the solution of the nonlinear system* (24). *If* $f : [a, b] \times I\!R \longrightarrow I\!R$ *satisfies Part (A4) of Assumption 2.1,* $\kappa \in H^p([a, b]^2)$, *and* $u \in H^p([a, b])$ *with* $p \leq 2N$, *then there exists* $h_0 > 0$ *such that, for* $h \leq h_0$,

$$\|u - u_h\|_{\mathrm{w}} \leq Ch^p|u|_p, \tag{39}$$

$$\|u - u_h\|_0 \leq Ch^s|u|_s, \quad s = \min(p, N+1). \tag{40}$$

Proof The proof is structured similarly as in [29, Theorem 4.1]. We have from (1) and (24) that

$$u(x_I) - u_I = S_I, \quad S_I = \langle \kappa(x_I, \cdot), f(\cdot, u)\rangle - \langle \kappa(x_I, \cdot), f(\cdot, u_h)\rangle_{\mathrm{w}}, \tag{41}$$

for $1 \leq I \leq n$. Note from (41) that

$$\|u - u_h\|_{\mathrm{w}} = \left(\sum_{I=1}^{n} w_I|S_I|^2\right)^{1/2}. \tag{42}$$

We must find an upper bound for the right-hand side of (42). We have from Lemma 5.2 and conditions (A3) and (A4) that

$$\begin{aligned}
|S_I| &= \left|\langle \kappa(x_I, \cdot), f(\cdot, u)\rangle - \langle \kappa(x_I, \cdot), f(\cdot, u_h)\rangle_{\mathrm{w}}\right| \\
&\leq \left|\langle \kappa(x_I, \cdot), f(\cdot, u)\rangle - \langle \kappa(x_I, \cdot), f(\cdot, u)\rangle_{\mathrm{w}}\right| + \left|\langle \kappa(x_I, \cdot), f(\cdot, u) - f(\cdot, u_h)\rangle_{\mathrm{w}}\right| \\
&\leq Ch^p|\kappa(x_I, \cdot)|_p|u|_p + \left|\sum_{J=1}^{n} w_J\kappa(x_I, x_J)[f(x_J, u(x_J)) - f(x_J, u_h(x_J))]\right| \\
&\leq \tau_2\left(\sum_{J=1}^{n} w_J\,|\kappa(x_I, x_J)|^2 \sum_{J=1}^{n} w_J\,|u(x_J) - u_J|^2\right)^{1/2} + Ch^p|\kappa(x_I, \cdot)|_p|u|_p \\
&= \tau_2\|u - u_h\|_{\mathrm{w}}\left(\sum_{J=1}^{n} w_J\,|\kappa(x_I, x_J)|^2\right)^{1/2} + Ch^p|\kappa(x_I, \cdot)|_p|u|_p.
\end{aligned}$$

It follows from Minkowski's inequality that

$$\left(\sum_{I=1}^{n} w_I |S_I|^2\right)^{1/2} \leq \tau_2 \|u - u_h\|_{\mathrm{w}} \left(\sum_{I,J=1}^{n} w_I w_J \, |\kappa(x_I, x_J)|^2\right)^{1/2} + Ch^p |u|_p \left(\sum_{I=1}^{n} w_I |\kappa(x_I, \cdot)|_p^2\right)^{1/2},$$

and using again Lemma 5.2, we have that there exists $\tilde{C} > 0$ such that

$$\left(\sum_{I=1}^{n} w_I |S_I|^2\right)^{1/2} \leq \tau_2 \|u - u_h\|_{\mathrm{w}} \left(\|\kappa\|_0 + \tilde{C}h^{p/2}|\kappa|_p\right) + Ch^p |u|_p \left(\left\|\frac{\partial^p \kappa}{\partial y^p}\right\|_0 + \tilde{C}h^{p/2}|\kappa|_p\right). \tag{43}$$

Thus, there exists a constant C depending on κ such that

$$\left(1 - \tau_2\|\kappa\|_0 - \tau_2\tilde{C}h^p|\kappa|_p\right) \|u - u_h\|_{\mathrm{w}} \leq Ch^p|u|_p, \tag{44}$$

and estimate (39) follows from part (A5) if $h \leq h_0$, where h_0 is such that $0 < \tau_2\tilde{C}h_0^p|\kappa|_p < 1 - \tau_2\|\kappa\|_0$. On the other hand, let $s = \min(p, N+1)$. It follows from the triangle inequality and Lemmas 5.1 and 5.3 that

$$\|u - u_h\|_0 \leq \|u - I_h u\|_0 + \|I_h u - u_h\|_0 \leq Ch^s|u|_s + C_2\|I_h u - u_h\|_{\mathrm{w}}. \tag{45}$$

Noting that since $\|I_h u - u_h\|_{\mathrm{w}} = \|u - u_h\|_{\mathrm{w}}$, we conclude that estimate (39) follows from (45) and (40). □

5.2 Convergence of Picard Iteration

In the following we study the error in matrix formulation (25) due to the iterative process (27). For this purpose, we introduce the following vector norm associated with the GLL quadrature:

$$\|\boldsymbol{v}\|_w^2 = \sum_{I=1}^{n} w_I |v_I|^2. \tag{46}$$

Theorem 5.2 *Under Assumption 2.1, the operator* $\boldsymbol{A} : I\!R^n \longrightarrow I\!R^n$ *defined by*

$$\boldsymbol{A}(\boldsymbol{u}) = \boldsymbol{g} + \boldsymbol{K}\boldsymbol{F}(\boldsymbol{u}). \tag{47}$$

is a contraction with respect to the norm $\|\cdot\|_w$.

Proof By definition of the operator **A** and norm $\|\cdot\|_w$, we have

$$\|\boldsymbol{A}(\boldsymbol{u}) - \boldsymbol{A}(\boldsymbol{v})\|_w^2 = \sum_{I=1}^{n} w_I \left| \sum_{J=1}^{n} w_J \kappa(x_I, x_J) f(x_J, u_J) - \sum_{J=1}^{n} w_J \kappa(x_I, x_J) f(x_J, v_J) \right|^2. \tag{48}$$

Using condition (A4), Hölder inequality, and Lemma 5.2, it follows that

$$\begin{aligned}
\|\boldsymbol{A}(\boldsymbol{u}) - \boldsymbol{A}(\boldsymbol{v})\|_w^2 &= \sum_{I=1}^{n} w_I \left| \sum_{J=1}^{n} w_J \kappa(x_I, x_J)(f(x_J, u_J) - f(x_J, v_J)) \right|^2 \\
&\leq \tau_2^2 \sum_{I=1}^{n} w_I \left(\sum_{J=1}^{n} w_J \, |\kappa(x_I, x_J)|^2 \right) \left(\sum_{J=1}^{n} w_J |u_J - v_J|^2 \right) \\
&= \tau_2^2 \left(\sum_{I,J=1}^{n} w_I w_J |\kappa(x_I, x_J)|^2 \right) \|\boldsymbol{u} - \boldsymbol{v}\|_w^2 \\
&= M \|\boldsymbol{u} - \boldsymbol{v}\|_w^2,
\end{aligned}$$

where

$$M = \tau_2^2 \left(\|\kappa\|_0^2 + C h^{p/2} |\kappa|_p \right). \tag{49}$$

Taking h sufficiently small and recalling from (A5) that $\tau_2 \|\kappa\|_0 < 1$, we have $M < 1$, i.e., **A** is a contraction. □

Corollary 5.1 *The sequence* $\{u_h^{(k)}\}_k$ *defined in* (31) *converges to the solution* u_h *of* (15) *for any initial guess* $u_h^{(0)}$. *Moreover,*

$$\|u_h - u_h^{(k)}\|_0 \leq C \frac{M^k}{1-M} r_0, \quad r_0 = \|A u_h^{(0)} - u_h^{(0)}\|_0, \tag{50}$$

where M *is given by* (49).

Proof From Banach Fixed Point Theorem, the sequence $\{\boldsymbol{u}^{(k)}\}_{k \in \mathbb{N}}$ generated by Picard iteration (27) converges to $\boldsymbol{u}$ in (25) for any $k \geq k_0$ and any initial guess $\boldsymbol{u}^{(0)}$. Moreover,

$$\|\boldsymbol{u} - \boldsymbol{u}^{(k)}\|_w \leq \frac{M^k}{1-M} \|A\boldsymbol{u}^{(0)} - \boldsymbol{u}^{(0)}\|_w. \tag{51}$$

Since $\|v_h\|_w = \|\boldsymbol{v}\|_w$ if $\boldsymbol{v} = [v_1, \ldots, v_n]^T$ and v_h are such that

$$v_h(x) = \sum_{I=1}^{n} v_I \phi_I(x),$$

the proof follows from Lemma 5.3. □

5.3 *Global Convergence of the Error*

We are now ready to state the main result of this section:

Theorem 5.3 *Let Assumption 2.1 hold and assume $u \in H^p([a, b])$ $(1 \le p \le N + 1)$ is the solution of* (1)*. There exists $h_0 > 0$ such that, for any $h \le h_0$, the total error $e_{h,k}$ defined in* (32) *satisfies*

$$e_{h,k} \le Ch^p|u|_p, \quad k \ge k_0 \tag{52}$$

for some $k_0 = k_0(h)$.

Proof Choosing $k_0 \le p \log_M h$, inequality (52) follows from (33), (39), and (50). □

Remark 5.1 Analogously to Theorem 5.3, if $u \in H^p([a, b])$ with $1 \le p \le 2N$, then $\|u - u_h^{(k)}\|_w \le Ch^p|u|_p$ for k sufficiently large.

6 Numerical Experiments

To show accuracy and efficiency of the method, we give some numerical examples of the nonlinear integral equation (1) that illustrate the proposed theory. To evaluate the convergence of the method, the error was computed using the following relative measures:

$$e_k = \frac{e_{h,k}}{\|u\|_0} = \frac{\|u - u_h^{(k)}\|_0}{\|u\|_0}, \quad E_k = \frac{\|u - u_h^{(k)}\|_w}{\|u\|_w}. \tag{53}$$

Furthermore, we employ the following stopping criteria:

$$k \le k_{max}, \; r_k = \|\boldsymbol{u}^{(k)} - \boldsymbol{u}^{(k-1)}\|_w > tol_1, \; \|\boldsymbol{u}^{(k-1)}\|_w < tol_2, \tag{54}$$

with $k_{max} = 1000$, $tol_1 = 10^{-15}$, and $tol_2 = 10\|\boldsymbol{u}^{(0)}\|_w$.

We approximate $\|\cdot\|_0$ by the GLL quadrature of degree 12 with 512 elements. We take as initial guess the constant function $u_h^{(0)}(x) = 2$, unless specified otherwise.

6.1 Smooth Initial Data

Consider the Hammerstein integral equation (1) with $[a, b] = [0, 2]$, $\kappa(x, y) = \sin(x + y)$, and $f(x, u) = |u|/10$. The function g is chosen such that the exact solution is $u(x) = \sin(x)$, i.e.,

$$g(x) = \sin(x) - (\cos(x)/10 - \cos(2 + x)\sin(2)/20)\,. \tag{55}$$

The functions f, g, and κ satisfy the conditions of Assumption 2.1. More precisely, g satisfies (A1), κ satisfies (A2), where $\|\kappa\|_0 \approx 1.604$, and f satisfies (A3) and (A4), with $\theta_1(x) = 0$ for all $x \in [0, 2]$, $\tau_1 = 1/10$, and $\tau_2 = 1/10$. Moreover, condition (A5) is also satisfied with $\tau_2\|\kappa\|_0 \approx 0.1604 < 1$. Therefore, by Theorem 3.1, we can ensure the existence and uniqueness of the solution to the Hammerstein integral equation considered.

Note that the exact solution u and the kernel κ are infinitely differentiable, in accordance with the conditions of Theorem 5.3. The code is validated by a comparison of the simulation results with the collocation method with piecewise linear basis functions [3, 28].

We verify in Fig. 1 the error dependence on the mesh refinement h and the polynomial degree N in semi-log scale. We can observe that the relative errors e_k and E_k decay with order $O(h^{N+1})$ and $O(h^{2N})$, respectively, according to slopes of the linear regression lines. The results show a good agreement with Theorem 5.3. A slight deviation occurs in the relative error E_k for $N = 4$, where the relative error E_k at the finer mesh is near the machine epsilon. Moreover, the relative errors e_k and E_k of the collocation method (Coll) are almost identical to the GLL method with $N = 1$.

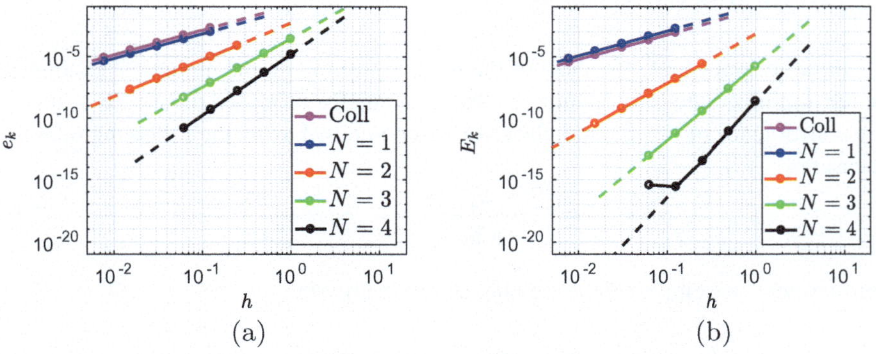

Fig. 1 Relative errors e_k (**a**) and E_k (**b**) in Sect. 6.1. The slopes of the dashed lines for e_k are 2.0 (Coll), 2.0 ($N = 1$), 3.0 ($N = 2$), 4.0 ($N = 3$) and 5.0 ($N = 4$), while those for E_k are 2.0 (Coll), 2.0 ($N = 1$), 4.0 ($N = 2$), 6.0 ($N = 3$) and 8.0 ($N = 4$)

6.2 Smooth Initial Data, Unknown Exact Solution

Now, we consider the Hammerstein integral equation (1), with $[a, b] = [0, \pi]$ and $\kappa(x, y)$ being a Gaussian kernel written as

$$\kappa(x, y) = e^{-\lambda^2(x-y)^2}, \tag{56}$$

where the parameter λ defines the length scale, while $f(x, u) = \exp(x) + \sin(u)/3$, and $g(x) = \sqrt{x^2 + \log(x + 1) + \exp(x)}$.

In this example, the exact solution is not known. However, such a solution exists and is unique, since all conditions of Theorem 3.1 are fulfilled. Note that conditions (A3) and (A4) are satisfied with $\theta_1(x) = \exp(x)$ for all $x \in [0, \pi]$, with $\tau_1 = 1/3$, and $\tau_2 = 1/3$. Furthermore, if $\lambda = 1$, for instance, condition (A5) is satisfied with $\tau_2\|\kappa\|_2 \approx 0.7124 < 1$, and if $\lambda = 10$, condition (A5) is satisfied with $\tau_2\|\kappa\|_2 \approx 0.2464 < 1$.

Here, the reference solution is computed by the GLL spectral element method of degree $N = 12$ using $N_e = 512$ elements.

In Fig. 2, we study how the length scale λ affects the relative error e_k. The error increases with λ regardless of the polynomial degree, which is a consequence of the steeper variation of the kernel κ. Nevertheless, an increase of the degree N renders the error e_k less sensitive to λ, as the lower-error plateaus become wider as N increases.

We proceed to the numerical convergence analysis of relative errors e_k and E_k for $\lambda = 1$ and $\lambda = 10$, shown in Figs. 3 and 4, respectively. Since the assumptions of Theorem 5.3 are satisfied, it is reasonable to expect that numerical approximations u_h properly converge to solution u. In the case of $\lambda = 1$ the numerical experiments suggest an order of convergence $\mathcal{O}(h^{N+1})$ in norm $\|\cdot\|_0$ and $\mathcal{O}(h^{2N})$ in norm $\|\cdot\|_w$, validating our theoretical results. For $\lambda = 10$, Fig. 4 indicates that the convergence rate begins to degrade, confirming the results of Fig. 2.

Fig. 2 Relative error e_k of Sect. 6.2 in terms of the parameter λ, with $N_e = 100$

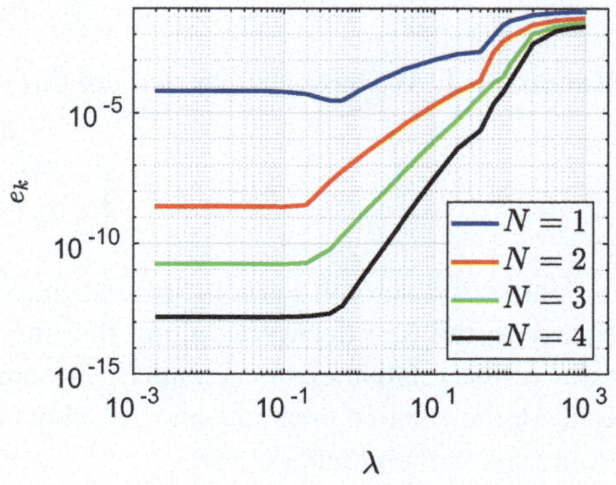

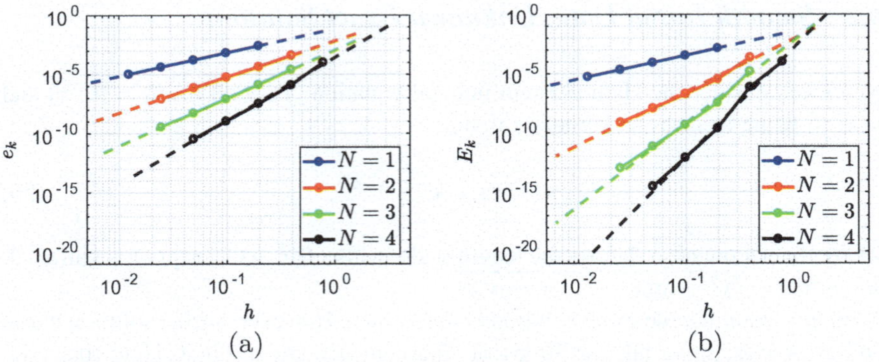

Fig. 3 Relative errors e_k (**a**) and E_k (**b**) in Sect. 6.2 with $\lambda = 1$. The slopes of the dashed lines for e_k are 2.0 ($N = 1$), 3.0 ($N = 2$), 4.0 ($N = 3$) and 5.3 ($N = 4$), while those for E_k are 2.0 ($N = 1$), 4.5 ($N = 2$), 7.0 ($N = 3$) and 9.2 ($N = 4$)

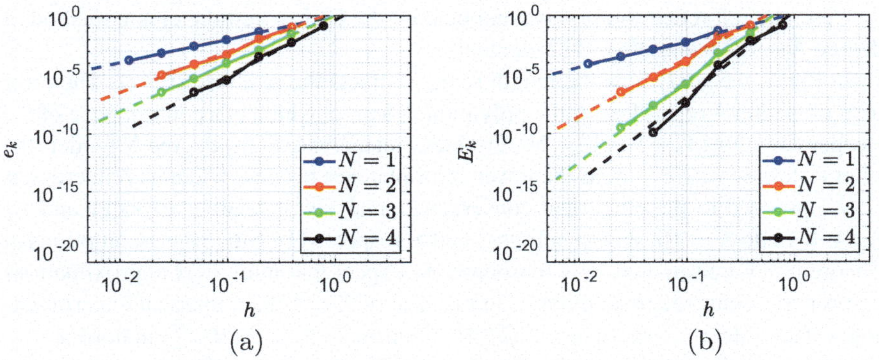

Fig. 4 Relative errors e_k (**a**) and E_k (**b**) in Sect. 6.2 with $\lambda = 10$. The slopes of the dashed lines for e_k are 2.0 ($N = 1$), 3.4 ($N = 2$), 4.0 ($N = 3$) and 4.8 ($N = 4$), while those for E_k are 2.2 ($N = 1$), 4.9 ($N = 2$), 6.8 ($N = 3$) and 7.2 ($N = 4$)

Remark 6.1 We have also carried out this experiment with the kernel

$$k(x, y) = \begin{cases} -y(1-x) & y \leq x, \\ -x(1-y) & y > x. \end{cases} \tag{57}$$

Unlike the smooth kernel (56), the function $\kappa(x, y)$ in (57) has a jump discontinuity in the first derivative along the line $x = y$. Consequently, the convergence rates of the relative errors e_k and E_k are not expected to be the same as in Sect. 5. Indeed, the relative errors e_k and E_k, shown in Fig. 5, have order $\mathcal{O}(h^2)$ regardless of the spectral element degree.

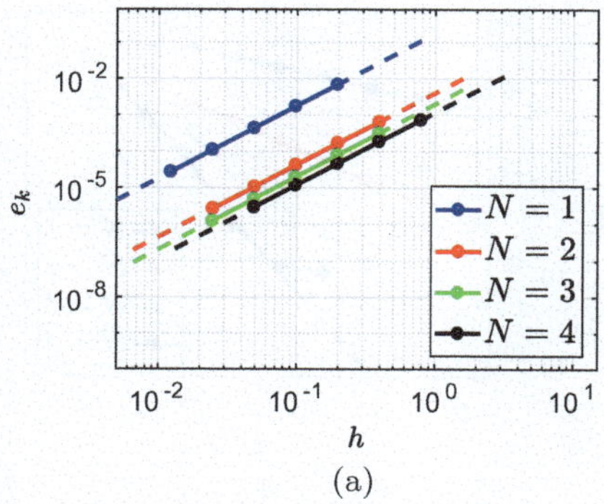

(a)

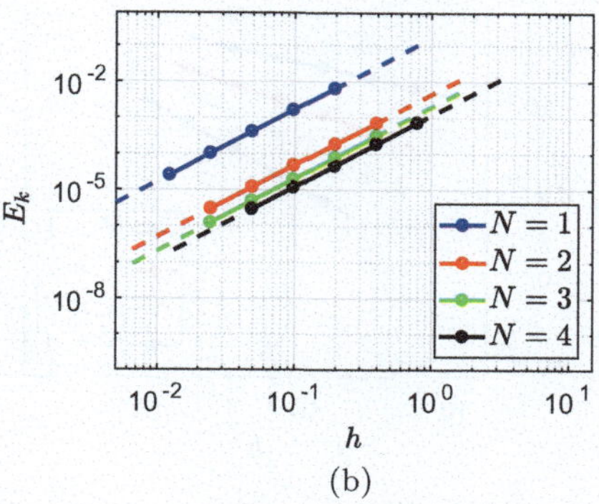

(b)

Fig. 5 Relative errors e_k (**a**) and E_k (**b**) in Sect. 6.2 with the non-smooth kernel (57). The slopes of the dashed lines for both e_k and E_k are nearly equal to 2.0

6.3 *Two-Dimensional Hammerstein Equation*

In order to consider an artificial example in the two-dimensional case, Assumption 2.1 must be rewritten as follows:

Assumption 6.1

(A1) $g \in L^2([a, b] \times [a, b])$.
(A2) $\kappa \in L^2([a, b]^2 \times [a, b]^2)$, i.e.,

$$\|\kappa\|_0 := \left(\int_a^b \int_a^b \int_a^b \int_a^b |k(x, y, t, s)|^2 \, dx \, dy \, dt \, ds \right)^{1/2} < \infty. \tag{58}$$

(A3) $f : [a, b] \times [a, b] \times \mathbb{R} \longrightarrow \mathbb{R}$ satisfies the Carathéodory conditions and there exist a positive constant $\tau_1 > 0$ and a nonnegative function $\theta_1 \in L^2([a, b] \times [a, b])$ such that

$$|f(\mathbf{x}, u)| \leq \theta_1(\mathbf{x}) + \tau_1|u|, \quad u \in \mathbb{R}, \ \mathbf{x} \in [a, b] \times [a, b]. \tag{59}$$

(A4) f is Lipschitz with respect to the second variable with Lipschitz constant $\tau_2 > 0$, that is,

$$|f(\mathbf{x}, u) - f(\mathbf{x}, v)| \leq \tau_2|u - v| \tag{60}$$

for all $\mathbf{x} \in [a, b] \times [a, b]$ and $u, v \in \mathbb{R}$.
(A5) The kernel κ and the constant τ_2 satisfy

$$\tau_2\|\kappa\|_0 < 1. \tag{61}$$

Similarly to [5], we consider the Hammerstein integral equation

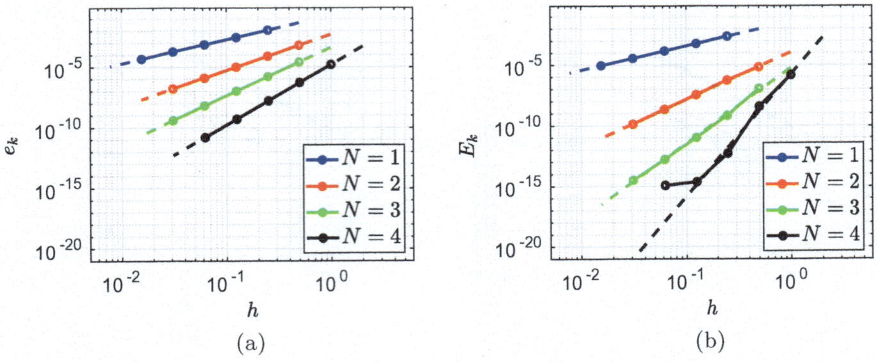

Fig. 6 Relative errors e_k (**a**) and E_k (**b**) in Sect. 6.3. The slopes of the dashed lines for e_k are 2.0 ($N = 1$), 3.0 ($N = 2$), 4.0 ($N = 3$) and 5.0 ($N = 4$), while those for E_k are 2.0 ($N = 1$), 3.9 ($N = 2$), 6.2 ($N = 3$) and 10.0 ($N = 4$)

$$u(x, y) = g(x, y) + \int_0^1 \int_0^1 \frac{\sin(t + s)}{(1 + x)(1 + y)} \frac{1}{1 + u(t, s)^2} \, dtds, \tag{62}$$

where the source term

$$g(x, y) = \sin(x + y) - \frac{c}{(1 + x)(1 + y)}, \quad c = 0.458543000997359. \tag{63}$$

is chosen so that the exact solution is $u(x, y) = \sin(x + y)$. Here, we have $\kappa(x, y, t, s) = \sin(t + s)/(1 + x)(1 + y)$ and $f(\mathbf{x}, u) = 1/(1 + u^2)$.

Note that f, g, and κ satisfy the conditions of Assumption 6.1, with $\|\kappa\|_0 \approx 0.608$, $\theta_1(\mathbf{x}) = 3/2$ for all $\mathbf{x} \in [0, 1] \times [0, 1]$, $\tau_1 = 1$, and $\tau_2 = 1$, so a two-dimensional version of Theorem 3.1 can be easily built to show that equation (62) admits a unique solution.

We employ a two-dimensional spectral element approximation as described in Remark 4.2. Figure 6 shows the relative errors e_k and E_k, using in both x and y directions the same polynomial degree as in the previous examples.

Note that convergence rates in the two-dimensional case are similar to those obtained in the one-dimensional case. In particular, since the initial data and the exact solution are smooth in the domain $[0, 1] \times [0, 1]$, the convergence rates are nearly the same as in Sect. 6.1.

6.4 *Application to Chemical Reactor Theory*

The least example concerns the model for an adiabatic tubular chemical reactor which processes an irreversible exothermic chemical reaction [23]. The model in steady state can be reduced to the boundary-value problem

Table 1 Numerical solution $u_h(x)$ of Sect. 6.2 obtained by the spectral element with $N = 1$ to $N = 3$. The parameters are $\lambda = 10$, $\beta = 3$, and $\mu = 0.02$ (cf. [23, Table 1])

x	$N = 1$	$N = 2$	$N = 3$
0.0	0.006079717752434	0.006048424440590	0.006048393273618
0.2	0.018224782011082	0.018193073184659	0.018192995033853
0.4	0.030456964483686	0.030424894946446	0.030424768822984
0.6	0.042701413585823	0.042669432192967	0.042669257503415
0.8	0.054400608254753	0.054372052651128	0.054371831831772
1.0	0.061459804251372	0.061459182257727	0.061458939882039

$$\begin{cases} u'' - \lambda u' + \mu(\beta - u)\exp(u) = 0, \\ u'(0) = \lambda u(0), \quad u'(1) = 0. \end{cases} \tag{64}$$

Applying the Green's function integral equation method to (64) leads to the Hammerstein integral equation

$$u(x) = \int_0^1 \kappa(x, y)\mu(\beta - u(y))\exp(u(y))\,dy, \quad x \in [0, 1]. \tag{65}$$

whose kernel is given by

$$k(x, y) = \begin{cases} e^{\lambda(x-y)}, & 0 \le x < y, \\ 1, & y \le x \le 1. \end{cases} \tag{66}$$

Let us first compare the numerical solutions obtained by the proposed method with the ones reported in [23], where the equation parameters were $\lambda = 10$, $\beta = 3$, and $\mu = 0.02$ and the element size was $h = 0.025$. Moreover, we employ the polynomial degrees $N = 1, 2$, and 3. As shown in Table 1, the results with $N = 1$ are identical to those obtained by the contraction mapping principle, while in degrees $N = 2$ and $N = 3$ the method converges rapidly to solution of the Adomian's and shooting methods (see [23] for details).

As in Sect. 6.2, integral equation (65) does not have a known analytical solution, and to obtain a reference solution we compute the GLL spectral element solution using $N = 12$ and $N_e = 512$ elements, considering the same equation parameters as in Table 1. According to the definition of the kernel, we have a discontinuity in the first derivative of $\kappa(x, y)$. As in Remark 6.1, the convergence rates were restricted to $\mathcal{O}(h^2)$ regardless of the degree N, as shown in Fig. 7. The relative errors E_k are nearly identical to e_k, with a slight difference at $N = 2$.

Finally, since in this example $f(x, y) = \mu(\beta - y)e^y$ does not satisfy condition (A4), an additional experiment is carried out to assess how convergence is affected by the choice of the initial guess.

We choose the parameters $\lambda = 10$, $\mu = 0.55$, $\beta = 0.57$ suggested in Feng et al. [13]. As in Table 1, the step size and the polynomial degree are $h = 0.025$ and $N = 1$,

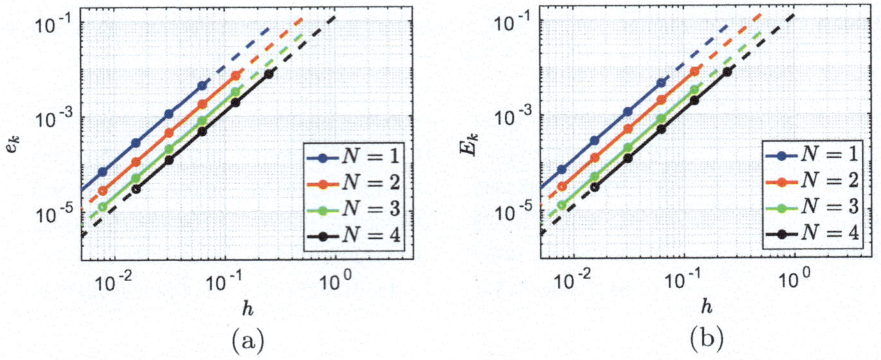

Fig. 7 Relative errors e_k (**a**) and E_k (**b**) in Sect. 6.4. The slopes of the dashed lines are nearly equal to 2.0 for all values of N

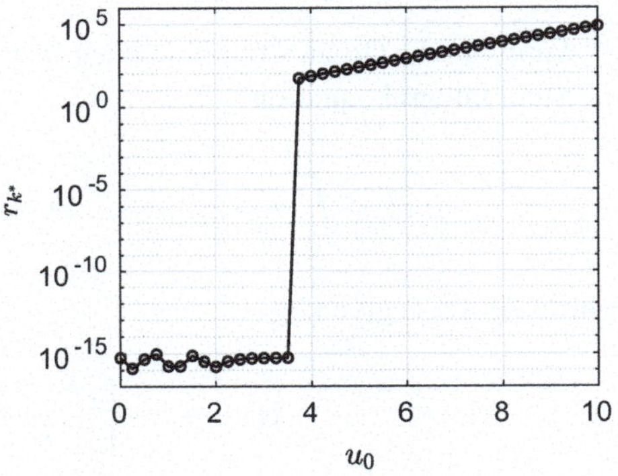

Fig. 8 Residual norm r_k at the final step k^* in Sect. 6.4 computed with initial guess $u_h^{(0)}(x) = u_0$, $u_0 \in [0, 10]$, and parameters $\lambda = 10$, $\mu = 0.55$, $\beta = 0.57$

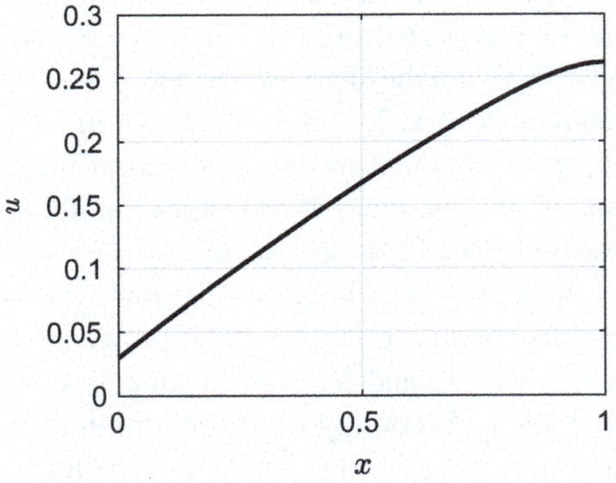

Fig. 9 Numerical solution in Sect. 6.4 with initial guess $u_h^{(0)}(x) = 2$ and parameters $\lambda = 10$, $\mu = 0.55$, $\beta = 0.57$

respectively. As shown in Fig. 8, the Picard iteration diverges when the constant initial guess $u_h^{(0)}(x) = u_0$ is such that $u_0 > 3.5$. It is worth noting that, when the Picard iteration converges (for instance with $u_0 = 2$, as in Fig. 9), the numerical solution satisfies the conditions of Theorem 3 in [13] (namely, $\min\{u_h(x) : x \in [0, 1]\} \leq \beta/2$ and $u_h(x) \leq \beta$ for $x \in [0, 1]$), unlike the solution provided in [13, Fig. 1].

7 Conclusions

We have proposed a spectral element method for nonlinear Hammerstein integral equations of type (1), which can be rewritten as a fixed-point equation $u = Au$, where A is a nonlinear operator defined by (8). We have established sufficient existence and uniqueness conditions in L^2-norm (Assumption 2.1), contributing to the theory developed by previous works that consider the L^2-norm, such as [12, 22].

From a numerical point of view, we discuss global convergence analysis of the Picard iteration for the spectral element discretization of equation (1) and prove that the global error has optimal order $O(h^{N+1})$ in the L^2 norm $\|\cdot\|_0$ and convergence rate $O(h^{2N})$ in $\|\cdot\|_w$, provided that the assumptions of Theorem 5.3 hold and the input data are sufficiently regular. This is confirmed by Sects. 6.2 and 6.4, showing that if the kernel is not sufficiently regular then the convergence rates are lower than the optimal ones. A similar result was obtained in [10] for a Fredholm-Hammerstein integral equation. It is important to stand out that convergence results obtained in the one-dimensional media can also be extended to higher dimensions, as suggested by Sect. 6.3.

The methodology used in this paper can be easily extended for the solution of nonlinear integro-differential equations in one- and two-dimensional media, as well as nonlinear integral equations of Hammerstein type considering singular kernel in the space $L^2([a, b])$. A revelant extension of this work would be the use of the Newton-Kantorovich method [14] rather than the Picard method in the solution of the nonlinear algebraic system.

Acknowledgements S. P. Oliveira is supported by CNPq under the grant 313100/2017-9.

References

1. Andreev, A., Kascieva, V., Vanmaele, M.: Some results in lumped mass finite-element approximation of eigenvalue problems using numerical quadrature formulas. J. Comput. Appl. Math. **43**(3), 291–311 (1992)
2. Atkinson, K.E.: A survey of numerical methods for solving nonlinear integral equations. J. Int. Eq. Appl. **4**(1), 15–46 (1992)
3. Atkinson, K.E., Flores, J.: The discrete collocation method for nonlinear integral equations. IMA J. Numer. Anal. **13**(2), 195–213 (1993)
4. Azevedo, J.S., Oliveira, S.P., Rocha, A.M.: Spectral element approximation of functional integral equations. Electron. T. Numer. Ana. **8**(2), 164–180 (2020)
5. Bazm, S.: Numerical solution of a class of nonlinear two-dimensional integral equations using Bernoulli polynomials. Sahand Commun. Math. Anal. **3**(1), 37–51 (2016)
6. Boffi, V.C., Spiga, G.: An equation of Hammerstein type arising in particle transport theory. J. Math. Phys. **24**(6), 1625–1629 (1983)
7. Canuto, C., Hussaini, M., Quarteroni, A., Zang, T.: Spectral Methods: Fundamentals in Single Domains. Springer, New York (2006)
8. Ciarlet, P.: The Finite Element Method for Elliptic Problems. SIAM, Philadelphia (2002)
9. Corduneanu, C.: Integral Equations and Applications. Cambridge University Press (1991)

10. Das, P., Nelakanti, G., Long, G.: Discrete Legendre spectral projection methods for Fredholm-Hammerstein integral equations. J. Comput. Appl. Math **278**, 293–305 (2015)
11. de Figueiredo, D.G., Gupta, C.P.: On the variational method for the existence of solutions of nonlinear equations of Hammerstein type. Proc. Amer. Math. Soc. **40**(2), 470–476 (1973)
12. Dolph, C.: Nonlinear integral equations of the Hammerstein type. Trans. Am. Math. Soc. **66**(2), 289–307 (1949)
13. Feng, W., Zhang, G., Chai, Y.: Existence of positive solutions for second order differential equations arising from chemical reactor theory. Discrete Contin. Dyn. Syst. 373–381 (2007)
14. Husam Hameed, H., Eshkuvatov, Z.K., Ahmedov, A., Nik Long, N.M.A.: On Newton-Kantorovich method for solving the nonlinear operator equation. In: Abstract and Applied Analysis. Hindawi (2015)
15. Kaneko, H., Xu, Y.: Superconvergence of the iterated Galerkin methods for Hammerstein equations. SIAM J. Numer. Anal. **33**(3), 1048–1064 (1996)
16. Karoui, A., Jawahdou, A.: Existence and approximate L^p and continuous solutions of nonlinear integral equations of the Hammerstein and Volterra types. Appl. Math. Comp. **216**(7), 2077–2091 (2010)
17. Komatitsch, D., Vilotte, J.-P.: The spectral element method: an efficient tool to simulate the seismic response of 2D and 3D geological structures. Bull. Seismol. Soc. Am. **88**(2), 368–392 (1998)
18. Kumar, S.: A discrete collocation-type method for Hammerstein equations. SIAM J. Numer. Anal. **25**(2), 328–341 (1988)
19. Kumar, S., Sloan, I.H.: A new collocation-type method for Hammerstein integral equations. Math. Comp. **48**(178), 585–593 (1987)
20. Lardy, L.: A variation of Nyström's method for Hammerstein equations. J. Int. Eq. 43–60 (1981)
21. Latif, A.: Banach contraction principle and its generalizations. In: Topics in Fixed Point Theory, pp. 33–64 (2014)
22. Li, F., Li, Y., Liang, Z.: Existence of solutions to nonlinear Hammerstein integral equations and applications. J. Math. Anal. Appl. **323**(1), 209–227 (2006)
23. Madbouly, N.M., McGhee, D.F., Roach, G.F.: Adomian's method for Hammerstein integral equations arising from chemical reactor theory. Appl. Math. Comput. **117**(2–3), 241–249 (2001)
24. Nahid, N., Das, P., Nelakanti, G.: Projection and multi projection methods for nonlinear integral equations on the half-line. J. Comput. Appl. Math **359**, 119–144 (2019)
25. Oliveira, S.P., Azevedo, J.S.: Spectral element approximation of Fredholm integral eigenvalue problems. J. Comput. Appl. Math. **257**, 46–56 (2014)
26. Oliveira, S.P., Leite, S.A.: Error analysis of the spectral element method with Gauss-Lobatto-Legendre points for the acoustic wave equation in heterogeneous media. Appl. Numer. Math. **129**, 39–57 (2018)
27. Patera, A.: A spectral element method for fluid dynamics: laminar flow in a channel expansion. J. Comput. Phys. **54**(3), 468–488 (1984)
28. Rocha, A.M., Azevedo, J.S., Oliveira, S.P., Correa, M.R.: Numerical analysis of a collocation method for functional integral equations. Appl. Numer. Math. **134**, 31–45 (2018)
29. Yang, Y., Chen, Y., Huang, Y., Yang, W.: Convergence Analysis of Legendre-collocation methods for nonlinear Volterra type integro equations. Adv. Appl. Math. Mech. **7**(1), 74–88 (2015)
30. Zakian, P., Khaji, N.: A stochastic spectral finite element method for wave propagation analyses with medium uncertainties. Appl. Math. Model. **63**, 84–108 (2018)

Approximate Methods for Solving Hypersingular Integral Equations

Ilya Boykov

Abstract The work is devoted to a review of analytical and numerical methods for solving linear hypersingular integral equations. Hypersingular integral equations of the first and second kind on closed and open integration intervals are considered. Particular attention is paid to equations with second-order singularities, since these equations are most in demand in applications. The proofs of the convergence of approximate methods are based on the general theory of approximate methods of analysis and on the continuous method for solving operator equations. Easily visible sufficient conditions for the solvability of computational schemes and estimates of the accuracy of the proposed methods are obtained.

Keywords Hypersingular integral equations · Collocation method · Mechanical quadrature method

MSC 2010 65R20

1 Introduction

The importance of solving hypersingular integral equations (HSIEs) is justified by numerous applications and intensive growing number of applications. HSIEs arise a variety of mixed boundary value problems in mathematical physics such as water wave scattering [25], problems associated with multilayer plates [42] and fracture mechanics [1], theory of elasticity and thermoelasticity [33], aerodynamics and electrodynamics [31] and many other fields.

A closed-form solution of HSIEs is only possible in exceptional cases. This circumstance stimulates the active development of approximate methods.

Galerkin and collocation methods for solving HSIEs of the first and second kind are developed and justified by imposing conditions on the kernels and the right-hand

I. Boykov (✉)
Penza State University, Penza, Russia
e-mail: i.v.boykov@gmail.com

H. Singh et al. (eds.), *Topics in Integral and Integro-Differential Equations*,
Studies in Systems, Decision and Control 340,
https://doi.org/10.1007/978-3-030-65509-9_3

sides of the equations in [6, 7, 9, 10, 17, 21, 22, 31, 35]. Projection method with Chebyshev polynomials were discussed to solve HSIEs of the first kind in [15, 27]. Spline-collocation method for solving linear and nonlinear HSIEs of the second kind with even singularity has been developed and justified in [9, 10].

In this work, we propose an approach to solving linear HSIEs. Using the collocation method and the method of mechanical quadrature HSIEs are approximated by a linear systems of algebraic equations. We demonstrate that this system of algebraic equations admits an iterative solution based on the procedure which we justify for solving nonlinear operator equations in Banach spaces. We associate the system of algebraic equations with a system of ordinary differential equations which generate trajectories converging to the solution of the system of algebraic equations.

Now will list several classes of functions, which will be used later.

Let γ be the unit circle: $\gamma = \{z : |z| = 1\}$.

Definition 1.1 A function f, defined on $\Delta = [a, b]$ or $\Delta = \gamma$, satisfies a Lipschits condition with constant M and exponent α, or belongs to the class $H_\alpha(M)$, $M \geq 0$, $0 < \alpha \leq 1$, if $|f(x') - f(x'')| \leq M|x' - x''|^\alpha$, $x', x'' \in \Delta$.

Definition 1.2 The class $W^r(M, \Delta)$, $r = 1, 2, \ldots$, $\Delta = [a, b]$ or $\Delta = \gamma$, consists of all functions $f \in C(\Delta)$, which have an absolutely continuous derivative $f^{(r-1)}(x)$ and piecewise derivative $f^{(r)}(x)$ with $|f^{(r)}(x)| \leq M$. Then $W^r H_\alpha$ is the set of all functions $f \in C(\Delta)$, which have continuous derivative $f^{(r)}(x) : |f^{(r)}(x)| \leq M$, $f^{(r)} \in H_\alpha(M_1)$, M_1- constant, on Δ.

Definition 1.3 [34]. A function f belongs to $W^{r,r} H_{\alpha,\alpha}$ if and only if all its partial derivatives $D^j f$ of order $j = 0, 1, \ldots, r$ exist and continuous and satisfy the following conditions: For each partial derivative $D^j f$ of order j, $\|D^j f\| \leq M_j$, $j = 0, 1, \ldots, r$, and in addition for each derivative of order r, $D^r f \in H_\alpha(M_{r+1})$.

If coefficient M is not essential we use designations H_α, $W^r H_\alpha$, $W^{rr} H_{\alpha,\alpha}$ instead of $H_\alpha(M)$, $W^r H_\alpha(M)$, $W^{rr} H_{\alpha,\alpha}(M)$ respectively.

Let's recall the Hadamard definition of hypersingular integrals [23].

Definition 1.4 [23]. The integral of the type $\int_a^b \frac{A(x)\,dx}{(b-x)^{p+\alpha}}$ for an integer p and $0 < \alpha < 1$, is defined as the limit of the sum

$$\int_a^b \frac{A(x)\,dx}{(b-x)^{p+\alpha}} = \lim_{x \to b} \left[\int_a^x \frac{A(t)\,dt}{(b-t)^{p+\alpha}} + \frac{B(x)}{(b-x)^{p+\alpha-1}} \right],$$

if one assumes that $A(x)$ has p derivatives in the neighborhood of the point b. Here $B(x)$ is any function that satisfies the following two conditions: (1) the above limit exists; (2) $B(x)$ has at least p derivatives in the neighborhood of the point $x = b$.

Chikin [12] introduced the definition of the Cauchy-Hadamard type integral that generalizes the notion of the singular integral in the Cauchy principal sense and in the Hadamard sense.

Definition 1.5 [12]. The Cauchy-Hadamard principal value of the integral $\int_a^b \frac{\varphi(\tau)\,d\tau}{(\tau-c)^p}$, $a < c < b$, is defined as the limit of the following expression

$$\int_a^b \frac{\varphi(\tau)\,d\tau}{(\tau-c)^p} = \lim_{v\to 0}\left[\int_a^{c-v} \frac{\varphi(\tau)\,d\tau}{(\tau-c)^p} + \int_{c+v}^{b} \frac{\varphi(\tau)\,d\tau}{(\tau-c)^p} + \frac{\xi(v)}{v^{p-1}}\right],$$

where $\xi(v)$ is a function constructed so that the limit exists.

The following notations are used in the work.

Let $f(t) \in H_\alpha, 0 < \alpha \le 1, t \in [a,b]$. The functional $\sup_{t_1 \ne t_2, t_1, t_2 \in [a,b]} |f(t_1) - f(t_2)|/|t_1 - t_2|^\alpha$ is indicated by $H(f, \alpha)$.

Recall [24] that the symbol $[X, Y]$ denotes the set linear bounded operators mapping the normed space X into normed space Y.

This chapter provides an overview of methods for solving HSIEs of the first and second kind with second-order singularities. Detailed proofs of the unique solvability and convergence of collocation and mechanical quadrature methods for HSIEs of the first and second kind, defined on closed and open circuits, are given. A number of the results presented below were previously published in Russian in untranslatable journals. In the author's opinion, it is of undoubted interest the acquaintance of the English-speaking reader with these results.

The chapter is organized as follows. The continuous method for linear and nonlinear operator equations is explained in Sect. 2. The analytical methods for solving hypersingular and polyhypersingular integral equations is presented in Sect. 3. The approximate methods for solving HSIEs of the first kind is given in Sect. 4. The numerical method for solving HSIEs of the second kind is presented in Sect. 5.

2 Continuous Method and Its Convergence Properties

Abstract. This section is devoted to a continuous method for solving operator equations, which makes it possible to reduce the solution of linear and nonlinear operator equations in a Banach space B to solving the Cauchy problem for an ordinary differential equation in the same space.

We will use the following notations: $B(a, r) = \{z \in B : \|z - a\| \le r\}$, $S(a, r) = \{z \in B : \|z - a\| = r\}$, $\Lambda(K) = \lim_{h \downarrow 0}(\|I + hK\| - 1)h^{-1}$.

Here B is a Banach space, $a \in B$, K is a linear and bounded operator on B, $\Lambda(K)$ is the logarithmic norm [13] of the operator K and I stands for the identity operator. Main properties of the logarithmic norm are given in [13].

The logarithmic norm is known for operators in the most frequently used spaces. Let $A = \{a_{ij}\}$, $i, j = 1, 2, \ldots, n$, be a real matrix. In the space R_n of vectors $x = (x_1, \ldots, x_n)$ the following norms are often used: octahedral—$\|x\|_1 = \sum_{i=1}^n |x_i|$; cubic—$\|x\|_2 = \max_{1 \le i \le n} |x_i|$; spherical (Euclidean)—$\|x\|_3 = (\sum_{i=1}^n x_i^2)^{1/2}$. Here are expressions of the logarithmic norm of a matrix $A = \{a_{ij}\}$, due to the above norms of the vectors: octahedral logarithmic norm $\Lambda_1(A) = \max_{1\le j\le n}(a_{jj} + \sum_{i\ne j} |a_{ij}|)$; cubic logarithmic norm $\Lambda_2(A) = \max_{1\le i\le n}(a_{ii} + \sum_{j\ne i} |a_{ij}|)$; spherical (Euclidean) logarithmic norm $\Lambda_3(A) = \lambda_{\max}\left(\frac{A+A^*}{2}\right)$, where A^* is the conjugate matrix for A.

Consider a nonlinear operator equation

$$A(x) - f = 0. \tag{1}$$

Here A is a nonlinear operator that acts from a Banach space B into B.

In a Banach space B, consider the Cauchy problem

$$\frac{dx(t)}{dt} = A(x(t)) - f, \tag{2}$$

$$x(0) = x_0. \tag{3}$$

The following theorem establish a connection between the solutions of the Eq. (1) and the Cauchy problem (2), (3).

Theorem 1 *[3]. Let the Eq. (1) has a solution x^*; and let inequality*

$$\lim_{t\to\infty} \frac{1}{t} \int_0^t \Lambda(A'(g(\tau)))d\tau \le -\alpha_g,\ \alpha_g > 0, \tag{4}$$

be true on each differentiable curve $g(t)$ lying in the ball $B(x^*, r)$, $0 < r \le \infty$. Then the solution of the Cauchy problem (2), (3) converges to the solution x^* of the Eq. (1) for any initial approximation $x(0) \in B(x^*, r)$.

3 Analytical Methods for Solving Hypersingular and Bihypersingular Integral Equations

Abstract. In this section we propose a method for transformating linear and nonlinear HSIEs into ordinary differential equations. Linear and nonlinear bihypersingular integral equations are transformed into partial differential equations. Many types of differential equations are solving in quadratures. So, we receive analytical solutions for linear and nonlinear hypersingular and bihypersingular integral equations.

3.1 Introduction

A closed-form solution of HSIEs are known only in exceptional cases. In the works [35–37] given the exact solution of the equation

$$Hx = \frac{d}{dt}\left(\int_{-1}^{1} \frac{x(\tau)}{\tau - t} d\tau\right) = f(t), \ -1 < t < 1, \ x(\pm 1) = 0.$$

The exact solution is obtained [40] for the non-linear hypersingular Peierls-Nabarro integral equation

$$\frac{1}{1-v}\int_{-\infty}^{\infty} \frac{x(\tau)}{(\tau - t)^2} d\tau + \sin\left(\frac{1 - x(t)}{b}\right) = 0.$$

3.2 Hypersingular Integral Equations

Let us consider HSIE

$$a(t)x(t) + \frac{b(t)}{\pi i}\int_{\gamma} \frac{x(\tau)d\tau}{(\tau - t)^p} = f(t), \ p = 2, 3, \ldots \tag{5}$$

where γ is a smooth closed contour in the plane of a complex variable.

Let us associate with a function $\varphi(z), z \in Z$, functions $\varphi^+(z)$ and $\varphi^-(z)$, which are analytical inside and outside of the contour γ respectively and connected with $\varphi(t), t \in \gamma$, by Sohotzky–Plemel formulas $\varphi^+(t) - \varphi^-(t) = \varphi(t), \quad \varphi^+(t) + \varphi^-(t) = \frac{1}{\pi i}\int_{\gamma} \frac{\varphi(\tau)d\tau}{\tau - t}$. Well known [19] that Sohotzky–Plemel formulas have place if the function $\varphi(t)$ belongs to common enough classes of functions.

We denote by $D^+(D^-)$ the inner (outer) domain with respect to the contour γ. We denote by $\bar{D}^+$ the closure of the domain D^+. Denote by G an open domain such that $\bar{D}^+ \subset G$.

We will seak a solution of the Eq. (5) in the class of functions $x(z), z \in \bar{D}^+$, which are analytical in D^+ and satisfied the condition: $x(t) \in W^{p-1}H_\alpha(M), t \in \gamma$.

Let the following conditions be satisfied for functions $a(t), b(t), f(t)$:
(1) functions $a(z), b(z), f(z)$ are analytical in G; (2) $a(z) \neq 0, z \in D^+$.

We will find, under these conditions, $p - 1$ linearly independent solutions of the Eq. (5).

Assume that the Eq. (5) has a solution $x^*(t) \in W^{p-1}H_\alpha(M)$.

As $a(z) \neq 0, z \in G$, that $x^*(t) = \frac{f(t)}{a(t)} - \frac{b(t)}{a(t)}\frac{1}{\pi i}\int_{\gamma} \frac{x^*(\tau)d\tau}{(\tau - t)^p} d\tau$.

From definition of hypersingular integral [12], we have

$$x^*(t) = \frac{f(t)}{a(t)} - \frac{b(t)}{a(t)}\frac{1}{\pi i(p-1)!}\int_{\gamma} \frac{x^{*(p-1)}(\tau)d\tau}{\tau - t} d\tau.$$

Function $\Psi(z) = \frac{1}{\pi i}\int_\gamma \frac{x^*(\tau)d\tau}{(\tau-z)^p} = \frac{1}{\pi i(p-1)!}\int_\gamma \frac{x^{*(p-1)}(\tau)d\tau}{\tau-z}$ is analytical for $z \notin \gamma$. So, the function $x^*(z) = \frac{f(z)}{a(z)} - \frac{1}{\pi i(p-1)!}\frac{b(z)}{a(z)}\frac{1}{\pi i}\int_\gamma \frac{x^{*(p-1)}(\tau)d\tau}{\tau-z}$ is analytical in D^+. In addition, $x^*(t) \in W^{p-1}H_\alpha(M)$ by assumption.

Then [12] $\frac{1}{\pi i}\int_\gamma \frac{x^*(\tau)d\tau}{(\tau-t)^p}d\tau = \frac{1}{\pi i(p-1)!}\frac{d^{p-1}}{dt^{p-1}}\int_\gamma \frac{x^*(\tau)d\tau}{\tau-t}d\tau$.

Well known [19], that for functions, analytical in D^+ and satisfying the condition $x^*(t) \in H_\alpha$, justly the equality $\frac{1}{\pi i}\int_\gamma \frac{x^*(\tau)}{\tau-t}d\tau = x^*(t)$.

So, we have $\frac{1}{\pi i}\int_\gamma \frac{x^*(\tau)}{(\tau-t)^p}d\tau = \frac{1}{(p-1)!}\frac{d^{p-1}}{dt^{p-1}}x^*(t)$.

Thus, the solution $x^*(t)$ of the Eq. (5) satisfies the equation

$$\frac{b(t)}{(p-1)!}\frac{d^{p-1}}{dt^{p-1}}x^*(t) + a(t)x^*(t) = f(t).$$

This is differential equation with $p-1$ linearly independent solutions.

Let us turn to nonlinear HSIE

$$£\left(\frac{1}{\pi i}\int\limits_L \frac{x(\tau)d\tau}{(\tau-t)^p}, \frac{1}{\pi i}\int\limits_L \frac{x(\tau)d\tau}{(\tau-t)^{p-1}}, \ldots, \frac{1}{\pi i}\int\limits_L \frac{x(\tau)d\tau}{\tau-t}, x(t)\right) = f(t), \quad (6)$$

where £ is non-linear operator, L is a smooth closed contour.

Let G—be a bounded closed domain in which the contour L is located.

If the Eq. (6) has an analytic solution in G, then this solution satisfies the differential equation

$$£_1(x^{(p-1)}(t), x^{(p-2)}(t), \ldots, x'(t), x(t)) = f(t), \quad (7)$$

where $£_1$ is the operator obtained from the operator £ by replacing the hypersingular operators and the singular operator into the corresponding differential operators.

If a solution of the Eq. (7) is analytic in G, then it is also a solution of the Eq. (6).

3.3 Bihypersingular Integral Equations

We will consider the bihypersingular integral equation

$$a(t_1,t_2)x(t_1,t_2) + b(t_1,t_2)S_1x + c(t_1,t_2)S_2x + d(t_1,t_2)S_{12}x = f(t_1,t_2), \quad (8)$$

$(t_1,t_2) \in (\gamma_1 \times \gamma_2)$, where γ_i is a smooth closed contour in the plane Z_i, $i = 1, 2$, of a complex variable z_i, $i = 1, 2$; $\gamma = \gamma_1 \times \gamma_2$; $S_1\varphi = \frac{1}{\pi i}\int_{\gamma_1} \frac{\varphi(\tau_1,t_2)d\tau_1}{\tau_1-t_1}$, $S_2\varphi = \frac{1}{\pi i}\int_{\gamma_2} \frac{\varphi(t_1,\tau_2)d\tau_2}{\tau_2-t_2}$, $S_{12}\varphi = -\frac{1}{\pi^2}\int_\gamma \frac{\varphi(\tau_1,\tau_2)d\tau_1 d\tau_2}{(\tau_1-t_1)(\tau_2-t_2)}$.

We denote by G_i a closed bounded domain in the plane z_i such that the contour γ_i lies inside G_i, $i = 1, 2$. Let $G = G_1 \times G_2$.

Suppose that the Eq. (8) has a solution $x^*(t_1, t_2)$ that is analytic in the domain G. Then $x^*(t_1, t_2)$ is a solution of the differential equation

$$a(t_1, t_2)x(t_1, t_2) + \frac{b(t_1,t_2)}{(p-1)!}\frac{\partial^{p-1}x(t_1,t_2)}{\partial t_1^{p-1}} + \frac{c(t_1,t_2)}{(p-1!)}\frac{\partial^{p-1}x(t_1,t_2)}{\partial t_2^{p-1}} + \\ + \frac{d(t_1,t_2)}{((p-1)!)^2}\frac{\partial^{2p-2}x(t_1,t_2)}{\partial t_1^{p-1}\partial t_2^{p-1}} = f(t_1, t_2),\ (t_1, t_2) \in (\gamma_1 \times \gamma_2).$$

For proof it we will use the analog of the Sokhotsky–Plemel formulas for multiple integrals of Cauchy type.

Let the contour γ_i divide the plane of the complex variable z_i, $i = 1, 2$, into two parts: the inner D_i^+ and the outer D_i^-.. Then $\gamma = \gamma_1 \times \gamma_2$ is the boundary of regular bicylindrical domains $D^{\pm\pm} = D_1^{\pm} \times D_2^{\pm}$.

Consider the double Cauchy-type integral

$$\Phi(z_1, z_2) = \frac{1}{(2\pi i)^2}\int\limits_{\gamma} \frac{\varphi(\tau_1, \tau_2)d\tau_1 d\tau_2}{(\tau_1 - z_1)(\tau_2 - z_2)}. \tag{9}$$

We denote by $\Phi^{\pm\pm}(t_1, t_2)$ the limit values of the integral (9) when a point $(z_1, z_2) \in D^{\pm\pm}$ tends to the point $(t_1, t_2) \in \gamma$.

We will need in the following multidimensional analogue of the Sokhotsky–Plemel formulas [19]:

$$\Phi^{++} + \Phi^{+-} + \Phi^{-+} + \Phi^{--} = S_{12}\varphi. \tag{10}$$

Using the definition of the bihypersingular integral, we have

$$-\frac{1}{\pi^2}\int\limits_{\gamma_1}\int\limits_{\gamma_2} \frac{x(\tau_1,\tau_2)d\tau_1 d\tau_2}{(\tau_1-t_1)^p(\tau_2-t_2)^p} = \\ = -\frac{1}{((p-1)!)^2}\frac{1}{\pi^2}\int\limits_{\gamma_1}\int\limits_{\gamma_2} \frac{x^{(p-1,p-1)}(\tau_1,\tau_2)d\tau_1 d\tau_2}{(\tau_1-t_1)(\tau_2-t_2)}. \tag{11}$$

It follows from (10) that

$$-\frac{1}{\pi^2}\int\limits_{\gamma_1}\int\limits_{\gamma_2} \frac{x^{(p-1,p-1)}(\tau_1,\tau_2)d\tau_1 d\tau_2}{(\tau_1-t_1)(\tau_2-t_2)} = \\ = X^{++}_{p-1,p-1}(t_1, t_2) + X^{+-}_{p-1,p-1}(t_1, t_2) + X^{-+}_{p-1,p-1}(t_1, t_2) + X^{--}_{p-1,p-1}(t_1, t_2), \tag{12}$$

where

$$X_{p-1,p-1}(z_1, z_2) = \frac{1}{(2\pi i)^2}\int\limits_{\gamma_1}\int\limits_{\gamma_2} \frac{x^{(p-1,p-1)}(\tau_1, \tau_2)d\tau_1 d\tau_2}{(\tau_1 - z_1)(\tau_2 - z_2)}.$$

Since the function $x^{(p-1,p-1)}(z_1, z_2)$ is assumed to be analytic in the domain G, then from the Cauchy integral formula [18] follows that $X^{++}_{p-1,p-1}(t_1, t_2) = x^{(p-1,p-1)}(t_1, t_2)$, $X^{\pm\mp}_{p-1,p-1}(t_1, t_2) = 0$, $X^{--}_{p-1,p-1}(t_1, t_2) = 0$.

From this and (11), (12) we have

$$-\frac{1}{\pi^2}\int\limits_{\gamma_1}\int\limits_{\gamma_2}\frac{x(\tau_1,\tau_2)d\tau_1 d\tau_2}{(\tau_1-t_1)^p(\tau_2-t_2)^p}=\frac{1}{((p-1)!)^2}x^{(p-1,p-1)}(t_1,t_2). \tag{13}$$

It was previously shown that

$$\frac{1}{\pi i}\int\limits_{\gamma_1}\frac{x(\tau_1,t_2)}{(\tau_1-t_1)^p}d\tau_1=\frac{1}{(p-1)!}\frac{\partial^{p-1}x(t_1,t_2)}{\partial t_1^{p-1}}, \tag{14}$$

$$\frac{1}{\pi i}\int\limits_{\gamma_2}\frac{x(t_1,\tau_2)}{(\tau_2-t_2)^p}d\tau_2=\frac{1}{(p-1)!}\frac{\partial^{p-1}x(t_1,t_2)}{\partial t_2^{p-1}}. \tag{15}$$

Substituting (13)–(15) into (8), we arrive at the differential equation

$$\begin{gathered}a(t_1,t_2)x(t_1,t_2)+\frac{b(t_1,t_2)}{(p-1)!}\frac{\partial^{p-1}x(t_1,t_2)}{\partial t_1^{p-1}}+\\+\frac{c(t_1,t_2)}{(p-1)!}\frac{\partial^{p-1}x(t_1,t_2)}{\partial t_2^{p-1}}+\frac{d(t_1,t_2)}{((p-1)!)^2}\frac{\partial^{2p-2}x(t_1,t_2)}{\partial t_1^{p-1}\partial t_2^{p-1}}=f(t_1,t_2).\end{gathered} \tag{16}$$

Thus, it is shown that if the bihypersingular integral equation (8) has a solution analytic in the domain G, then it is also a solution of the differential equation (16). The converse is also true—if a differential equation (16) has a solution analytic in the domain G, then it is also a solution of the Eq. (8).

Remark 1 Since for many classes of ODE and PDE exact solutions are known, thus exact solutions for hypersingular and bihypersingular integral equations are obtained.

4 Approximate Solution of the First Kind HSIEs

Abstract. The section is devoted to approximate methods for solving HSIEs of the first kind. The main attention in the section is paid to HSIEs of the first kind with singularities of the second order. This is due to the fact that at present these equations are most often found in applications. Constructed numerical schemes of collocation and mechanical quadrature methods and given its justifications.

4.1 Introduction

This section gives a brief overview of approximate methods for solving HSIEs of the first kind.

We study approximate methods for solving HSIEs of the form

$$Kx \equiv \int_{-1}^{1} \frac{x(\tau)}{(\tau - t)^2} d\tau + \int_{-1}^{1} h(t, \tau)x(\tau)d\tau = f(t), \ -1 < t < 1, \tag{17}$$

where $h(t, \tau) \in W^{r,r} H_{\alpha,\alpha}(M)$, $f(t) \in W^r H_\alpha(M), r = 1, 2 \ldots, 0 < \alpha \leq 1$.

Some of the first works devoted to approximate methods for solving HSIEs of the first kind of the form (17) were articles [21, 22]. In these articles was constructed computational schemes of Galerkin and collocation methods for solving equations of the form (17) under assumptions that $h(t, \tau)$, $f(t)$ are smooth functions and the Eq. (17) has a solution of the form $x(t) = \omega(t)\varphi(t)$, where $\omega(t) = (1 - t^2)^{1/2}$, $\varphi(t)$ is a smooth function.

In the paper [14], a numerical—analytical method for solving hypersingular integral equations of the first kind of the form

$$\begin{gathered}(Ax)t + \alpha(Lx)(t) + \beta(Kx)(t) \equiv \tfrac{1}{\pi}\tfrac{\partial}{\partial t} \int_{-1}^{1} x(\tau)\tfrac{\partial}{\partial \tau} \ln \tfrac{1}{|\tau - t|} d\tau + \\ + \alpha \int_{-1}^{1} x(\tau) \ln \tfrac{1}{|\tau - t|} d\tau + \beta \int_{-1}^{1} h(t, \tau)x(\tau)d\tau = f(t), -1 \leq t \leq 1,\end{gathered} \tag{18}$$

was investigated. Here α and β be constants, $h(t, \tau)$ is a continuous function.

The method is based on the following property of the operator $(Ax)(t)$: on the class of functions $x(t) = \sqrt{1 - t^2}\varphi(t)$, where $\varphi(t)$ is a smooth function, the operator $(Ax)(t)$ has the continuous inverse operator

$$(A^{-1} f)(t) = \frac{1}{\pi} \int_{-1}^{1} f(\tau) \ln \left| \frac{\tau - t}{1 - \tau t + \sqrt{1 - \tau^2}\sqrt{1 - t^2}} \right| d\tau.$$

Applying the operator A^{-1} to the Eq. (18), we arrive at the Fredholm equation

$$x(t) + \alpha(A^{-1}Lx)(t) + \beta(A^{-1}Kx)(t) = A^{-1} f. \tag{19}$$

Equation (19) is solved by the reduction method with basis functions $\varphi_n(t) = \sqrt{\frac{2}{\pi n}}\sqrt{1 - t^2}U_n(t), n = 1, 2, \ldots$, where $U_n(t), n = 0, 1, \ldots$, are the Chebyshev polynomials of the second kind.

In [30], in the class of generalized functions, investigated the equation

$$\begin{gathered}\tfrac{a}{4\pi} \int_{0}^{2\pi} \tfrac{x(\tau)d\sigma}{\sin^2 \frac{\sigma - s}{2}} d\tau + \tfrac{b}{2\pi} \int_{0}^{2\pi} x(\tau)ctg \tfrac{\sigma - s}{2} d\sigma + \\ + \tfrac{c}{\pi} \int_{0}^{2\pi} \ln \left|\sin \tfrac{\sigma - s}{2}\right| d\sigma + \int_{0}^{2\pi} h(s, \sigma)x(\sigma)d\sigma = f(s).\end{gathered} \tag{20}$$

Here a, b, c are constants, $h(s, \sigma)$, $f(s)$ are smooth functions.

Given the explicit solutions of Eq. (20) in several special cases. In the general case, Eq. (20) is solved by the method of moments.

In the work [17] built an approximate method for solving the equation

$$\frac{1}{4\pi}\int\limits_0^{2\pi}\frac{x(\sigma)}{\sin^2\frac{\sigma-s}{2}}d\sigma = f(s),\ s\in(0,2\pi) \tag{21}$$

under the additional condition $\int_0^{2\pi} f(\sigma)d\sigma = 0$.

Two node systems are introduced $t_k = 2\pi k/n$, $k = 0, 1, \ldots, n$, and $\hat{t}_k = t_{k-1} + h/2$, $k = 1, 2, \ldots, n$, $h = \pi/n$.

An approximate solution is sought in the form of a piecewise constant function $x_n(t) = \sum_{i=1}^n x(\hat{t}_i)\varphi_i(t)$, where

$$\varphi_i(t) = \begin{cases} 1,\ t\in[t_{i-1}, t_i], \\ 0,\ t\in[0, 2\pi]\backslash[t_{i-1}, t_i]. \end{cases}$$

The coefficients $\{\alpha_i\}_{i=1}^n$ are found from the system of equations

$$\begin{cases} \gamma_{0n} + \frac{1}{2\pi}\sum\limits_{l=1}^n\left(ctg\frac{\hat{t}_k-t_l}{2} - ctg\frac{\hat{t}_k-t_{l-1}}{2}\right)\alpha_l = f(\hat{t}_k),\ k = 1, 2, \ldots, n, \\ \sum\limits_{l=1}^n \alpha_l = 0, \end{cases} \tag{22}$$

where $\gamma_{0n} + \frac{1}{2\pi}\sum_{k=1}^n f(\hat{t}_k)h = 0$, $h = \frac{2\pi}{n}$.

It is shown in [17] that if the solution x^* of the Eq. (21) belongs to the class of functions $W^{3+\alpha}[0, 2\pi]$, then for sufficiently large n, the system (22) is uniquely solvable and the estimate $\max_{1\le i\le n}|x^*(\hat{t}_i) - x_n(\hat{t}_i)| \le Ch^2$ is valid.

Remark 2 Throughout this work, we denote the constants that do not depend on n by C.

In [28], the Eq. (21) was solved by the method of discrete vortices.

The discrete vortex method, originally developed for solving aerodynamic problems, has been widely used in solving singular integral equations [29]. In [31, 32], the method of discrete vortices and its generalizations are used to solve a number of hypersingular integral equations of the first kind.

For solving hypersingular integral equations of the first kind, among others, the homotopy method is used. A detailed review of the results obtained in this method can be found in [16].

Let's consider the compound singular integral equation

$$Kx \equiv \frac{a}{\pi}\int_{-1}^{1}\frac{x(\tau)d\tau}{(\tau-t)^2} + \frac{b}{\pi}\int_{-1}^{1}\frac{x(\tau)d\tau}{\tau-t} + \frac{c}{\pi}\int_{-1}^{1}\ln|\tau-t|x(\tau)d\tau +$$
$$+\frac{1}{\pi}\int_{-1}^{1}h(t,\tau)d\tau = f(t), t \in (-1,1). \tag{23}$$

Here a, b, c are constants.

A series of applied problems lead to equations of the form (23): the problem of calculating the input resistance of a wire antenna, the problem of calculating the electromagnetic field in a coaxial gyrotron for the case of a TM wave [20].

To solve the Eq. (23), the following approximate method was studied in [20]. An approximate solution is sought as a function $x_{n-2}(t) = \sqrt{1-t^2}\varphi_{n-2}(t)$, where $\varphi_{n-2}(t) = \sum_{k=0}^{n-2}\alpha_k t^k$. Substituting $x_{n-2}(t)$ into the Eq. (23), accurately calculating the singular integrals and approximately calculating the regular integral, they obtain a computational scheme.

Assuming that the operator K is continuously invertible in the Holder space, the unique solvability of the computational scheme follows from the general theory of approximate methods [24].

The modeling of many problems of physics and technology leads to nonlinear hypersingular integral equations of the first kind.

In particular, equations of the form

$$\int_{-\infty}^{\infty}\frac{x(\tau)}{(\tau-t)^2}d\tau + \int_{-\infty}^{\infty}h(t-\tau)x(\tau)d\tau - f(x(t)) = 0, \tag{24}$$

where $f(x)$ is a nonlinear function, arises in the theory of dislocations and in the theory of waves on water.

Approximate methods for solving the Peierls–Nabarro equation

$$\frac{1}{1-v}\int_{-\infty}^{\infty}\frac{x(\tau)}{(\tau-t)^2}d\tau + \sin\left(\frac{1-x(t)}{b}\right) = 0$$

are investigated in [26].

In the article [26] the hypersingular integral $\int_{-\infty}^{\infty}\frac{x(\tau)}{(\tau-t)^2}d\tau$ was approximated by integral $\int_{-A}^{A}\frac{x(\tau)}{(\tau-t)^2}d\tau$; the last integral was determined by the expression

$$\int_{-A}^{A}\frac{x(\tau)}{(\tau-t)^2}d\tau = \int_{-A}^{A}\frac{x(\tau)-x(t)-(\tau-t)x'(t)}{(\tau-t)^2}d\tau -$$
$$-\frac{2Ax(t)}{A^2-t^2} + x'(t)\ln\left|\frac{A-t}{A+t}\right|.$$

Equation (24) is solved with the stabilization method

$$\gamma\frac{\partial x}{\partial u}-\int\limits_{-\infty}^{\infty}K(t-\tau)x(\tau)d\tau-f(x(t))=0.$$

Here γ is a complex constant, $K(\zeta)=-(\zeta^{-2}+h(\zeta))$.

To solve the last equation, the Euler method was used.

Nonlinear hypersingular integral equations of the form

$$\int\limits_{-1}^{1}\frac{x(\tau)}{(\tau-t)^2}d\tau+f(x(t))=0 \tag{25}$$

find applications in aerodynamics problems.

In the papers [11] investigated an approximate method for solving of the Eq. (25) under the assumption that the solution has the form $x(t)=\sqrt{1-t^2}\varphi(t)$, where $\varphi(t)$ is a smooth function.

4.2 An Approximate Solution of HSIEs of the First Kind by the Mechanical Quadrature Method

4.2.1 Weight Function $(1-t^2)^{1/2}$

In this section, we study approximate methods for solving HSIEs of the form

$$Kx\equiv\int\limits_{-1}^{1}\frac{x(\tau)}{(\tau-t)^2}d\tau+\int\limits_{-1}^{1}h(t,\tau)x(\tau)d\tau=f(t),\ -1<t<1, \tag{26}$$

where $h(t,\tau)\in W^{r,r}H_{\alpha,\alpha}(M),\ f(t)\in W^rH_\alpha(M), r=1,2\ldots,0<\alpha\le 1.$

An approximate solution of the Eq. (26) will be sought in the form of the function

$$x_n(t)=\sqrt{1-t^2}\sum_{k=0}^{n}\alpha_k t^k.$$

Coefficients $\{\alpha_k\}$ are determined from the system of linear algebraic equations, which is written in operator form as

$$K_nx_n\equiv T_n\left[\int\limits_{-1}^{1}\frac{x_n(\tau)d\tau}{(\tau-t)^2}+\int\limits_{-1}^{1}\sqrt{1-t^2}U_n^\tau[h(t,\tau)\varphi_n(\tau)]d\tau\right]=T_n[f(t)], \tag{27}$$

where $\varphi_n(t) = \sum_{k=0}^{n} \alpha_k t^k$, $T_n[f]$ is the projection operator on the set of interpolating polynomials of n-th order over nodes of Chebyshev orthonormal polynomials of the first kind; $U_n^T[f]$-operator of projection onto the set of interpolating polynomials of n-th order over nodes of Chebyshev orthonormal polynomials of the second kind.

We introduce the space X of functions of the form $x(t) = \sqrt{1-t^2}\varphi(t)$ with the norm $\|x(t)\| = \max_{-1\le t\le 1} |\varphi(t)| + \max_{-1\le t\le 1} |\varphi'(t)| + H(\varphi', \beta)$ and the space Y of continuous functions $y(t) \in H_\beta, t \in [-1, 1]$, with the norm $\|y(t)\| = \max_{-1\le t\le 1} |y(t)| + H(y, \beta), \beta > 1/2$.

Denote by X_n the subspace of the space X consisting of functions $x_n(t) = \sqrt{1-t^2}\varphi_n(t)$, $\varphi_n(t) = \sum_{k=0}^{n} \alpha_k t^k$. Let Y_n be the subspace of the space Y consisting of polynomials $\sum_{k=0}^{n} \gamma_k t_n^k$. The norms in the subspaces X_n and Y_n are induced by the norms of the spaces X and Y, respectively.

We will show that the operator K maps the space X to Y. To do this, it suffices to restrict ourselves to considering the operator

$$K^0 x \equiv \int_{-1}^{1} \frac{x(\tau)}{(\tau - t)^2} d\tau, \ -1 < t < 1.$$

From the definition of hypersingular integral it follows that

$$\int_{-1}^{1} \frac{x(\tau)}{(\tau-t)^2} d\tau = \int_{-1}^{1} \frac{x'(\tau)d\tau}{\tau-t} =$$
$$= \int_{-1}^{1} \frac{\sqrt{1-\tau^2}\varphi'(\tau)}{\tau-t} d\tau - \int_{-1}^{1} \frac{\tau\varphi(\tau)}{\sqrt{1-\tau^2}(\tau-t)} d\tau = J_1(t) - J_2(t).$$

Since the numerator in the integral $J_1(t)$ vanishes in the points ± 1, the integral $J_1(t)$ can be extended by zero to any smooth curve connecting the points ± 1. From Privalov theorem [19] it follows that if $\varphi'(t) \in H_\beta$, then the function $J_1(t) \in H_\gamma$, $\gamma = \min(1/2, \beta)$.

Let us study the integral $J_2(t)$. Let $\psi(t) = t\varphi(t)$. Then

$$|J_2(t)| = \left| \int_{-1}^{1} \frac{\psi(\tau) - \psi(t)}{\sqrt{1-\tau^2}(\tau - t)} d\tau \right| \le C = \text{const};$$

$$\left|\frac{d}{dt} J_2(t)\right| = \left| \int_{-1}^{1} \frac{\psi(\tau)d\tau}{\sqrt{1-\tau^2}(\tau-t)^2} \right| = \left| \int_{-1}^{1} \frac{\psi(\tau)-\psi(t)-\psi'(t)(\tau-t)}{\sqrt{1-\tau^2}(\tau-t)^2} d\tau \right| \le$$
$$\le \int_{-1}^{1} \frac{|(\psi'(t+\theta(t,\tau)(\tau-t))-\psi'(t)(\tau-t)|}{\sqrt{1-\tau^2})|\tau-t|^2} d\tau \le \int_{-1}^{1} \frac{C|\tau-t|^{\beta-1}}{\sqrt{1-\tau^2}} d\tau = C = \text{const}.$$

Thus, the function $J_2(t)$ belongs to the class of Golder functions H_1 and, therefore, in conjunction with the inclusion $J_1(t) \in H_\gamma$, proved above, we have $K^0 \in [X, Y]$. Therefore, $K \in [X, Y]$.

Let's return to the Eq. (27). It is known [27] that

$$\frac{1}{\pi}\int_{-1}^{1}\frac{\sqrt{1-\tau^2}\tau^n}{(\tau-t)^2}d\tau=\sum_{k=0}^{n}c_k t^k,\ n=0,1,\dots, \tag{28}$$

From (28) it follows that the Eq. (27) is equivalent to the equation

$$K_n x_n \equiv \int_{-1}^{1}\frac{x_n(\tau)d\tau}{(\tau-t)^2}+T_n[\int_{-1}^{1}\sqrt{1-\tau^2}U_n^{\tau}[h(t,\tau)\varphi_n(\tau)]d\tau]=T_n[f(t)]. \tag{29}$$

Let's prove the unique solvability of the Eq. (29). The assumption was made above that the operator $K\in[X,Y]$ is continuously invertible.

Therefore, the operator K is left invertible on the subspace X_n.

Let's introduce the operator

$$\tilde{K}x\equiv\int_{-1}^{1}\frac{x(\tau)}{(\tau-t)^2}d\tau+T_n[\int_{-1}^{1}\sqrt{1-\tau^2}h(t,\tau)\varphi(\tau)d\tau].$$

Now we will appreciate $\|(K-\tilde{K})x\|$. Obviously

$$\begin{aligned}\|(K-\tilde{K})x\|_{C[-1,1]}&=\|D_n[\int_{-1}^{1}\sqrt{1-\tau^2}h(t,\tau)\varphi(\tau)d\tau]\|_{C[-1,1]}=\\&=\|\int_{-1}^{1}\sqrt{1-\tau^2}D_n^t[h(t,\tau)]\varphi(\tau)d\tau\|_{C[-1,1]}\le\frac{C\ln n}{n^{r+\alpha}}\|x\|.\end{aligned} \tag{30}$$

Here $D_n=I-T_n$, D_n^t means that the operator D_n acts in the variable t; I is the identity operator.

Let's evaluate the functional $H((K-\tilde{K})x;\beta)$.

For simplify we introduce the following notations:

$$\psi(t)=\int_{-1}^{1}\sqrt{1-\tau^2}h(t,\tau)\varphi(\tau)d\tau,\ \psi_n(t)=\int_{-1}^{1}\sqrt{1-\tau^2}T_n^t[h(t,\tau)]\varphi(\tau)d\tau.$$

It was shown above that $|\psi(t)-\psi_n(t)|\le Cn^{-r-\alpha}\ln n\|x\|$.

Using the method of proving the S. N. Bernstein inverse theorems [38], we come to the estimate

$$H((\psi(t)-\psi_n(t));\beta)\le Cn^{-r-\alpha+\beta}\ln n\|x\|. \tag{31}$$

From (30)–(31) we have $\|(K-\tilde{K})x\|\le Cn^{-r-\alpha+\beta}\ln n\|x\|$.

Obviously

$$\|\tilde{K}x_n\| = \|Kx_n - (K - \tilde{K})x_n\| \geq \|Kx_n\| - \|(K - \tilde{K})x_n\| \geq \\ \geq q\|x_n\|, \quad q = \|K^{-1}\|^{-1} - Cn^{-r-\alpha+\beta}\ln n. \tag{32}$$

From the inequality (32) and the left inverse operator theorem [24] follows that for n such that $q > 0$, the operator $\tilde{K}$ has a left inverse operator on X_n. Since the left invertible operator $\tilde{K}$ maps the finite-dimensional subspace X_n to the finite-dimensional subspace Y_n, the operator $\tilde{K} \in [X_n, Y_n]$ is continuously invertible and the inequality $\|\tilde{K}^{-1}\| \leq \|K^{-1}\|/(1-q)$ is valid.

Let's estimate the norm of the following difference

$$\|(\tilde{K} - K_n)x_n\| = \|T_n[\int_{-1}^{1} \sqrt{1-\tau^2}(h(t,\tau)\varphi_n(\tau) - U_n^{\tau}[h(t,\tau)\varphi_n(\tau)])d\tau]\| \leq$$
$$\leq \|T_n[\int_{-1}^{1} \sqrt{1-\tau^2}(h(t,\tau)\varphi_n(\tau) - P_{n-1}^{\tau}[h(t,\tau)]\varphi_n(\tau))d\tau]\| +$$
$$+\|T_n[\int_{-1}^{1} \sqrt{1-\tau^2}(U_n^{\tau}[(P_{n-1}^{\tau}[h(t,\tau)] - h(t,\tau))\varphi_n(\tau)])d\tau]\| = J_1 + J_2. \tag{33}$$

Here $P_n^{\tau}[h(t,\tau)]$ is the polynomial of the best uniform approximation of degree n to the function $h(t,\tau)$ by the variable τ; the equality

$$\int_{-1}^{1} \sqrt{1-\tau^2} P_{n-1}^{\tau}[h(t,\tau)]\varphi_n(\tau)d\tau = \int_{-1}^{1} \sqrt{1-\tau^2} U_n^{\tau}[P_{n-1}^{\tau}[h(t,\tau)]\varphi_n(\tau)]d\tau$$

follows from the properties of the Gauss quadrature formula.

We will evaluate the expressions $\|J_1\|$ and $\|J_2\|$ in $C[-1,1]$.

It is known [38, 39, 41] that $\|T_n\|_{C[-1,1]} \leq C\ln n, \quad \|U_n\|_{C[-1,1]} \leq Cn$.

Obviously

$$\|J_1\|_{C[-1,1]} \leq C\|(h(t,\tau) - P_{n-1}^{\tau}[h(t,\tau)])\varphi_n(\tau)\|_{C[-1,1]}\ln n \leq \\ \leq Cn^{-r-\alpha}\ln n\|\varphi_n\|_{C[-1,1]} \leq Cn^{-r-\alpha}\ln n\|x_n\|, \tag{34}$$

$$\|J_2\| \leq Cn\ln n\|(h(t,\tau) - P_{n-1}^{\tau}[h(t,\tau)])\varphi_n(\tau)\|_{C[-1,1]} \leq \\ \leq Cn^{-r-\alpha+1}\ln n\|\varphi_n\|_{C[-1,1]} \leq Cn^{-r-\alpha+1}\ln n\|x_n\|. \tag{35}$$

From (33)–(35) we have $\|(\tilde{K} - K_n)x_n\|_{C[-1,1]} \leq Cn^{-r-\alpha+1}\ln n\|x_n\|$.

Using the method of proving the S. N. Bernstein inverse theorems [38, 39], we come to the estimate $\|(\tilde{K} - K_n)x_n\|_{X_n} \leq Cn^{-r-\alpha+1+\beta}\ln n\|x_n\|$.

From this and the inequality (32) it follows that the inequality

$$\|K_n x_n\| \geq \|\tilde{K} x_n\| - \|(\tilde{K} - K_n)x_n\| \geq q\|x_n\| - Cn^{-r-\alpha+1+\beta}\ln n\|x_n\| \geq q_1\|x_n\|$$

is valid for sufficiently large n. Here $q_1 = q - Cn^{-r-\alpha+1+\beta}\ln n > 0$.

From this inequality and the theorem on the left inverse operator [24] it follows that there is a left inverse operator $K_{n,l}^{-1}$ with the norm $\|K_{n,l}^{-1}\| \leq 1/q_1$.

Since the operator $K_n \in [X_n, Y_n]$ is finite-dimensional, the invertibility on the left implies continuous invertibility and the estimate $\|K_n^{-1}\| \leq 1/q_1$.

Thus, the unique solvability of the system of equations (27) is proved.

Let's estimate the error value of the proposed computational scheme. Denote by x^* the solution of the Eq. (26).

Let's introduce the intermediate equation

$$\tilde{K}x = f. \tag{36}$$

It was shown above that $\|(K - \tilde{K})x\| \leq Cn^{-r-\alpha+\beta}\ln n\|x\|$.

So, as it follows from Banach's theorem, that for n such that $q_2 = C\|K^{-1}\|n^{-r-\alpha+\beta}\ln n < 1$, the operator $\tilde{K}$ is continuously invertible and the estimate $\|\tilde{K}^{-1}\| \leq \|K^{-1}\|/(1 - q_2)$ is valid.

Denote by $\tilde{x}^*$ the solution of the Eq. (36).

It follows from the Banach theorem that

$$\|x^* - \tilde{x}^*\| \leq Cn^{-r-\alpha+\beta}\ln n\|x_n\|. \tag{37}$$

Let's introduce the equation

$$\tilde{K}x = f_n, \ f_n = T_n[f]. \tag{38}$$

Obviously, $\|f - f_n\| \leq Cn^{-r-\alpha+\beta}\ln n$.

Denote by $\bar{x}^*$ the solution of the Eq. (38).

It follows from the Banach theorem that

$$\|\tilde{x}^* - \bar{x}^*\| \leq Cn^{-r-\alpha+\beta}\ln n. \tag{39}$$

Note that $\bar{x}^* \in X_n$. This follows from the structure of the Eq. (38). In fact, the expression $T_n[\int_{-1}^{1}\sqrt{1-\tau^2}h(t,\tau)\bar{\varphi}(\tau)d\tau] - f_n(t)$ is a polynomial of n-th order.

Let's consider the equation

$$\int_{-1}^{1}\frac{x(\tau)}{(\tau - t)^2}d\tau = g_n, \tag{40}$$

where g_n is a polynomial of the n-th order. We will show that its solution is also a polynomial of the n-th order.

Below we will need in the famous [27] equality

$$\frac{1}{\pi}\int_{-1}^{1}\frac{\sqrt{1-\tau^2}U_n(\tau)}{(\tau-t)^2}d\tau = -(n+1)U_n(t),\ n=0,1,\ldots,$$

where $U_n(t)$ is the orthonormal Chebyshev polynomial of the second kind.

Let's show that the polynomial $g_n(t)=\sum_{k=0}^{n}\alpha_k t^k$ can be represented as $g_n(t)=\sum_{k=0}^{n}\beta_k U_k(t)$. It is known [38, 39] that for the Chebyshev polynomials of the second kind $\tilde{U}_n(t)$ with leading coefficient 1, the recursion formula $\tilde{U}_{n+2}(t)=t\tilde{U}_{n+1}(t)-\frac{1}{4}\tilde{U}_n(t)$ is valid. Here $\tilde{U}_0(t)=1,\ \tilde{U}_1(t)=t,\ \tilde{U}_2(t)=t^2-\frac{1}{4}$. It is easy to see that $g_n(t)=\sum_{k=0}^{n}\alpha_k t^k=\alpha_n U_n(t)+\alpha_{n-1}^1 U_{n-1}(t)+\cdots+\alpha_1^{n-1}U_1(t)+\alpha_0^n U_0(t)$.

It was shown above that the Eq. (40) has a unique solution. It is easy to see that this solution is the function $x(t)=-\sum_{k=0}^{n}\frac{\alpha_k^{n-k}}{k+1}\sqrt{1-t^2}U_k(t)$, where α_n^0 denotes the constant α_n.

Thus, the inclusion $\bar{x}^*\in X_n$ is valid.

It follows from this inclusion that Eq. (29) can be considered in the subspace X_n and there exists the linear operator $K_n^{-1}\in[Y_n,X_n]$ with the norm $||K^{-1}||/(1-q_2),q_2=C||K^{-1}||n^{-r-\alpha+1+\beta}$.

Let's estimate the norm of the difference

$$\|\tilde{K}x_n-K_nx_n\|=\|T_n[\int_{-1}^{1}\sqrt{1-\tau^2}(h(t,\tau)\varphi_n(\tau)-U_n^\tau[h(t,\tau)\varphi_n(\tau)])d\tau]\|\leq$$
$$\leq Cn^{-r-\alpha+1+\beta}\ln n\|x_n\|.$$

Here we used the estimates obtained above (in proving the existence of the operator K_n^{-1}).

From this estimate and the Banach theorem, it follows for $q_4=C\|\tilde{K}^{-1}\|n^{-r-\alpha+1+\beta}\ln n<1$, that the inequality

$$\|\bar{x}^*(t)-x_n^*(t)\|\leq Cn^{-r-\alpha+1+\beta}\ln n \tag{41}$$

is valid. Here x_n^* is the solution of the Eq. (27).

Collecting estimates (37), (39), (41), we have $\|x^*(t)-x_n^*(t)\|\leq Cn^{-r-\alpha+1+\beta}\ln n$.

Thus, the following statement is proved.

Theorem 2 *[6]. Let be $h(t,\tau)\in W^{r,r}H_{\alpha\alpha}(M)$, $f(t)\in W^rH_\alpha(M)$, $r=2,3,\ldots,$ $0<\alpha\leq1$, $M=$ const. Let be $K\in[X,Y]$ and there exists the linear inverse operator $K^{-1}\in[Y,X]$. Then, for n such that $q=Cn^{-r-\alpha+1+\beta}\ln n<1$, system of equations (27) is uniquely solvable and the inequality $\|x^*(t)-x_n^*(t)\|\leq Cn^{-r-\alpha+1+\beta}\ln n$ is valid. Here x^* and x_n^* solutions of equations (26) and (27), respectively.*

4.2.2 Weight Function $(1 - t^2)^{-1/2}$

Let's study the solvability of characteristic HSIEs of the form

$$K^0 x \equiv \frac{1}{\pi} \int_{-1}^{1} \frac{x(\tau)d\tau}{\sqrt{1-\tau^2}(\tau - t)^2} = f(t) \tag{42}$$

on the class of functions satisfying the following condition: the solution $x^*(t)$ of the Eq. (42) expands in powers of t: $x^*(t) = c_0 + c_1 t + \sum_{k=2}^{\infty} x_k t^k$; where c_0 and c_1 are predefined constants.

Let's introduce the space V_n of polynomials of the form $x_n(t) = c_0 + c_1 t + \sum_{k=2}^{n} \alpha_k t^k$ with the norm $\|x_n\| = \sum_{k=2}^{n} |\alpha_k|$ and the space W_n of polynomials of the form $y_n(t) = \sum_{k=0}^{n-2} \beta_k t^k$ with the norm $\|y_n\| = \sum_{k=0}^{n-2} |\beta_k|$.

We will show that the operator K^0 one-to-one maps the space V_n to the space W_n. To this end, we show that the Eq. (42) has a unique solution $x^*(t) \in V_n$ for any right-hand side $f(t) \in W_n$.

The solution of the Eq. (42) will be sought in the form

$$x_n(t) = c_0 + c_1 t + \sum_{k=2}^{n} \alpha_k t^k. \tag{43}$$

It is known [27] that

$$\frac{1}{\pi} \int_{-1}^{1} \frac{\tau^n d\tau}{\sqrt{1-\tau^2}(\tau - t)^2} = \begin{cases} 0, & n = 0, 1, \\ \sum\limits_{k=0}^{n-2} e_k^n t^k, & n = 2, 3, \ldots, \end{cases} \tag{44}$$

where

$$e_k^n = \begin{cases} 0, & n - k = 2j - 1, \\ \frac{(k+1)}{\sqrt{\pi}} \frac{\Gamma\left(\frac{n-k-1}{2}\right)}{\Gamma\left(\frac{n-k}{2}\right)}, & n - k = 2j \ (n \geq 2, k \leq n-2), \end{cases}$$

Γ is the gamma function, $j = 1, 2, \ldots$.

Using the formula (44), we have

$$\frac{1}{\pi} \int_{-1}^{1} \frac{x_n(\tau)d\tau}{\sqrt{1-\tau^2}(\tau - t)^2} = \sum_{k=2}^{n} \alpha_k \sum_{l=0}^{k-2} e_l^k t^l. \tag{45}$$

Since $f(t) \in W_n$, then

$$f(t) = \sum_{k=0}^{n-2} f_k t^k. \tag{46}$$

The equality (45) can be written as

$$\frac{1}{\pi}\int\limits_{-1}^{1}\frac{x_n(\tau)d\tau}{\sqrt{1-\tau^2}(\tau-t)^2}=\sum_{k=2}^{n}\alpha_k\begin{cases} e^k_{k-2}t^{k-2}+e^k_{k-4}t^{k-4}+\cdots+e^2_0,\, k=2j,\\ e^k_{k-2}t^{k-2}+e^k_{k-4}t^{k-4}+\cdots+e^k_1 t,\, k=2j-1,\end{cases}$$

$j=1,2,\ldots,$ because $e^k_l=0$ if $k-l$ is odd.

Substituting (45), (46) into (42), we have

$$\sum_{k=2}^{n}\alpha_k\sum_{l=0}^{k-2}e^k_l t^l=\sum_{k=0}^{n-2}f_k t^k. \tag{47}$$

Equating the coefficients in the left and right sides of equality (47) for the same powers of t, we obtain two systems of recurrence formulas.

Let n be an even number. Then $\alpha_n e^n_{n-2}=f_{n-2}$, $\alpha_n e^n_{n-4}+\alpha_{n-2}e^{n-2}_{n-4}=f_{n-4}$, $\cdots$, $\alpha_n e^n_0+\alpha_{n-2}e^{n-2}_0+\cdots+\alpha_2 e^2_0=f_0$.

A similar system is obtained when n is an odd number: $\alpha_{n-1}e^{n-1}_{n-3}=f_{n-3}$, $\alpha_{n-1}e^{n-1}_{n-5}+\alpha_{n-3}e^{n-3}_{n-5}=f_{n-5}$, $\cdots$, $\alpha_{n-1}e^{n-1}_1+\alpha_{n-3}e^{n-3}_1+\cdots+\alpha_1 e^3_1=f_1$.

It follows from two last systems that the coefficients α_k, $k=2,\ldots,n$, are uniquely determined. Thus, it is shown that the operator K^0 one-to-one maps the space V_n to the space W_n.

Denote by X the space of functions $x(t)$, satisfying the following conditions: functions $x(t)$ are expanded on the segment $[-1,1]$ in series $x(t)=c_0+c_1t+\sum_{k=2}^{\infty}x_k t^k$, where c_0 and c_1 are predefined numbers; derivative $x''(t)\in H_\alpha$, $0<\alpha\le 1$.

The norm in the space X is determined by the expression $\|x(t)\|=\sum_{j=0}^{2}\max_{t\in[-1,1]}|x^{(j)}(t)|+H(x'';\beta),\quad 0<\beta<\alpha.$

Denote by Y the space of functions $y(t)\in H_\beta$, $t\in[-1,1]$, with the norm $\|y(t)\|=\max_{-1\le t\le 1}|y(t)|+H(y;\beta)$.

Denote by X_n the subspace of the space X, and by Y_n the subspace of the space Y.

Let's show that the operator K maps the space X to the space Y.

Obviously

$$\begin{aligned}\psi(t)&=\int\limits_{-1}^{1}\frac{x(\tau)d\tau}{\sqrt{1-\tau^2}(\tau-t)^2}=\\&=\int\limits_{-1}^{1}\frac{1}{\sqrt{1-\tau^2}}\left[x(\tau)-x'(t)\frac{\tau-t}{1!}-x''(t)\frac{(\tau-t)^2}{2!}\right]\frac{d\tau}{(\tau-t)^2}+\frac{x''(t)}{2!}\int\limits_{-1}^{1}\frac{d\tau}{\sqrt{1-\tau^2}}=\\&=\int\limits_{-1}^{1}\frac{1}{\sqrt{1-\tau^2}}\frac{1}{(\tau-t)^2}\left[\int\limits_{t}^{\tau}(x''(v)-x''(t))(\tau-v)dv\right]d\tau+\frac{\pi}{2}x''(t).\end{aligned} \tag{48}$$

From the representation (48) it follows that

$$\| \int_{-1}^{1} \frac{1}{\sqrt{1-\tau^2}} \frac{1}{(\tau-t)^2} x(\tau) d\tau \|_{C[-1,1]} \leq C\|x\|.$$

Let's evaluate the functional $H(\psi, \beta)$. From (48) we have

$$|\psi(t_1) - \psi(t_2)| \leq \frac{\pi}{2}|x''(t_1) - x''(t_2)| +$$
$$+ \left| \int_{-1}^{1} \frac{1}{\sqrt{1-\tau^2}} \frac{1}{(\tau-t_1)^2} \left(\int_{t_1}^{\tau} (x''(v) - x''(t_1))(\tau - v) dv \right) d\tau - \right.$$
$$\left. - \int_{-1}^{1} \frac{1}{\sqrt{1-\tau^2}} \frac{1}{(\tau-t_2)^2} \left(\int_{t_2}^{\tau} (x''(v) - x''(t_2)(\tau - v) dv \right) d\tau \right| = |I_1| + |I_2|.$$

Obviously $|I_1| \leq \frac{\pi}{2} H(x'', \beta)|t_1 - t_2|^{\beta}$.

Let's proceed to evaluate the expression $|I_2|$. Let's $\rho = |t_2 - t_1|$, $t_2 > t_1$, $-1 < t_1 < 0$, $\Delta_1 = [-1, 1] \cap [t_1 - \rho, t_2 + \rho]$, $\Delta_2 = [-1, 1] \setminus \Delta_1$.

Let's imagine the expression $|I_2|$ as follows:

$$|I_2| \leq \left| \int_{\Delta_1} \frac{1}{\sqrt{1-\tau^2}} \frac{1}{(\tau-t_1)^2} \left(\int_{t_1}^{\tau} (x''(v) - x''(t_1))(\tau - v) dv \right) d\tau \right| +$$
$$+ \left| \int_{\Delta_1} \frac{1}{\sqrt{1-\tau^2}} \frac{1}{(\tau-t_2)^2} \left(\int_{t_2}^{\tau} (x''(v) - x''(t_2))(\tau - v) dv \right) d\tau \right| +$$
$$+ \left| \int_{\Delta_2} \frac{1}{\sqrt{1-\tau^2}} \frac{1}{(\tau-t_1)^2} \left(\int_{t_1}^{\tau} (x''(v) - x''(t_1))(\tau - v) dv \right) d\tau - \right.$$
$$\left. - \int_{\Delta_2} \frac{1}{\sqrt{1-\tau^2}} \frac{1}{(\tau-t_2)^2} \left(\int_{t_2}^{\tau} (x''(v) - x''(t_2))(\tau - v) dv \right) d\tau \right| = |I_{21}| + |I_{22}| + |I_{23}|.$$

Let's evaluate the expression $|I_{21}|$ (the expression $|I_{22}|$ is evaluated in the same way). Obviously
$|I_{21}| \leq C \int_{\Delta_1} \frac{1}{\sqrt{1-\tau^2}} |\tau - t_1|^{\beta} d\tau \leq C\rho^{\beta+1/2} \leq C|t_2 - t_1|^{\beta+1/2}$.

Let's evaluate the expressions I_{23} :

$$|I_{23}| \leq \left| \int_{\Delta_2} \left[\frac{1}{(\tau-t_1)^2} - \frac{1}{(\tau-t_2)^2} \right] \left(\int_{t_1}^{\tau} (x''(v) - x''(t_1))(\tau - v) dv \right) \frac{d\tau}{\sqrt{1-\tau^2}} \right| +$$
$$+ \left| \int_{\Delta_2} \frac{1}{\sqrt{1-\tau^2}} \frac{1}{(\tau-t_2)^2} \left[\int_{t_1}^{\tau} (x''(v) - x''(t_1))(\tau - v) dv) d\tau - \right. \right.$$
$$\left. \left. - \int_{t_2}^{\tau} (x''(v) - x''(t_2))(\tau - v) dv \right] d\tau \right| = |I_{231}| + |I_{232}|.$$

Let's evaluate each expression individually.

After simple calculations, we have

$$|I_{231}| \le C \int\limits_{\Delta_2} \frac{1}{\sqrt{1-\tau^2}} \frac{|t_2-t_1||2\tau-t_1-t_2||\tau-t_1|^{\beta}}{|\tau-t_2|^2} d\tau \le$$
$$\le C|t_2-t_1| \int\limits_{\Delta_2} \frac{1}{\sqrt{1-\tau^2}} \frac{1}{|\tau-t_2|^{1-\beta}} d\tau \le C|t_2-t_1|^{\beta}.$$

The following inequality is used here:

$$\frac{|2\tau-t_1-t_2|}{|\tau-t_2|} = \frac{2|\tau-(t_1+t_2)/2|}{|\tau-t_2|} =$$
$$= \frac{2|\tau-t_2+t_2-(t_1+t_2)/2|}{|\tau-t_2|} \le 2\left(\frac{|\tau-t_2|+|t_2-t_1|/2}{|\tau-t_2|}\right) \le 3.$$

It is correct, because $|\tau - t_2| \ge \rho$, $|t_2 - t_1| = \rho$.

More difficult to evaluate the expression I_{232}:

$$|I_{232}| \le \left| \int\limits_{\Delta_2} \frac{1}{\sqrt{1-\tau^2}} \frac{1}{(\tau-t_2)^2} [\int\limits_{t_1}^{t_2} (x''(v) - x''(t_1))(\tau - v) dv \right] d\tau| +$$
$$+ |\int\limits_{\Delta_2} \frac{1}{(\tau-t_2)^2} [\int\limits_{t_2}^{\tau} ((x''(t_1)) - x''(t_2)))(\tau - v) dv] \frac{d\tau}{\sqrt{1-\tau^2}}| = |I_{2321}| + |I_{2322}|.$$

Obviously

$$|I_{2321}| \le C \int\limits_{\Delta_2} \frac{1}{\sqrt{1-\tau^2}} \frac{|t_2-t_1|^{\beta}}{(\tau-t_2)^2} [(\tau - t_2)^2 - (\tau - t_1)^2] d\tau \le C|t_2 - t_1|^{\beta}.$$
$$|I_{2322}| \le C \int\limits_{\Delta_2} \frac{1}{\sqrt{1-\tau^2}} ||x''(t_1) - x''(t_2)| d\tau \le C||x''(t_1) - x''(t_2)|| \le C|t_1 - t_2|^{\beta}.$$

Collecting the obtained estimates, we are convinced of the validity of the inclusion $K^0 x \in H_\beta$. Thus, it is proved that $K \in [X, Y]$.

Let's assume that the operator $K^0 \in [X, Y]$ is continuously invertible.

We will approximate the Eq. (42) as follows

$$\bar{K}x \equiv \frac{1}{\pi} \int\limits_{-1}^{1} \frac{x(\tau)d\tau}{\sqrt{1-\tau^2}(\tau-t)^2} = f_{n-2}(t), \tag{49}$$

where $f_{n-2}(t)$ is the interpolation polynomial of degree $n - 2$ at nodes of the Chebyshev polynomial of the first kind.

Since $f_{n-2}(t)$ is a polynomial of degree $n - 2$, then, as was proved above, the Eq. (42) has a unique solution in the space X_n.

Denote respectively by $x_*(t)$ and $\bar{x}_*(t)$ the solutions of equations (42) and (49), satisfying the conditions $x_*(0) = c_0$, $x'_*(0) = c_1$ and $\bar{x}_*(0) = c_0$, $\bar{x}'_*(0) = c_1$. Then $\|x_*(t) - \bar{x}_*(t)\| \le Cn^{-r-\alpha+\beta} \ln n$.

Thus, the following statement is proved.

Theorem 3 *[6]. Let the operator $K^0 \in [X, Y]$ be continuously invertible. Let an approximate solution of the Eq. (42) be sought in the form of a polynomial (43) whose coefficients $\{\alpha_k\}$ are found from the system of equations (49). Then the system of equations (49) is uniquely solvable for any $n = 3, 4, \dots$, and the estimate $\|x_*(t) - \bar{x}_*(t)\| \le Cn^{-r-\alpha+\beta} \ln n$ is valid. Here x_* and $\bar{x}_*$ are the solutions of equations (42) and (49), respectively.*

Let's consider a hypersingular integral equation of the first kind

$$Kx \equiv \frac{1}{\pi} \int_{-1}^{1} \frac{x(\tau)d\tau}{\sqrt{1-\tau^2}(\tau - t)^2} + \int_{-1}^{1} \frac{h(t,\tau)x(\tau)d\tau}{\sqrt{1-\tau^2}} = f(t) \tag{50}$$

on the class of functions satisfying the following condition: the solution $x^*(t)$ of the Eq. (50) is expanded in a series by powers of t: $x^*(t) = c_0 + c_1 t + \sum_{k=2}^{\infty} x_k t^k$, where c_0 and c_1 are predefined constants.

An approximate solution of the Eq. (50) will be sought in the form of a polynomial $x_n(t) = c_0 + c_1 t + \sum_{k=2}^{n} \alpha_k t^k$, the coefficients $\{\alpha_k\}$ of which are determined from the system of equations

$$K_n x_n \equiv T_{n-2} \left[\frac{1}{\pi} \int_{-1}^{1} \frac{x_n(\tau)d\tau}{\sqrt{1-\tau^2}(\tau - t)^2} + \int_{-1}^{1} \frac{T_n^{\tau}[h(t,\tau)x_n(\tau)]}{\sqrt{1-\tau^2}} d\tau - f(t) \right] = 0, \tag{51}$$

where T_n is the operator of projection onto the set of interpolating polynomials of n-th order over the nodes of Chebyshev polynomials of the first kind; the superscript on the operator T_n^{τ} means that the interpolation is performed with respect to the variable τ.

From (45) it follows that Eq. (51) can be written in the form

$$\frac{1}{\pi} \int_{-1}^{1} \frac{x_n(\tau)d\tau}{\sqrt{1-\tau^2}(\tau - t)^2} + T_{n-2} \left[\int_{-1}^{1} \frac{T_n^{\tau}[h(t,\tau)x_n(\tau)]d\tau}{\sqrt{1-\tau^2}} - f(t) \right] = 0. \tag{52}$$

We assume that the operator $K \in [X, Y]$ has a linear inverse operator. It follows that on the subspace X_n the operator K has the left inverse operator.

Let's introduce the operator

$$\bar{K} x_n \equiv \frac{1}{\pi} \int_{-1}^{1} \frac{x_n(\tau)d\tau}{\sqrt{1-\tau^2}(\tau - t)^2} + T_{n-2} [\int_{-1}^{1} \frac{P_{n-1}^{\tau}[h(t,\tau)]x_n(\tau)d\tau}{\sqrt{1-\tau^2}}], \tag{53}$$

where P_{n-1}^{τ} is the projector onto the set of $n-1$ order polynomials of best uniform approximation. Assuming that $h(t,\tau) \in H_{\alpha,\alpha}$ and repeating the arguments used to justify the mechanical quadrature method in the theory of second-kind singular integral equations [4, 5], we have $\|Kx_n - \bar{K}x_n\| \le Cn^{-(\alpha-\beta)} \|x_n\| \ln n$.

It follows from the theorem of the left inverse operator [24] that for n such that $q = Cn^{-\alpha+\beta}\ln n < 1$, the operator $\bar{K}$ has the left invertible operator in the subspace X_n. Since the expression $P^{\tau}_{n-1}[h(t,\tau)]x_n(\tau)$ is a polynomial of degree $2n-1$ by the variable τ and the operator T_n is constructed on the nodes of the Chebyshev polynomial of the first kind, then

$$\int_{-1}^{1} \frac{1}{\sqrt{1-\tau^2}} P^{\tau}_{n-1}[h(t,\tau)]x_n(\tau)d\tau \equiv \int_{-1}^{1} \frac{1}{\sqrt{1-\tau^2}} T^{\tau}_n[P^{\tau}_{n-1}[h(t,\tau)]x_n(\tau)]d\tau.$$

Let's introduce the operator $\bar{K}_n$

$$\bar{K}_n x_n \equiv \frac{1}{\pi}\int_{-1}^{1} \frac{x_n(\tau)d\tau}{\sqrt{1-\tau^2}(\tau-t)^2} + T_{n-2}[\int_{-1}^{1} \frac{1}{\sqrt{1-\tau^2}} T_n[P^{\tau}_{n-1}[h(t,\tau)]x_n(\tau)]d\tau].$$

It is easy to see that for n such that $q_1 = Cn^{-\alpha+\beta}\ln^2 n < 1$, the left invertibility of the operator $\bar{K}$ implies the left invertibility of the operator $\bar{K}_n$. The transition from the operator $\bar{K}_n$ to the operator K_n is carried out by the Banach theorem. Since the operator K_n is finite-dimensional, the left invertibility implies its continuous invertibility.

Thus, the system of equations (51) is uniquely solvable.

It follows from the common theory of approximate methods [24] that if $f \in H_\alpha$, then for n such that $Cn^{-\alpha+\beta}\ln^2 n < 1$, the estimate $\|x^*(t)-x^*_n(t)\|_X \le Cn^{-\alpha+\beta}\ln^2 n$ is valid.

Thus, the following statement is proved.

Theorem 4 *[6]. Let the operator $K \in [X,Y]$ be continuously invertible and $h(t,\tau) \in H_{\alpha,\alpha}$, $f(t) \in H_\alpha$. Then for n such that $Cn^{-\alpha+\beta}\ln^2 n < 1$, the system of equations (51) is uniquely solvable and the estimate $\|x^* - x^*_n\|_X \le Cn^{-\alpha+\beta}\ln^2 n$ is valid. Here x^* and x^*_n are solutions of equations (50) and (51), respectively.*

4.2.3 Weight Functions $((1+t)/(1-t))^{\pm 1/2}$

In this section, we study approximate methods for solving hypersingular integral equations of the form

$$Kx \equiv \frac{1}{\pi}\int_{-1}^{1} \frac{x(\tau)}{(\tau-t)^2}d\tau + \int_{-1}^{1} h(t,\tau)x(\tau)d\tau = f(t), \ -1 < t < 1, \tag{54}$$

whose solution is sought in the class of functions $x(t) = ((1+t)/(1-t))^{\pm 1/2}\varphi(t)$. Here $\varphi(t)$ is a smooth function.

For definiteness, we restrict ourselves to considering the case when the weight function is equal to $((1+t)/(1-t))^{1/2}$.

An approximate solution of Eq. (54) will be sought in the form of a function

$$x_n(t) = \sqrt{\frac{1+t}{1-t}}\varphi_n(t) = \sqrt{\frac{1+t}{1-t}}\sum_{k=0}^{n}\alpha_k t^k,$$

with a predetermined coefficient $\alpha_0 = \text{const}$. This corresponds to the fact that the value $x^*(0) = \alpha_0$ is taken as an additional condition imposed on the solution of the Eq. (54).

Coefficients $\{\alpha_k\}$, $k = 1, 2, \ldots, n$, of function $x_n(t)$ are found from the following system of algebraic equations

$$K_n x_n \equiv T_{n-1}\left[\frac{1}{\pi}\int_{-1}^{1}\sqrt{\frac{1+\tau}{1-\tau}}\frac{\varphi_n(\tau)d\tau}{(\tau-t)^2} + \int_{-1}^{1}\frac{T_n^\tau[(1+\tau)h(t,\tau)\varphi_n(\tau)]}{\sqrt{1-\tau^2}}d\tau\right] = \\ = T_{n-1}[f(t)]. \quad (55)$$

Here T_n is the projection operator onto the set of interpolating polynomials of n-th order by the nodes of Chebyshev polynomials of the first kind.

We will transform the first integral in the Eq. (55):

$$\frac{1}{\pi}\int_{-1}^{1}\sqrt{\frac{1+\tau}{1-\tau}}\frac{\varphi_n(\tau)}{(\tau-t)^2}d\tau = \frac{1}{\pi}\int_{-1}^{1}\frac{1+\tau}{\sqrt{1-\tau^2}}\frac{\varphi_n(\tau)}{(\tau-t)^2}d\tau = \frac{1}{\pi}\sum_{k=0}^{n-1}\beta_k' t^k.$$

So,

$$T_{n-1}\left[\frac{1}{\pi}\int_{-1}^{1}\sqrt{\frac{1+\tau}{1-\tau}}\frac{\varphi_n(\tau)}{(\tau-t)^2}d\tau\right] = \frac{1}{\pi}\int_{-1}^{1}\sqrt{\frac{1+\tau}{1-\tau}}\frac{\varphi_n(\tau)}{(\tau-t)^2}d\tau$$

and the Eq. (55) can be represented as

$$K_n x_n \equiv \frac{1}{\pi}\int_{-1}^{1}\sqrt{\frac{1+\tau}{1-\tau}}\frac{\varphi_n(\tau)}{(\tau-t)^2}d\tau + \\ +T_{n-1}\left[\int_{-1}^{1}\frac{1}{\sqrt{1-\tau^2}}T_n^\tau[(1+\tau)h(t,\tau)\varphi_n(\tau)]d\tau\right] = T_{n-1}[f(t)]. \quad (56)$$

Let's introduce the space X of functions of the form $x(t) = \sqrt{\frac{1+\tau}{1-\tau}}\varphi(t)$, where $\varphi(t)$ is a function included in the class $W^r H_\alpha(M)$, $r = 2, 3, \ldots$, $0 < \alpha \le 1$, and satisfying the condition $\varphi(0) = c_0$. The norm in the space X is determined by the formula $\|x(t)\| = \sum_{j=0}^{2}\|\varphi^{(j)}(t)\|_{C[-1,1]} + H(\varphi^{(2)}, \beta)$, $\beta < \alpha$. Denote by X_n the subspace of the space X, consisting of functions $x_n(t) = \sqrt{\frac{1+\tau}{1-\tau}}\varphi_n(\tau)$, where $\varphi_n(\tau) = \sum_{k=0}^{n}\alpha_k t^k$, $\alpha_0 = c_0$, with the norm induced by the space X.

We denote by Y the space of functions $y(t) \in H_\beta$ with the norm $\|y(t)\| = \|y(t)\|_{C[-1,1]} + H(y, \beta)$. Denote by Y_n the subspace of the space Y, consisting of polynomials of the form $y_n(t) = \sum_{k=0}^{n-1} \beta_k t^k$. The norm in Y_n is induced by the norm of the space Y.

Let's show that the operator K maps X to Y. For this, it is sufficient to consider the characteristic equation $K^0 x \equiv \frac{1}{\pi}\int_{-1}^{1} \frac{x(\tau)}{(\tau-t)^2} d\tau = f(t)$.

Using the results of the previous section, we estimate the module $K^0 x$:

$$\left|\frac{1}{\pi}\int_{-1}^{1} \frac{x(\tau)}{(\tau-t)^2} d\tau\right| = \left|\frac{1}{\pi}\int_{-1}^{1} \frac{1+\tau}{\sqrt{1-\tau^2}} \frac{\varphi(\tau)}{(\tau-t)^2} d\tau\right| \leq C\|x\|.$$

Let's introduce the notation

$$\Psi(t) = \frac{1}{\pi}\int_{-1}^{1} \frac{x(\tau)}{(\tau-t)^2} d\tau = \frac{1}{\pi}\int_{-1}^{1} \frac{(1+\tau)\varphi(\tau)}{\sqrt{1-\tau^2}(\tau-t)^2} d\tau.$$

Let's estimate the difference

$$|\Psi(t+h) - \Psi(t)| = \left|\frac{1}{\pi}\int_{-1}^{1} \frac{(1+\tau)\varphi(\tau)d\tau}{\sqrt{1-\tau^2}(\tau-(t+h))^2} - \frac{1}{\pi}\int_{-1}^{1} \frac{(1+\tau)\varphi(\tau)d\tau}{\sqrt{1-\tau^2}(\tau-t))^2}\right| \leq$$

$$\leq \left|\frac{1}{\pi}\int_{-1}^{1} \frac{\varphi(\tau)d\tau}{\sqrt{1-\tau^2}(\tau-(t+h))^2} - \frac{1}{\pi}\int_{-1}^{1} \frac{\varphi(\tau)d\tau}{\sqrt{1-\tau^2}(\tau-t))^2}\right| +$$

$$+ \left|\frac{1}{\pi}\int_{-1}^{1} \frac{\tau\varphi(\tau)d\tau}{\sqrt{1-\tau^2}(\tau-(t+h))^2} - \frac{1}{\pi}\int_{-1}^{1} \frac{\tau\varphi(\tau)d\tau}{\sqrt{1-\tau^2}(\tau-t))^2}\right| = |J_3(t)| + |J_4(t)|.$$

Since both expressions are evaluated equally, we restrict ourselves to the consideration of the first of them:

$$|I_3(t)| \leq \left|\frac{1}{\pi}\int_{-1}^{1} \frac{\varphi(\tau)-\varphi(t+h)-\frac{\varphi'(t+h)}{1!}(\tau-(t+h))-\frac{\varphi''(t+h)}{2!}(\tau-(t+h))^2 d\tau}{\sqrt{1-\tau^2}(\tau-(t+h))^2} -\right.$$

$$\left.-\frac{1}{\pi}\int_{-1}^{1} \frac{\varphi(\tau)-\varphi(t)-\frac{\varphi'(t)}{1!}(\tau-t)-\frac{\varphi''(t)}{2!}(\tau-t)^2 d\tau}{\sqrt{1-\tau^2}(\tau-t)^2}\right| + \frac{1}{2}|\varphi''(t+h) - \varphi''(t)| = |I_{31}(t)| + |I_{32}(t)|.$$

Obviously $|I_{32}(t)| \leq Ch^\beta$.

To estimate $|I_{31}|$, we use the Taylor formula with the remainder term in integral form. We have

$$|I_{31}(t)| = \left|\frac{1}{\pi}\int_{-1}^{1} \frac{\int_{t+h}^{\tau} (\varphi''(v)-\varphi''(t+h))(\tau-v)dv}{\sqrt{1-\tau^2}(\tau-(t+h))^2} d\tau -\right.$$

$$\left.-\frac{1}{\pi}\int_{-1}^{1} \frac{\int_{t}^{\tau} (\varphi''(v)-\varphi''(t))(\tau-v)dv}{\sqrt{1-\tau^2}(\tau-t)^2} d\tau\right| \leq Ch^\beta.$$

Collecting the obtained estimates, we verify the inclusion $K^0 \in [X, Y]$.

Let's show that if the operator $K \in [X, Y]$ has a linear inverse operator $K^{-1} \in [Y, X]$, then for n such that $q = Cn^{-r-\alpha+\beta} \ln n < 1$, the operator $K_n \in [X_n, Y_n]$ continuously reversible and the estimate $\|x^*(t) - x_n^*(t)\| \le Cn^{-r-\alpha+\beta} \ln n$ is valid. Here x^* and x_n^* are solutions of equations (54) and (55), respectively.

Let's introduce the intermediate equation

$$\bar{K}x \equiv \frac{1}{\pi}\int_{-1}^{1} \sqrt{\frac{1+\tau}{1-\tau}} \frac{\varphi(\tau)d\tau}{(\tau - t)^2} + T_{n-1}\left[\int_{-1}^{1} \frac{(1+\tau)h(t,\tau)\varphi(\tau)}{\sqrt{1-\tau^2}} d\tau\right] = f(t). \quad (57)$$

It is easy to see that $\Psi(t) = \int_{-1}^{1} \frac{1}{\sqrt{1-\tau^2}}[(1+\tau)h(t,\tau)\varphi(\tau)]d\tau \in W^r H_\alpha(M)$.

Consequently,

$$\|(K - \bar{K})x\| = \|D_{n-1}[\Psi(t)]\| \le Cn^{-r-\alpha+\beta} \ln n \|\varphi\|_{C[-1,1]} \le Cn^{-r-\alpha+\beta} \ln n \|x\|,$$

where $D_{n-1} = I - T_{n-1}$.

From this inequality and the Banach theorem it follows that for n such that $q = C\|K^{-1}\| n^{-r-\alpha+\beta} \ln n < 1$, the Eq. (57) is uniquely solvable and estimates $\|\bar{K}^{-1}\| \le \|K^{-1}\|/(1-q)$ and $\|x^* - \bar{x}^*\| \le Cn^{-r-\alpha+\beta} \ln n$ are valid. Here $\bar{x}^*$ is the solution of the Eq. (57).

Let's consider the equation

$$\bar{K}x_n = f_{n-1} = T_{n-1}[f]. \quad (58)$$

Obviously $\|x_n^* - \bar{x}_n^*\| \le Cn^{-r+\beta} \ln n$, where $\bar{x}_n^*$ is the solution of the Eq. (58).

Let us prove that the operator $\bar{K}$ maps the space X_n to Y_n.

To do this, just consider the equation

$$K^0 x \equiv \frac{1}{\pi}\int_{-1}^{1} \sqrt{\frac{1+\tau}{1-\tau}} \varphi_n(\tau) \frac{d\tau}{(\tau - t)^2} = f_{n-1}(t). \quad (59)$$

In the space X the Eq. (59) has a unique solution.

Let us show that this solution belongs to the subspace X_n.

To do this, we represent the Eq. (59) in the form

$$\frac{1}{\pi}\int_{-1}^{1} \frac{1}{\sqrt{1-\tau^2}}(1+\tau)\frac{\varphi(\tau)}{(\tau - t)^2} d\tau = f_{n-1}(t). \quad (60)$$

We will seek a solution of the Eq. (60) in the form of a polynomial $\tilde{\varphi}(t) = \sum_{k=0}^{n} v_k t^k$.

Then, using the formulae (44), we receive

$$\frac{1}{\pi}\int_{-1}^{1}\frac{1}{\sqrt{1-\tau^2}}(1+\tau)\frac{\tilde{\varphi}(\tau)}{(\tau-t)^2}d\tau=\sum_{k=1}^{n}v_k\sum_{l=0}^{k-1}e_l^{k+1}t^l+\sum_{k=2}^{n}v_k\sum_{l=0}^{k-2}e_l^k t^l.$$

Here $e_l^k=0$, if $k-l$ be odd.

Substituting this expression into the Eq. (60), we have

$$\sum_{k=1}^{n}v_k\sum_{l=0}^{k-1}e_l^{k+1}t^l+\sum_{k=2}^{n}v_k\sum_{l=0}^{k-2}e_l^k t^l=\sum_{k=0}^{n-1}c_k t^k. \tag{61}$$

From the Eq. (61) follows the system of equations: $v_n e_{n-1}^{n+1}=c_{n-1}$, $v_{n-1}e_{n-2}^{n}+v_n e_{n-2}^{n}=c_{n-2}$, $v_{n-2}e_{n-3}^{n-1}+v_{n-1}e_{n-3}^{n-1}=c_{n-3}$, $\ldots$.

It follows from this system that the coefficients v_k, $k=0,1,\ldots,n$, are determined uniquely.

Therefore, Eqs. (59) and (60) have solutions in the subspace X_n.

Since the Eq. (60) has a unique solution in the space X, it follows that the operator K^0 acts from X_n to Y_n and has a linear inverse operator $[K^0]^{-1}\in[Y_n,X_n]$.

From the above considerations it follows that the operator $K\in[X_n,Y_n]$ has the linear inverse operator $\bar{K}^{-1}\in[Y_n,X_n]$.

Let's introduce the intermediate equation

$$\bar{K}_n x_n\equiv\frac{1}{\pi}\int_{-1}^{1}\sqrt{\frac{1+\tau}{1-\tau}}\frac{\varphi_n(\tau)}{(\tau-t)^2}d\tau+$$
$$+T_{n-1}\left[\frac{1}{\pi}\int_{-1}^{1}\frac{1}{\sqrt{1-\tau^2}}(1+\tau)P_{n-2}^{\tau}[h(t,\tau)]\varphi_n(\tau)d\tau\right]=f_{n-1}(t).$$

Here, $P_n^{\tau}[h(t,\tau)]$ denotes the polynomial of the best uniform approximation of degree n to the function $h(t,\tau)$ (by the variable τ).

Easy to see that $\|\bar{K}x_n-\bar{K}_n x_n\|_{C[-1,1]}\le Cn^{-r-\alpha}\ln n\|x_n\|_X$.

Using the method of proving the S. N. Bernstein inverse theorems [38, 39], we come to the estimate $\|\bar{K}x_n-\bar{K}_n x_n\|\le Cn^{-r-\alpha+\beta}\ln n\|x_n\|$.

Since the operator $\bar{K}\in[X_n,Y_n]$ is continuously invertible, then from the last estimate and Banach's theorem it follows that for n such that $q_1=C\|\bar{K}^{-1}\|n^{-r-\alpha+\beta}\ln n<1$, the operator $\bar{K}_n\in[X_n,Y_n]$ continuously reversible and $\|\bar{K}_n^{-1}\|\le\|\bar{K}^{-1}\|/(1-q_1)$.

Let's estimate the norm of the difference $\|\bar{K}x_n-K_n x_n\|$. We have

$$\|\bar{K}x_n-K_n x_n\|_{C[-1,1]}=\|T_{n-1}\left[\int_{-1}^{1}T_n^{\tau}[(1+\tau)D_{n-2}^{\tau}[h(t,\tau)]\varphi_n(\tau)]\frac{d\tau}{\sqrt{1-\tau^2}}\right]\|\le$$
$$\le Cn^{-r-\alpha}\ln^2 n\|\varphi_n\|\le Cn^{-r-\alpha}\ln^2 n\|x_n\|,\quad D_{n-2}=I-T_{n-2}.$$

Using the method of proving the S. N. Bernstein inverse theorems [38], we come to the estimate $\|\bar{K}_n x_n-K_n x_n\|\le Cn^{-r-\alpha+\beta}\ln^2 n\|x_n\|$.

From this estimate and the Banach theorem it follows that, for n such that $q_2 = C\|\bar{K}_n^{-1}\| n^{-r-\alpha+\beta} \ln^2 n < 1$, the operator $K_n \in [X_n, Y_n]$ is continuously reversible with estimate $\|K_n^{-1}\| \le \|\bar{K}_n^{-1}\|/(1-q_2)$.

Thus, it is proved that the Eq. (55) is uniquely solvable.

Denote by x^* and x_n^* the solutions of equations (54) and (55), respectively. It is easy to see that $\|x^* - x_n^*\| \le Cn^{-r-\alpha+\beta} \ln^2 n$.

Theorem 5 *[7]. Let the following conditions be satisfied:*
(1) $h(t, \tau) \in W^{r,r} H_{\alpha\alpha}(M)$, $f(t) \in W^r H_\alpha$;
(2) operator $K \in [X, Y]$ *is continuously reversible.*
Then for n such that $q = Cn^{-r-\alpha+\beta} \ln n < 1$, *the system of equations (55) is uniquely solvable and the estimate* $\|x^* - x_n^*\| \le Cn^{-r-\alpha+\beta} \ln^2 n$ *is valid. Here* x^* *and* x_n^* *solutions of equations (54) and (55), respectively.*

5 An Approximate Solution of Second Kind HSIEs

Abstract. The method of mechanical quadrature for the approximate solution of second kind HSIEs is studied. Estimates of the error and speed of convergence are given. Note that the results obtained in this section are also valid for HSIEs of the first kind.

5.1 An Approximate Solution of Linear HSIEs on Closed Circuits

Let's consider the hypersingular integral equation

$$Kx \equiv a(t)x(t) + \frac{1}{2\pi i} \int\limits_\gamma \frac{h(t,\tau)x(\tau)}{(\tau - t)^p} d\tau = f(t), \quad p = 2, 3, \ldots, \tag{62}$$

where γ is the unit circle centered at the origin.

We assume that the coefficients and the right-hand side of the Eq. (62) satisfy to one of the following conditions:

(1) $a(t)$, $f(t) \in H_\alpha$, function $h(t, \tau)$ satisfies the Holder condition H_α in first variable and belongs to the class of functions $W^{p-1} H_\alpha$ in second variable, $0 < \alpha \le 1$,

(2) $a(t)$, $f(t) \in W^r H_\alpha$, function $h(t, \tau)$ belongs to the class of functions $W^r H_\alpha$ in first variable and to the class of functions $W^{r+p-1} H_\alpha$ in second variable, $0 < \alpha \le 1$.

By X we denote the Banach space of functions $W^{p-1}H_\beta$, $0 < \beta < \alpha \leq 1$. The norm in the space X is introduced by the formula $\|x(t)\|_X = \sum_{k=0}^{p-1} \max_{t\in\gamma} |x^{(k)}(t)| + H(x^{(p-1)}, \beta)$.

By Y we denote the Banach space of functions $y(t) \in H_\beta$ $(0 < \beta < \alpha)$ with the norm $\|y(t)\|_Y = \|y(t)\|_{H_\beta} = \max_{t\in\gamma} |y(t)| + H(y, \beta)$.

An approximate solution of the Eq. (62) we will search in the form of a polynomial

$$x_n(t) = \sum_{k=0}^{n} \alpha_k t^{k+p-1} + \sum_{k=-n}^{-1} \alpha_k t^k, \tag{63}$$

which coefficients $\{\alpha_k\}$ are determined from the system

$$K_n x_n \equiv \bar{P}_n \left[a(t)x_n(t) + \frac{1}{2\pi i} \int_\gamma P_n \left[\frac{h(t,\tau)x_n(\tau)}{(\tau - t)^p} \right] d\tau \right] = \bar{P}_n[f(t)]. \tag{64}$$

Here $P_n(\bar{P}_n)$ is the projector onto the set of interpolation trigonometric polynomials by nodes $t_k = e^{is_k}$, $s_k = 2k\pi/(2n+1)$ $(\bar{t}_k = e^{i\bar{s}_k}, \bar{s}_k = (2k+1)\pi/(2n+1))$, $k = 0, 1, \ldots, 2n$.

Theorem 6 *[8]. Let the Eq. (62) be uniquely solvable and conditions (1) are satisfied. Then for n such that $q = C \ln^3 n/n^{\alpha-\beta}$, the system of equations (64) is uniquely solvable and the estimate $\|x^* - x_n^*\|_{H_\beta} \leq C \ln^3 n/n^{\alpha-\beta}$ is valid. Here x^* and x_n^* are solutions of equations (62) and (64), respectively.*

Proof. At first, we will study the solvability of collocation method for the Eq. (62). Collocation method in operator form is written as

$$\tilde{K}_n x_n \equiv \bar{P}_n \left[a(t)x_n(t) + \frac{1}{2\pi i} \int_\gamma \frac{h(t,\tau)x_n(\tau)}{(\tau - t)^p} d\tau \right] = \bar{P}_n[f(t)]. \tag{65}$$

Using the definition of a hypersingular integral, we transform the equations (62) and (65) to the equivalent singular integro-differential equation

$$Kx \equiv a(t)x(t) + \frac{1}{2\pi i} \frac{(-1)^{p-1}}{(p-1)!} \int_\gamma \frac{(h(t,\tau)x(\tau))_\tau^{(p-1)}}{\tau - t} d\tau = f(t) \tag{66}$$

and to the system of equations approximating it:

$$\tilde{K}_n x_n \equiv \bar{P}_n \left[a(t)x_n(t) + \frac{(-1)^{p-1}}{2\pi i (p-1)!} \int_\gamma \frac{(h(t,\tau)x(\tau))_\tau^{(p-1)} d\tau}{\tau - t} \right] = \bar{P}_n[f(t)]. \tag{67}$$

Here by $(h(t,\tau)x(\tau))_\tau^{(p-1)}(\tau)$ is denoted the $(p-1)$-order derivative of function $(h(t,\tau)x(\tau))$ with respect to variable τ.

Let's continue the transformation of Eqs. (66)–(67). Obviously

$$
\begin{aligned}
Kx &\equiv a(t)x(t) + \frac{1}{2\pi i}\frac{(-1)^{p-1}}{(p-1)!}\int_\gamma \frac{h(t,\tau)x^{(p-1)}(\tau)}{\tau-t}d\tau + \\
&+\sum_{k=0}^{p-2}\frac{1}{2\pi i}\int_\gamma \frac{h_k^*(t,\tau)x^{(k)}(\tau)}{\tau-t}d\tau = f(t),
\end{aligned}
\tag{68}
$$

$$
\begin{aligned}
\tilde{K}_n x_n &\equiv \bar{P}_n\Bigg[a(t)x_n(t) + \frac{1}{2\pi i}\frac{(-1)^{p-1}}{(p-1)!}\int_\gamma \frac{h(t,\tau)x_n^{(p-1)}(\tau)}{\tau-t}d\tau + \\
&+\sum_{k=0}^{p-2}\frac{1}{2\pi i}\int_\gamma \frac{h_k^*(t,\tau)x_n^{(k)}(\tau)}{\tau-t}d\tau\Bigg] = \bar{P}_n[f(t)].
\end{aligned}
\tag{69}
$$

Here, $h_k^*(t,\tau)$, $k=0,1,\ldots,p-2$, denotes functions, construction which are obvious.

It is convenient to present Eqs. (68) and (69) in the following form

$$
\begin{aligned}
Kx &\equiv a(t)x(t) + \sum_{k=0}^{p-1}\Bigg[b_k(t)\frac{1}{2\pi i}\int_\gamma \frac{x^{(k)}(\tau)d\tau}{\tau-t} + \\
&+\int_\gamma g_k(t,\tau)x^{(k)}(\tau)d\tau\Bigg] = f(t),
\end{aligned}
\tag{70}
$$

and

$$
\begin{aligned}
\tilde{K}_n x_n &\equiv \bar{P}_n\Bigg[a(t)x_n(t) + \sum_{k=0}^{p-1}\Bigg[b_k(t)\frac{1}{2\pi i}\int_\gamma \frac{x_n^{(k)}(\tau)d\tau}{\tau-t} + \\
&+\int_\gamma g_k(t,\tau)x_n^{(k)}(\tau)d\tau\Bigg]\Bigg] = \bar{P}_n[f(t)].
\end{aligned}
\tag{71}
$$

Here the coefficient $b_{p-1}(t) = \frac{h(t,t)(-1)^{p-1}}{(p-1)!}$; the construction of the coefficients $b_k(t)$, $k=0,1,\ldots,p-2$, $g_k(t,\tau)$, $k=0,1,\ldots,p-1$, is obvious.

It is easy to see that Eqs. (62) and (70), as well as (64) and (71) are identical. Thus, the problem of substantiation of the collocation method for the Eq. (62) reduced to justification of the collocation method for singular integro-differential equation (70).

Collocation method for linear singular integro-differential equations substantiated in the work [5]. From results of this work it follows that, for n such that $q = C\ln n/n^{\alpha-\beta} < 1$, the system of equations (71) is uniquely solvable and the estimates $\|[\tilde{K}_n]^{-1}\| \le C/(1-q)$ and $\|x^* - x_n^{**}\| \le C\ln n/n^{\alpha-\beta}$ are valid. Here x^* and x_n^{**} are solutions of equations (62) and (71), respectively.

Since the system of equations (71) is obtained from the system of Eq. (65) with identical transformations, then for the above values of n the system of equations (65) is uniquely solvable and the function x_n^{**} is it solution.

Thus, it is proved that, if the operator $K \in [X, Y]$ is continuously invertible, for n such that $q = C \ln n / n^{\alpha-\beta} < 1$, the operator $\tilde{K}_n \in [X_n, Y_n]$ is continuously invertible too.

Here, by $X_n \subset X$ we denoted the set of polynomials of the form (63) and by Y_n we denoted the set of polynomials of the form $\sum_{k=-n}^{n} \alpha_k t^k$.

To justify the mechanical quadrature method (64), we estimate the norm of the following difference:

$$\|K_n x_n - \tilde{K}_n x_n\| = \|\bar{P}_n \left[\int_\gamma R_n \left[\frac{h(t,\tau)x_n(\tau)}{(\tau-t)^p} \right] d\tau \right] \|, \tag{72}$$

where $R_n = I - P_n$, I is the identity operator.

Consider the polynomial $\bar{P}_n \left[\int_\gamma \frac{h(t,\tau)x_n(\tau)}{(\tau-t)^p} d\tau \right]$.

Using the definition of hypersingular integrals, we have

$$\begin{aligned} &\bar{P}_n \left[\int_\gamma \frac{h(t,\tau)x_n(\tau)}{(\tau-t)^p} d\tau \right] = \bar{P}_n \left[\frac{(-1)^{p-1}}{(p-1)!} \int_\gamma \frac{(h(t,\tau)x_n(\tau))_\tau^{(p-1)}}{\tau-t} d\tau \right] = \\ &= \frac{(-1)^{p-1}}{(p-1)!} \bar{P}_n \left[\int_\gamma \frac{h(t,\tau)x_n^{(p-1)}(\tau)}{\tau-t} d\tau + \cdots + \int_\gamma \frac{[h(t,\tau)]_\tau^{(p-1)} x_n(\tau)}{\tau-t} d\tau \right]. \end{aligned} \tag{73}$$

Denote by $\tilde{h}(t,\tau)$ the trigonometric polynomial of the best uniform approximation of degree $n-p+1$ in each variable to the function $h(t,\tau)$. Then

$$\begin{aligned} J = &\|\bar{P}_n \left[\frac{(-1)^{p-1}}{(p-1)!} \int_\gamma \left[\left[\frac{h(t,\tau)x_n^{(p-1)}(\tau)}{\tau-t} \right] - P_n \left[\frac{h(t,\tau)x_n^{(p-1)}(\tau)}{\tau-t} \right] \right] d\tau \right] \|_{Y_n} \leq \\ &\leq \|\bar{P}_n \left[\frac{(-1)^{p-1}}{(p-1)!} \int_\gamma \frac{h(t,\tau)-\tilde{h}(t,\tau)}{\tau-t} x_n^{(p-1)}(\tau) d\tau \right] \|_{Y_n} + \\ &+ \|\bar{P}_n \left[\frac{(-1)^{p-1}}{(p-1)!} \int_\gamma \frac{\tilde{h}(t,\tau)x_n^{(p-1)}(\tau) - P_n[\tilde{h}(t,\tau)x_n^{(p-1)}(\tau)]}{\tau-t}]d\tau \right] \|_{Y_n} + \\ &+ \|\bar{P}_n \left[\frac{(-1)^{p-1}}{(p-1)!} \int_\gamma \frac{P_n[(\tilde{h}(t,\tau)-h(t,\tau))x_n^{(p-1)}(\tau)]}{\tau-t} d\tau \right] \|_{Y_n} + \\ &+ \|\bar{P}_n \left[\frac{(-1)^{p-1}}{(p-1)!} \int_\gamma \left[\frac{P_n[(h(t,\tau)x_n^{(p-1)}(\tau)]}{\tau-t} - P_n \left[\frac{h(t,\tau)x_n^{(p-1)}(\tau)}{\tau-t} \right] \right] d\tau \right] \|_{Y_n} = \\ &= J_1 + J_2 + J_3 + J_4. \end{aligned} \tag{74}$$

Let us evaluate each of the expressions $J_1 - J_4$ separately.

To estimate J_1, we note, by the well-known Bernstein theorem [2], that $\|h_l(t,\tau)-\tilde{h}_l(t,\tau)\|_C \le C\ln n \max\{E^t_{n-p+1}[h(t,\tau)], E^\tau_{n-p+1}[h(t,\tau)]\}$.

It follows that the function $\eta(t,\tau)=h(t,\tau)-\tilde{h}(t,\tau)$ enters into the Holder class of functions (with respect to the variable τ) with exponent $1/\ln n$ and with the coefficient $A\ln n E^*_n(h)$, where $E^*_n(h)=\ln n\max\{E^t_n[h(t,\tau)], E^\tau_n[h(t,\tau)]\}$.

Then

$$J_1 \le C\|x_n^{(p-1)}(t)\|_{H_\beta} n^\beta \ln n E^*_n(h). \tag{75}$$

We now note that the function $\psi_l(t,\tau)=\frac{\tilde{h}(t,\tau)-\tilde{h}(t,t)}{\tau-t}x_n^{(p-1)}(\tau)$ is a polynomial of degree $2n-1$ in the variable τ.

Below we will need in the Hilbert formula [19]:

$$\frac{1}{\pi}\int_\gamma \frac{\varphi(\tau)}{\tau-t} = \frac{1}{2\pi}\int_0^{2\pi}\varphi(e^{i\sigma})ctg\frac{\sigma-s}{2}d\sigma + \frac{i}{2\pi}\int_0^{2\pi}\varphi(e^{i\sigma})d\sigma.$$

Using the Hilbert formula and properties of Gauss quadrature formulas, we will prove the following formula

$$\|\bar{P}_n\left[\int_\gamma \frac{P_n^\tau[h(t,\tau)x_n^{(p-1)}(\tau)]}{\tau-t}d\tau\right]\|_{Y_n} = \|\bar{P}_n\left[\int_\gamma P_n^\tau\left[\frac{h(t,\tau)x_n^{(p-1)}(\tau)}{\tau-t}\right]d\tau\right]\|_{Y_n}. \tag{76}$$

Then

$$\frac{1}{\pi}\bar{P}_n\left[\int_\gamma \frac{P_n^\tau[h(t,\tau)x_n^{(p-1)}(\tau)]}{\tau-t}d\tau\right] = \frac{1}{2\pi}\bar{P}_n[\int_0^{2\pi} P_n^\sigma[h(e^{is},e^{i\sigma})x_n^{(p-1)}(e^{i\sigma})]ctg\tfrac{\sigma-s}{2}d\sigma +$$
$$+\frac{i}{2\pi}\int_0^{2\pi} P_n^\sigma[h(e^{is},e^{i\sigma})x_n^{(p-1)}(e^{i\sigma})]d\sigma] = \frac{1}{2\pi}\bar{P}_n\int_0^{2\pi}\{P_n^\sigma[h(e^{is},e^{i\sigma})x_n^{(p-1)}(e^{i\sigma})](\sigma)-$$
$$-P_n^\sigma[h(e^{is},e^{i\sigma})x_n^{(p-1)}(e^{i\sigma})](s)\}ctg\tfrac{\sigma-s}{2}d\sigma]+$$
$$+\bar{P}_n^s\left[\frac{i}{2\pi}\int_0^{2\pi} P_n^\sigma[h(e^{is},e^{i\sigma})x_n^{(p-1)}(e^{i\sigma})]d\sigma\right] =$$
$$=\frac{1}{2\pi}\bar{P}_n^s\left[\int_0^{2\pi} P_n^\sigma[(P_n^\sigma[h(e^{is},e^{i\sigma})]x_n^{(p-1)}(e^{i\sigma})(\sigma)-\right.$$
$$\left.-P_n^\sigma[h(e^{is},e^{i\sigma})]x_n^{(p-1)}(e^{i\sigma})(s)\Big)\,ctg\tfrac{\sigma-s}{2}]d\sigma\right]+$$
$$+\frac{i}{2\pi}\bar{P}_n^s\left[\int_0^{2\pi} P_n^\sigma[h(e^{is},e^{i\sigma})x_n^{(p-1)}(e^{i\sigma})]d\sigma\right] =$$

$$= \bar{P}_n^s \left[\int_0^{2\pi} P_n^\sigma \left[\frac{1}{2\pi} h(e^{is}, e^{i\sigma}) x_n^{(p-1)}(e^{i\sigma}) ctg \frac{\sigma - s}{2} + \right.\right.$$

$$\left.\left. + \frac{i}{2\pi} h(e^{is}, e^{i\sigma}) x_n^{(p-1)}(e^{i\sigma}) \right] d\sigma \right] = \bar{P}_n^t \left[\frac{1}{\pi} \int_\gamma P_n^\tau \left[\frac{h(t,\tau) x_n^{(l+p-1)}(\tau)}{\tau - t} \right] d\tau \right].$$

Returning to the formula (76), we have

$$\|\bar{P}_n \left[\int_\gamma \frac{\tilde{h}(t,\tau) x_n^{(p-1)}(\tau)}{\tau - t} d\tau \right] \|_{Y_n} = \|\bar{P}_n \left[\int_\gamma P_n \left[\frac{\tilde{h}(t,\tau) x_n^{(p-1)}(\tau)}{\tau - t} \right] d\tau \right] \|_{Y_n} =$$
$$\|\bar{P}_n \left[\int_\gamma \frac{P_n[\tilde{h}(t,\tau) x_n^{(p-1)}(\tau)]}{\tau - t} d\tau \right] \|_{Y_n}. \quad (77)$$

From formula (77) it follows that

$$J_2 = 0. \quad (78)$$

Let's estimate the norm

$$\|\bar{P}_n \left[\int_\gamma \frac{P_n[(\tilde{h}(t,\tau) - h(t,\tau)) x_n^{(p-1)}(\tau)]}{\tau - t} d\tau \right] \|_Y \le C n^\beta \ln^2 n E_n^*(h). \quad (79)$$

From inequalities (74)–(79) it follows that

$$J \le C n^\beta E_n^*(h_l) \ln^2 n \|x_n\|. \quad (80)$$

Repeating the above reasoning, we have

$$\|\bar{P}_n \left[\int_\gamma R_n^\tau \left[\frac{(h(t,\tau))_\tau^{(p-1)} x_n^{(l)}(\tau)}{\tau - t} \right] d\tau \right] \|_{Y_n} \le C n^\beta E_n^*(h_s) \ln^2 n \|x_n\|_{X_n}. \quad (81)$$

From (72), (80), (82) we obtain the estimate

$$\|K_n x_n - \tilde{K}_n x_n\|_{Y_n} \le C n^{-(\alpha - \beta)} \ln^2 n \|x_n\|_{X_n}. \quad (82)$$

It was shown above that, for n such that $q < 1$, the operator $\tilde{K}_n$ is continuously invertible and the estimate $\|[\tilde{K}_n]^{-1}\| \le C/(1-q)$ is valid. It follows from this inequality, estimate (82) and the Banach theorem that, for n such that $q_1 = Cn^{-(\alpha-\beta)} \ln^2 n < 1$, the operator K_n has the inverse operator K_n^{-1} with the norm $\|K_n^{-1}\| \le C/(1-q_1)$.

The proximity of the solutions of equations (62) and (64) is estimated in the same way as in studying the proximity of solutions of singular integral equations [4, 5].

The Theorem is proved.

Theorem 7 *[8]. Suppose that the Eq. (62) is uniquely solvable and condition (2) holds. Then for n such that* $C\ln^3 n/n^{r-p+1+\alpha-\beta} < 1$, *the system of equations (64) is uniquely solving and the estimate* $\|x^* - x_n^*\|_{H_\beta} \le C\ln^3 n/n^{r-p+1+\alpha-\beta}$ *is valid.*

The proof of Theorem 7 is similar to the proof of Theorem 6 and therefore it is omitted.

5.2 An Approximate Solution of Linear HSIEs with Even Order Singularity

Consider a one-dimensional HSIE equation of the type

$$Kx \equiv a(t)x(t) + b(t)\int_{-1}^{1}\frac{x(\tau)d\tau}{(\tau-t)^p} + \int_{-1}^{1} h(t,\tau)x(\tau)d\tau = f(t), \tag{83}$$

where $p = 2, 4, \ldots$. It is assumed that $b(t) \ne 0$, $t \in [-1, 1]$.

Let us introduce the nodes $t_k = -1 + 2k/N$, $k = 0, 1, \ldots, N$, $\bar{t}_k = t_k + 1/N$, $k = 0, 1, 2, \ldots, N-1$, and the corresponding intervals $\Delta_k = [t_k, t_{k+1})$, $k = 0, 1, \ldots, N-2$, $\Delta_{N-1} = [t_{N-1}, t_N]$.

We will seek for an approximate solution of Eq. (83) in the form of a piecewise constant function

$$x_N(t) = \sum_{k=0}^{N-1} \alpha_k \psi_k(t), \tag{84}$$

where $\psi_k(t) = 1$ for $t \in \Delta_k$ and $\psi_k(t) = 0$ for $t \in [-1, 1] \setminus \Delta_k$.

The coefficients $\{\alpha_k\}$, $k = 0, 1, \ldots, N-1$, are determined from the following system of linear algebraic equations

$$a(\bar{t}_k)\alpha_k + b(\bar{t}_k)\sum_{l=0}^{N-1}\alpha_l\int_{\Delta_l}\frac{d\tau}{(\tau-\bar{t}_k)^p} + \frac{2}{N}\sum_{l=0}^{N-1}\alpha_l h(\bar{t}_k,\bar{t}_l) = f(\bar{t}_k),\ k = 0, 1, \ldots, N-1. \tag{85}$$

Calculating hypersingular integrals in the Eq. (85), we arrive at the system

$$a(\bar{t}_k)\alpha_k - \frac{N^{p-1}}{p-1}b(\bar{t}_k)\sum_{l=0}^{N-1}\alpha_l\left(\frac{1}{(2l-2k+1)^{p-1}} - \frac{1}{(2l-2k-1)^{p-1}}\right) +$$

$$+\frac{2}{N}\sum_{l=0}^{N-1}\alpha_l h(\bar{t}_k,\bar{t}_l) = f(\bar{t}_k),\ k = 0, 1, \ldots, N-1. \tag{86}$$

The system (86) is equivalent to the system

$$\mathrm{sgn} b(\bar{t}_k)\left(a(\bar{t}_k)\alpha_k - \frac{N^{p-1}}{p-1} b(\bar{t}_k)\sum_{l=0}^{N-1}\alpha_l\left(\frac{1}{(2l-2k+1)^{p-1}} - \frac{1}{(2l-2k-1)^{p-1}}\right) + \right.$$
$$\left. + \frac{2}{N}\sum_{l=0}^{N-1}\alpha_l h(\bar{t}_k,\bar{t}_l)\right) = \mathrm{sgn} b(\bar{t}_k)\left(f(\bar{t}_k)\right),\ k = 0, 1, \ldots, N-1. \tag{87}$$

We represent the system of equations (87) in the matrix form as $CX = F$, where $C = \{c_{kl}\}, k, l = 0, 1, \ldots, N-1, X = (x_0, x_1, \ldots, x_{N-1}), F = (f_0, f_1, \ldots, f_{N-1})$. The values $\{c_{kl}\}$, $\{x_k\}$, and $\{f_k\}$ are obvious.

The cubic logarithmic norm $\Delta_2(C)$ of the matrix C is equal to

$$\Delta_2(C) = \max_{0\le k\le N-1}\left(\mathrm{sgn} b(\bar{t}_k)(a(\bar{t}_k)) - \frac{2N^{p-1}}{p-1}|b(\bar{t}_k)| + \right.$$
$$+ \sum_{l=0, l\ne k}^{N}\frac{N^{p-1}}{p-1}|b(\bar{t}_k)|\left[\frac{1}{(2k+1)^{p-1}} + \frac{1}{(2N-2k-1)^{p-1}}\right] +$$
$$\left. + \frac{2}{N}\left[\mathrm{sgn} b(\bar{t}_k)(h(\bar{t}_k,\bar{t}_k)) + \sum_{l=0, l\ne k}^{N-1}|h(\bar{t}_k,\bar{t}_l)|\right]\right).$$

Under the condition $\Delta_2(C) < 0$ and using the Theorem 1, we see that the system (86) has a unique solution.

Theorem 8 *[9]. Assume the following conditions have been fulfilled:* $p = 2, 4, \ldots$; $|b(t)| \ge b_0 = const > 0, t \in [-1, 1]$; *the function* $h(t,\tau)$ *is bounded on the square* $[-1, 1]^2$; $\Delta_2(C) < 0$.
Then the system of equations (87) (and the system of equations (86)) has the unique solution $x_N^*(t)$.

So, we can use the continuous method proposed in Sect. 2 for operator equations to solve hypersingular integral equations.

According to the results from Sect. 2, solution of the system of ordinary differential equations

$$\frac{d\alpha_k(t)}{dt} = \mathrm{sgn}(b(\bar{t}_k))\left(a(\bar{t}_k)\alpha_k(t) - \frac{N^{p-1}b(\bar{t}_k)}{p-1}\sum_{l=0}^{N-1}\alpha_l(t)\left(\frac{1}{(2l-2k+1)^{p-1}} - \frac{1}{(2l-2k-1)^{p-1}}\right) + \right.$$
$$\left. + \frac{2}{N}\sum_{l=0}^{N-1}h(\bar{t}_k,\bar{t}_l)\alpha_l(t) - f(\bar{t}_k)\right),\ k = 0, 1, \ldots, N-1$$

converges to the solution of the system of algebraic equations (86), (87) when $t \to \infty$.

Let us consider the case when $p = 2$.

Divide the interval $[-1, 1]$ into $2N$ subintervals at the points $t_k = -1 + k/N$, $k = 0, 1, \ldots, 2N$.

We shall seek for an approximate solution of (83) in the form of continuous function

$$x_N(t) = \sum_{k=0}^{2N}\alpha_k\varphi_k(t), \tag{88}$$

where $\varphi_k(t), k = 0, 1, \ldots, 2N$, is a family of basis functions.

For nodes t_k, $k = 1, \ldots, 2N-1$, the basis elements are determined by

$$\varphi_k(t) = \begin{cases} 0, & t_{k-1} \le t \le t_{k-1} + \frac{1}{N^2}, \\ \frac{N^2}{N-2}(t - t_{k-1}) - \frac{1}{N-2}, & t_{k-1} + \frac{1}{N^2} \le t \le t_k - \frac{1}{N^2}, \\ 1, & t_k - \frac{1}{N^2} \le t \le t_k + \frac{1}{N^2}, \\ -\frac{N^2}{N-2}(t - t_{k+1}) - \frac{1}{N-2}, & t_k + \frac{1}{N^2} \le t \le t_{k+1} - \frac{1}{N^2}, \\ 0, & t_{k+1} - \frac{1}{N^2} \le t \le t_{k+1}, \\ 0, & t \in [-1, 1] \backslash [t_{k-1}, t_{k+1}]. \end{cases} \tag{89}$$

For boundary nodes t_0, t_{2N} the basis elements are defined as

$$\varphi_0(t) = \begin{cases} 1, & -1 \le t \le -1 + \frac{1}{N^2}, \\ -\frac{N^2}{N-2}(t - t_1) - \frac{1}{N-2}, & -1 + \frac{1}{N^2} \le t \le t_1 - \frac{1}{N^2}, \\ 0, & t_1 - \frac{1}{N^2} \le t \le t_1, \\ 0, & [-1, 1] \backslash [t_0, t_1]; \end{cases} \tag{90}$$

$$\varphi_{2N}(t) = \begin{cases} 0, & -1 \le t \le t_{N-1} + \frac{1}{N^2}, \\ \frac{N^2}{N-2}(t - t_{N-1}) - \frac{1}{N-2}, & t_{N-1} + \frac{1}{N^2} \le t \le 1 - \frac{1}{N^2}, \\ 1, & 1 - \frac{1}{N^2} \le t \le 1. \end{cases} \tag{91}$$

The coefficients α_k in (88) are determined from the following system of linear algebraic equations

$$a(t_k)\alpha_k + \sum_{l=0}^{2N} h(t_k, t_l)\alpha_l \int_{-1}^{1} \frac{\varphi_l(\tau)}{(\tau - t_k)^2} d\tau = f(t_k), \quad k = 0, 1, \ldots, 2N. \tag{92}$$

The system (92) is equivalent to the system

$$\begin{aligned} &(\operatorname{sgn} h(t_k, t_k)) \left(a(t_k)\alpha_k - h(t_k, t_k) \frac{2N^2 \ln(N-1)}{N-2} \alpha_k + \alpha_0 h(t_k, t_0) \int_{t_0}^{t_1} \varphi_0(\tau) \frac{d\tau}{(\tau - t_k)^2} + \right. \\ &\left. + \sum_{l=1}^{2N-1}{}' \alpha_l h(t_k, t_l) \int_{t_{l-1}}^{t_{l+1}} \varphi_l(\tau) \frac{d\tau}{(\tau - t_k)^2} + \alpha_{2N} h(t_k, t_{2N}) \int_{t_{2N-1}}^{1} \varphi_{2N}(\tau) \frac{d\tau}{(\tau - t_k)^2} \right) = \\ &= (\operatorname{sgn} h(t_k, t_k)) f(t_k), k = 1, \ldots, 2N-1, \\ &(\operatorname{sgn} h(t_0, t_0)) \left(a(t_0)\alpha_0 - h(t_0, t_0) \frac{N^2 \ln(N-1)}{N-2} \alpha_0 + \sum_{l=1}^{2N-1} \alpha_l h(t_0, t_l) \int_{t_{l-1}}^{t_{l+1}} \frac{\varphi_l(\tau) d\tau}{(\tau+1)^2} + \right. \\ &\left. + \alpha_{2N} h(t_0, t_{2N}) \int_{t_{2N-1}}^{1} \varphi_{2N}(\tau) \frac{d\tau}{(\tau+1)^2} \right. = (\operatorname{sgn} h(t_0, t_0)) f(t_0), \\ &(\operatorname{sgn} h(t_{2N}, t_{2N})) \left(a(t_{2N})\alpha_{2N} - h(t_{2N}, t_{2N}) N^2 \frac{\ln(N-1)}{N-2} \alpha_{2N} + \right. \\ &+ \sum_{l=1}^{2N-1} \alpha_l h(t_{2N}, t_l) \int_{t_{l-1}}^{t_{l+1}} \varphi_l(\tau) \frac{d\tau}{(\tau+1)^2} + \\ &\left. + \alpha_0 h(t_{2N}, t_0) \int_{-1}^{t_1} \varphi_0(\tau) \frac{d\tau}{(\tau-1)^2} \right) = (\operatorname{sgn} h(t_{2N}, t_{2N})) f(t_{2N}). \end{aligned} \tag{93}$$

Let us write the system (93) in a matrix form $DX = F$, where $D = \{d_{kl}\}$, $k, l = 0, 1, \ldots, 2N$, $X = (x_0, x_1, \ldots, x_{2N})$, $F = (f_0, f_1, \ldots, f_{2N})$. The values $\{d_{kl}\}$, $\{x_k\}$, and $\{f_k\}$ are obvious.

If the cubic logarithmic norm $\Lambda_2(D)$ of the matrix D is negative (for all $t \in [-1, 1]$), than the systems (92) and (93) have a unique solution $x_N^*(t)$ and $\|D^{-1}\| \leq 1/|\Lambda_2(D)|$.

Let $x^*(t)$ and x_N^* be solutions of (83) and (92), respectively. In the paper [9] received the estimate $\|x^* - x_N^*\| \leq CN^{-1} \ln N$.

Theorem 9 *[9] Let the following conditions be fulfilled:* $p = 2$; *Equation (83) has a unique solution* $x^*(t) \in W^2(M)$, $M = const$; *(2) the function* $|h(t, t)| \geq b_0 = const > 0$ *for all* $t \in [-1, 1]$; *(3) the cubic logarithmic norm* $\Lambda_2(D)$ *of the matrix* D *is negative (for all* $t \in [-1, 1]$*).*
Then the system of equations (92) has a unique solution $x_N^*(t)$ *and the following estimate holds:* $\|x^* - x_N^*\|_2 \leq CN^{-1} \ln N$.

Remark 3 In the paper [10], approximate methods for solving HSIEs of the form (83) with $p = 2$ are constructed and justified in the case when the right-hand sides have singularities that are not Riemann integrable.

6 Summary and Discussion

Methods of collocation and mechanical quadratures for the approximate solution of HSIEs of the first and second kind are constructed and justified.

HSIEs of the first kind are investigated with a singularity $p = 2$. If the right-hand sides of the equations are smooth functions, then the solutions of the equations belong to one of the following types: $x(t) = (\sqrt{1 - t^2})^{\pm 1}\varphi(t)$, $x(t) = ((1 - t)/(1 + t))^{\pm 1/2}\varphi(t)$, where $\varphi(t)$ is a smooth function. Approximate solutions are constructed and justified in all cases. If the right-hand sides of the equations have singularities, then the method described in Sect. 5 is used to solve HSIEs of the first kind.

A mechanical quadrature method is constructed and justified for an approximate solution of HSIEs of the second kind that have a singularity $p = 2, 3, \ldots$, and are defined on smooth closed contours. To solve HSIEs of the second kind, defined on the segment $[-1, 1]$, the method of mechanical quadratures is presented, the justify of which is carried out on the basis of a continuous method for solving operator equations.

What are the advantages of the presented method?

(1) The method is applicable for solving linear HSIEs, which right-hand sides contain non Riemann integrable functions. It is shown [10] that for linear HSIEs the method converges for sufficiently large N and for $|b(t)| \geq B_0 > 0$, $t \in [-1, 1]$;
(2) it allows us to evaluate the norms of the matrix inverses and to establish the stability boundaries of the computational scheme;

(3) the variety of the logarithmic norms employed makes it possible to produce a number of different solvability criteria;
(4) the requirements for the equation coefficients, kernels and the inhomogeneous terms are rather weak.

References

1. Ang, W.-T.: Hypersingular Integral Equations in Fracture Analysis, 212 pp. Woodhead Publishing (2014)
2. Bernshtein, S.N.: On the best approximation of functions of several variables by means of polynomials or trigonometric sums. Collected papers. Dedicated to Academician Ivan Matveevich Vinogradov on his 60th birthday, Tr. Steklov Mathematical Institute of the USSR, 38, Publishing house of the Academy of Sciences of the USSR, Moscow, pp. 24–29 (1951). (in Russian)
3. Boikov, I.V.: On a continuous method for solving nonlinear operator equations. Differ. Equ. **48**(9), 1308–1314 (2012)
4. Boykov, I.V.: Approximate Methods of Solution of Singular Integral Equations. The Penza State University, Penza (2004). (in Russian)
5. Boykov, I.V.: Numerical methods for solutions of singular integral equations, 182 pp. arXiv:1610.09611 [math. NA]
6. Boykov, I.V., Boykova, A.I.: Approximate methods for solving hypersingular integral equations of the first kind with second-order singularities on classes of functions with weights $(1-t^2)^{-1/2}$. Univ. Proc. Volga Reg. Phys. Math. Sci. Math. **2**, 79–90 (2017). (in Russian)
7. Boykov, I.V., Boykova, A.I.: An approximate solution of hypersingular integral equations of the first kind with singularities of the second order on the class of functions with weight $((1-t)/(1+t))^{\pm 1/2}$. Univ. Proc. Volga Reg. Phys. Math. Sci. Math. **3**, 76–92 (2019). (in Russian)
8. Boykov, I.V., Zakharova, Yu.F.: An approximate solution to hypersingular integro-differential equations. Univ. Proc. Volga Reg. Phys. Math. Sci. Math. **1**, 80–90 (2010). (in Russian)
9. Boykov, I.V., Roudnev, V.A., Boykova, A.I., Baulina, O.A.: New iterative method for solving linear and nonlinear hypersingular integral equations. Appl. Numer. Math. **127**, 280–305 (2018)
10. Boykov, I.V., Roudnev, V.A., Boykova, A.I.: Approximate methods for solving linear and nonlinear hypersingular integral equations. Axioms **9**(74), 2–18 (2020). https://doi.org/10.3390/axioms9030074
11. Capobiano, M.R., Criscuolo, G., Junghanns, P.: On the numerical solution of a nonlinear integral equation of Prandtl's type. In: Operator Theory: Advances and Applications, vol. 160, pp. 53–79. Birkhauser Verlag, Basel (2005)
12. Chikin, L.A.: Special cases of the Riemann boundary value problems and singular integral equations. Scientific Notes of the Kazan State University 113 **10**, 53–105 (1953). (in Russian)
13. Daletskii, Y.L., Krein, M.G.: Stability of solutions of differential equations in Banach space. Transl. Math. Monographs **43**. Amer. Math. Soc. Providence. RI (1974)
14. Eminov, S.I., Eminova, V.S.: Justification of the Galerkin method for hypersingular equations. Comput. Math. Math. Phys. **56**, 417–425 (2016). https://doi.org/10.1134/S0965542516030039
15. Eshkuvatov, Z.K., Nik Long, N.M.A., Abdulkawi, M.: Approximate solution of singular integral equations of the first kind with Cauchy kernel. Appl. Math. Lett. **22**(5), 651–657 (2009)
16. Eshkuvatov, Z.K., Zulkarnain, F.S., Nik Long, N.M.A., Muminov, Z.: Modified homotopy perturbation method for solving hypersingular integral equations of the first kind. SpringerPlus **5**, 1473 (2016)
17. Feng, H., Zhang, X., Li, J.: Numerical solution of a certain hypersingular equation of the first kind. BIT Numer. Math. **51**, 609–630 (2011)

18. Fuks, B.A.: Introduction to the Theory of Analytic Functions of Several Complex Variables, 388 pp. American Mathematical Society (1963)
19. Gakhov, F.D.: Boundary Value Problems, p. 561. Dover Publication, USA (1990)
20. Gandel', Yu.V., Zaginailov, G.I., Steshenko, S.A.: Rigorous electrodynamic analysis of resonator systems of coaxial gyrotrons. Tech. Phys. **49**(7), 887–894 (2004)
21. Golberg, M.A.: The convergence of several algorithms for solving integral equations with finite-part integrals I. J. Integr. Equ. **5**(4), 329–340 (1983)
22. Golberg, M.A.: The convergence of several algorithms for solving integral equations with finite-part integrals II. J. Integr. Equ. **9**(3), 267–275 (1985)
23. Hadamard, J.: Lectures on Cauchy's Problem in Linear Partial Differential Equations. Dover Publ. Inc., New York (1952)
24. Kantorovich, L.V., Akilov, G.P.: Functional Analysis, 604 pp. Elsevier (2014)
25. Kanoria, M., Mandal, B.N.: Water wave scattering by a submerged circular-arc-shaped plate. Fluid Dyn. Res. **31**, 317–331 (2002)
26. Karlin, V., Maz'ya, V.G., Movchan, A.B., Willis, J.R., Bullough, R.: Numerical solution of nonlinear hypersingular integral equations of the Peierls type in dislocation theory. SIAM J. Appl. Math. **60**(2), 664–678 (2000). https://doi.org/10.1137/S003613999631162X
27. Kaya, A.C., Erdogan, F.: On the solution of integral equations with strongly singular kernels. Quart. Appl. Math. **95**, 105–122 (1987)
28. Lifanov, I.K.: Numerical solution of singular integral Hilbert equations with a strong singularity. Optimal methods computing and their application. Interuniversity Collection of Scientific Papers, vol. 7, pp. 38–45. Penza Polytechnic Institute, Penza (1985). (in Russian)
29. Lifanov, I.K.: Singular Integral Equations and Discrete Vortices. VSP, Utrecht, The Netherlands (1996)
30. Lifanov, I.K.: To the solution of composite integral equations. Successes of modern radio electronics **8**, 62–67 (2006). (in Russian)
31. Lifanov, I.K., Poltavskii, L.N., Vainikko, G.M.: Hypersingular Integral Equations and their Applications. Chapman Hall/CRC, CRC Press Company, London, New York, Washington, D.C, Boca Raton (2004)
32. Lifanov, I.K., Nenashev, A.S.: Analysis of some computational schemes for a hypersingular integral equation on an interval. Differ. Equ. **41**, 1343–1348 (2005). https://doi.org/10.1007/s10625-005-0283-2
33. Linkov, A.M., Mogilevskaya, S.G.: Complex hypersingular integrals and integral equations in plane elasticity. Acta Mech. **105**, 189–205 (1994)
34. Lorentz, G.G.: Approximation of Functions. Chelsea Publishing Company, New York (1986)
35. Mandal, B.N., Chakrabarti, A.: Applied Singular Integral Equations, 260 pp. CRC Press (2011)
36. Martin, P.A.: End-Point behavior of solutions to hypersingular integral equations. Proc. R. Soc. Lond. A. **432**, 301–320 (1991)
37. Martin, P.A.: Exact solution of a simple hypersingular integral equation. J. Integr. Equ. Appl. **4**, 197–204 (1992)
38. Natanson, I.P.: Constructive Function Theory. Volume 1. Uniform Approximation. Frederick Ungar Publishing Co., NY (1965)
39. Natanson, I.P.: Constructive Function Theory. Volume II. Approximation in Mean. Frederick Ungar Publishing Co., NY (1965)
40. Peierls, R.: The size of a dislocations. Proc. Phys. Soc. **52**, 34 (1940)
41. Szego, G.: Orthogonal Polynomials. NY (1959)
42. Ventsel, E.S.: A boundary element method applied to sandwich plates of arbitrary plan form. Eng. Anal. Bound. Elem. **27**, 597–601 (2003)

Solutions of Integral Equations by Reproducing Kernel Hilbert Space Method

Ali Akgül

Abstract The theory of reproducing kernels was considered for the first time at the beginning of the 20th century by Zaremba. Reproducing kernel theory has valuable implementations in numerical analysis, differential equations, probability and statistics. Some authors discussed fractional differential equations, nonlinear oscillators with discontinuity, singular nonlinear two-point periodic boundary value problems and nonlinear partial differential equations by the reproducing kernel Hilbert space method recently. In this chapter, we apply the reproducing kernel Hilbert space method to the integral equations. We give the solutions in the form of a series in the reproducing kernel Hilbert space. We demonstrate some numerical examples to show the accuracy of the technique.

1 Introduction

The study of the reproducing kernel Hilbert space method for Volterra integral equations of the second kind is one of most researched topic in numerical analysis; the significance of integral equations in modeling biological phenomenon is well-known since many years. It is possible to obtain many techniques based on different interpolation or functions approximation method that have been applied in order to solve Volterra integral equations [1]. Brunner [2] has investigated the collocation methods for Volterra integral and related functional equations and he has also studied the theory and applications of the Volterra integral equations [3]. Calvetti et al. [4] have researched the computation of Gauss-Kronrod quadrature rules. Cao et al. [5–7] have worked a hybrid collocation method for Volterra integral equations with weakly singular kernels, the approximation operators with sigmoidal functions and the construction and approximation of a class of neural networks operators with ramp functions. Coroianu [8] has investigated the approximation by truncated max-product operators of Kantorovich-type based on generalized kernels.

A. Akgül (✉)
Department of Mathematics, Siirt University Art and Science Faculty, 56100 Siirt, Turkey
e-mail: aliakgul00727@gmail.com

H. Singh et al. (eds.), *Topics in Integral and Integro-Differential Equations*,
Studies in Systems, Decision and Control 340,
https://doi.org/10.1007/978-3-030-65509-9_4

Costarelli [9] has worked the interpolation by neural network operators activated by ramp functions.

Lü et al. [10] have investigated the solutions of a singular system of two nonlinear ordinary differential equations. Tahmasbi et al. [11] have obtained the numerical solution of linear Volterra integral equations system of the second kind. Wang et al. [12] have used the new algorithm for second-order boundary value problems of integro-differential equation. Wang et al. [13, 14] have applied the reproducing kernel method for solving a class of singular weakly nonlinear boundary value problems and have applied the reproducing kernel method for investigating a class of singularly perturbed problems. Yang et al. [15] have implemented the reproducing kernel method for solving the system of the linear Volterra integral equations with variable coefficients. For more details see [16–18].

In this paper, we present the reproducing kernel Hilbert space method to solve the problem. Cui et al. [19] have studied the nonlinear numerical analysis in the reproducing kernel space. In this work there are many applications of the reproducing kernel method. Aronszajn [20] has investigated the theory of reproducing kernels. Akgül et al. [21–24] have worked on many problems by the reproducing kernel Hilbert space method: Boundary layer flow of a Powell-Eyring non-Newtonian fluid, new reproducing kernel functions, solutions of variable-order fractional differential equations and numerical solutions of fractional differential equations of Lane-Emden type. For more details see [25–33].

We organize the manuscript as follow: We give the main definitions of the reproducing kernel functions in Sect. 2. We present the reproducing kernel Hilbert space method in Sect. 3. We construct some useful reproducing kernel Hilbert spaces in this section. We have also obtained the reproducing kernel functions in these spaces. We apply the reproducing kernel Hilbert space method to the integral equations in Sect. 4. We present the numerical simulations in Sect. 5. We give some conclusion in the last section.

2 Preliminaries

We describe the concept of a reproducing kernel Hilbert space, exhibit some particular instances of these spaces which will play roles in this work, and describe some well-known properties of these spaces. Convenient references for this section are [19], [20], and [34]. For more details see [35].

Definition 2.1 Let $(H, \langle \cdot, \cdot \rangle)$ be a Hilbert space of real functions defined on a nonempty set E. A function $K : E \times E \to \mathbb{R}$ is called a *reproducing kernel* for H if and only if

(a) $K(\cdot, z) \in H$ for each fixed $z \in E$,
(b) $\langle \varphi, K(\cdot, z) \rangle = \varphi(z)$ for all $z \in E$ and all $\varphi \in H$.

We will refer to such a Hilbert space H for which there exists a reproducing kernel function K as a *reproducing kernel Hilbert space*.

Condition (b) is called "the reproducing property" of the kernel K because the value of an arbitrary function $\varphi \in H$ at an arbitrary point $z \in E$ is reproduced by the inner product of φ with $K(\cdot, z)$. For brevity we will freely use RKHS instead of the term reproducing kernel Hilbert space. Furthermore, we will usually adhere to the standard convention $K_z(\cdot) = K(\cdot, z)$ for denoting the reproducing kernel in such spaces.

We now list some RKHSs on the closed, bounded interval $[a, b]$. In what follows we will use the symbol $AC[a, b]$ to denote the vector space of real, absolutely continuous functions on the interval $[a, b]$ and $L^2[a, b]$ to denote the real, Lebesgue square integrable functions on [a,b].

Definition 2.2 Let m be a positive integer. The space $V_2^m[a, b]$ consists of the functions $u : [a, b] \to \mathbb{R}$ such that $u^{(m-1)} \in AC[a, b]$ and $u^{(m)} \in L^2[a, b]$. Equip $V_2^m[a, b]$ with the inner product

$$\langle u, v\rangle_{V_2^m} = \sum_{i=0}^{m-1} u^{(i)}(a)v^{(i)}(a) + \int_a^b u^{(m)}(x)v^{(m)}(x)dx.$$

Lemma 2.3 ([19], p. 8) *If m is a positive integer then $V_2^m[a, b]$ is a RKHS.*

One particular instance is

$$V_2^1[0, 1] = \{u \in AC[0, 1] : u' \in L^2[0, 1]\}$$

equipped with the inner product

$$\langle u, v\rangle_{V_2^1} = u(0)v(0) + \int_0^1 u'(x)v'(x)dx.$$

The reproducing kernel R of $V_2^1[0, 1]$ is given by ([19], pp. 10 and 17)

$$R_y(x) = \begin{cases} 1 + x, & 0 \le x \le y \le 1, \\ 1 + y, & 0 \le y < x \le 1. \end{cases} \tag{2.1}$$

3 Reproducing Kernel Hilbert Space Method

We take into consideration the following Volterra-Hamanerstein integral equation [1];

$$u(t) = 1 + \sin^2(t) - 3\int_0^t \sin(t - x)(u(x))^2 dx, \quad t \in [0, 1]$$

We know that the kernel has the convolution form.The exact solution of the above equation has been obtained as:

$$u(t) = \cos(t).$$

We use the above integral equation and obtain:

$$u'(t) = 2\sin(t)\cos(t) - 3\int_0^t \cos(t-x)u(x)^2 dx$$

$$u''(t) = 2\cos(2t) + 3\int_0^t \sin(t-x)u(x)^2 dx - 3u(t)^2$$

We know that

$$3\int_0^t \sin(t-x)(u(x))^2 dx = 1 + \sin^2(t) - u(t)$$

Then, we get

$$u''(t) = 2\cos(2t) + 1 + \sin^2(t) - u(t) - 3(u(t))^2$$

Thus, we reach the following initial value problem:

$$\begin{cases} u''(t) + u(t) = 1 + sin^2(t) + 2cos(2t) - 3(u(t))^2, \\ u(0) = 1. \end{cases}$$

We need to homogenize the condition to apply the reproducing kernel Hilbert space method. Therefore, we use the following transformation:

$$v(t) = u(t) - 1$$

Then, we obtain

$$\begin{aligned} u(t) &= v(t) + 1, \\ u'(t) &= v'(t), \\ u''(t) &= v''(t). \end{aligned}$$

If we use the above equations in the problem, we will get

$$\begin{cases} v''(t) + 7v(t) = sin^2(t) + 2cos(2t) - 3(v(t))^2 - 3, \\ v(0) = 0. \end{cases}$$

Now we can apply the reproducing kernel Hilbert space method to the above problem. Firstly we need to construct the reproducing kernel Hilbert spaces. Then, we need to obtain the reproducing kernel functions in these reproducing kernel Hilbert spaces. The first reproducing kernel Hilbert space that we need is $V_2^3[0,1]$. We define this space as:

$$V_2^3[0,1] = \{v : v, v', v'' \text{ are absolutely continuous functions, } v''' \in L^2[0,1], \quad v(0) = 0\}$$

We have the inner product and the norm of this space as:

$$\langle u, v\rangle_{V_2^3[0,1]} = u(0)v(0) + u'(0)v'(0) + u''(0)v''(0) + \int_0^1 u'''(x)v'''(x)dx$$

and

$$\|u\|_{v_2^3[0,1]} = \sqrt{\langle u, v\rangle_{V_2^3[0,1]}}$$

Let $A_z(t)$ be the reproducing kernel function of the reproducing kernel Hilbert space $V_2^3[0,1]$. Then, we have

$$\langle u, A_z\rangle_{V_2^3[0,1]} = u(0)A_z(0) + u'(0)A_z'(0) + u''(0)A_z''(0) + \int_0^1 u'''(t)A_z'''(t)dt$$

We know that

$$(1)\ A_z(0) = 0$$

by the initial condition. Then, we get

$$\langle u, A_z\rangle_{V_2^3[0,1]} = u'(0)A_z'(0) + u''(0)A_z''(0) + \int_0^1 u'''(t)A_z'''(t)dt$$

We use integration by parts and obtain

$$\begin{aligned}\langle u, A_z\rangle_{V_2^3[0,1]} = {} & u'(0)A_z'(0) + u''(0)A_z''(0) + u''(1)A_z'''(1) - u''(0)A_z'''(0) \\ & -u'(1)A_z^{(4)}(1) + u'(0)A_z^{(4)}(0) + u(1)A_z^{(5)}(1) - u(0)A_z^{(5)}(0) \\ & -\int_0^1 u(t)A_z^{(6)}(t)dt\end{aligned}$$

If we have the following equations

$$(2)\ A_z'(0) + A_z^{(4)}(0) = 0$$
$$(3)\ A_z''(0) - A_z'''(0) = 0$$
$$(4)\ A_z'''(1) = 0$$
$$(5)\ A_z^{(4)}(1) = 0$$
$$(6)\ A_z^{(5)}(1) = 0$$

thus we can get

$$\langle u, A_z \rangle_{V_2^3[0,1]} = -\int_0^1 u(t) A_z^{(6)}(t) dt$$

By reproducing property, we know

$$\langle u, A_z \rangle_{V_2^3[0,1]} = u(z)$$

Then, we will have

$$-\int_0^1 u(t) A_z^{(6)}(t) dt = u(z)$$

This will give us the Dirac-Delta function as:

$$A_z^{(6)}(t) = -\delta(z - t).$$

When $z \neq t$, we get

$$A_z^{(6)}(t) = 0.$$

Therefore, we obtain

$$A_z(t) = \begin{cases} \sum_{k=1}^{6} a_k t^{k-1}, & t \leq z, \\ \sum_{k=1}^{6} b_k t^{k-1}, & t > z. \end{cases}$$

We have twelve unknown coefficients and six equations. We need six more equations to get these coefficients. We use the properties of Dirac-Delta function and obtain

$$(7)\ A_{z^+}(z) = A_{z^-}(z)$$
$$(8)\ A'_{z^+}(z) = A'_{z^-}(z)$$
$$(9)\ A''_{z^+}(z) = A''_{z^-}(z)$$
$$(10)\ A'''_{z^+}(z) = A'''_{z^-}(z)$$
$$(11)\ A^{(4)}_{z^+}(z) = A^{(4)}_{z^-}(z)$$
$$(12)\ A^{(5)}_{z^+}(z) - A^{(5)}_{z^-}(z) = -1$$

We use these twelve equations and obtain the unknown coefficients as:

$$a_1 = 0, a_2 = z, a_3 = \frac{z^2}{4}, a_4 = \frac{z^2}{12}, a_5 = \frac{-z}{24}, a_6 = \frac{1}{120}$$
$$d_1 = \frac{z^5}{120}, d_2 = \frac{-z^4}{24}, d_3 = \frac{z^3}{12}, d_4 = 0, d_5 = 0, d_6 = 0$$

Then, we obtain the reproducing kernel function as

$$A_z(t) = \begin{cases} tz + \frac{t^2z^2}{4} + \frac{t^3z^2}{12} - \frac{t^4z}{24} + \frac{t^5}{120}, & t \le z, \\ zt + \frac{z^2t^2}{4} + \frac{z^3t^2}{12} - \frac{z^4t}{24} + \frac{z^5}{120}, & t > z. \end{cases}$$

Theorem 3.1 *The reproducing kernel function $A_z(t)$ of the reproducing kernel Hilbert space $V_2^3[0, 1]$ has been obtained as:*

$$A_z(t) = \begin{cases} tz + \frac{t^2z^2}{4} + \frac{t^3z^2}{12} - \frac{t^4z}{24} + \frac{t^5}{120}, & t \le z, \\ zt + \frac{z^2t^2}{4} + \frac{z^3t^2}{12} - \frac{z^4t}{24} + \frac{z^5}{120}, & t > z. \end{cases}$$

Proof We need to show that

$$\langle u, A_z \rangle_{V_2^3[0,1]} = u(z).$$

We have

$$\langle u, A_z \rangle_{V_2^3[0,1]} = u(0)A_z(0) + u'(0)A'_z(0) + u''(0)A''_z(0) + \int_0^1 u'''(t)A'''_z(t)dt$$

We use integration by parts and obtain

$$\langle u, A_z\rangle_{V_2^3[0,1]} = u(0)A_z(0) + u'(0)A_z'(0) + u''(0)A_z''(0) + u''(1)A_z'''(1)$$
$$-u''(0)A_z'''(0) - u'(1)A_z^{(4)}(1) + u'(0)A_z^{(4)}(0)$$
$$+\int_0^1 u'(t)A_z^{(5)}(t)dt$$

Thus, we get

$$\langle u, A_z\rangle_{V_2^3[0,1]} = u(0)A_z(0) + u'(0)\Big(A_z'(0) + A_z^{(4)}(0)\Big)$$
$$+u''(0)\Big(A_z''(0) - A_z'''(0)\Big) + u''(1)A_z'''(1) - u'(1)A_z^{(4)}(1)$$
$$+\int_0^1 u'(t)A_z^{(5)}(t)dt$$

We need to compute $A_z'(0)$, $A_z''(0)$, $A_z'''(0)$, $A_z^{(4)}(0)$, $A_z'''(1)$ and $A_z^{(4)}(1)$. We have

$$A_z(0) = 0$$
$$A_z'(0) = y$$
$$A_z''(0) = \frac{y^2}{2}$$
$$A_z'''(0) = \frac{y^2}{2}$$
$$A_z^{(4)}(0) = -y$$
$$A_z'''(1) = 0$$
$$A_z^{(4)}(1) = 0$$

If we use the above equations, we will get

$$\langle u, A_z\rangle_{V_2^3[0,1]} = u(0).0 + u'(0)(y - y) + u''(0)\left(\frac{y^2}{2} - \frac{y^2}{2}\right) + u''(1).0$$
$$-u'(1).0 + \int_0^1 u'(t)A_z^{(5)}(t)dt$$

Thus, we have

$$\begin{aligned}\langle u, A_z\rangle_{V_2^3[0,1]} &= \int_0^1 u'(t)A_z^{(5)}(t)dt \\ &= \int_0^z u'(t)A_z^{(5)}(t)dt + \int_z^1 u'(t)A_z^{(5)}(t)dt \\ &= \int_0^z u'(t)1dt + \int_z^1 u'(t)0dt \\ &= \int_0^z u'(t)dt = u(z) - u(0).\end{aligned}$$

Since $u(0) = 0$, we obtain

$$\langle u, A_z\rangle_{V_2^3[0,1]} = u(z).$$

This completes the proof. □

The second reproducing kernel Hilbert space that we need is $V_2^1[0, 1]$. We define it as:

$$V_2^1[0, 1] = \{v : v \quad \text{is absolutely continuous functions,} \quad v' \epsilon L^2[0, 1]\}$$

We have the inner product and the norm of this space as:

$$\langle u, v\rangle_{V_2^1[0,1]} = u(0)v(0) + \int_0^1 u'(t)v'(t)dt$$

and

$$\|u\|_{V_2^1[0,1]} = \sqrt{\langle u, u\rangle_{V_2^1[0,1]}}$$

We obtain the reproducing kernel function $B_z(t)$ of the reproducing kernel Hilbert space $V_2^1[0, 1]$ as:

$$\langle u, B_z\rangle_{V_2^1[0,1]} = u(0)B_z(0) + \int_0^1 u'(t)B_z'(t)dt$$

We use the integration by parts and obtain

$$\langle u, B_z\rangle_{V_2^1[0,1]} = u(0)B_z(0) + u(1)B_z'(1) - u(0)B_z'(0) - \int_0^1 u(t)B_z''(t)dt$$

If we have the following equations:

$$(1)\ B_z(0) - B_z'(0) = 0$$
$$(2)\ B_z'(1) = 0$$

Then, we will reach

$$\langle u, B_z\rangle_{V_2^1[0,1]} = -\int_0^1 u(t)B_z''(t)dt$$

By the reproducing property, we know that

$$\langle u, B_z\rangle_{V_2^1[0,1]} = u(z)$$

Therefore, we get

$$-\int_0^1 u(t)B_z''(t)dt = u(z)$$

This will give us the Dirac-delta function as:

$$B_z''(t) = -\delta(z-t)$$

When $z \neq t$, we get

$$B_z''(t) = 0$$

Therefore, we reach

$$B_z(t) = \begin{cases} a_1 + a_2 t, & t \leq z, \\ b_1 + b_2 t, & t > z. \end{cases}$$

We have four unknown coefficients and two equations. We need two more equations to get there coefficients. We use the properties of the Dirac-Delta function and obtain:

$$(3)\ B_{z^+}(z) = B_{z^-}(z)$$
$$(4)\ B_{z^+}'(z) - B_{z^-}'(z) = -1$$

Thus, we reach

$$(1)\ B_z(0) - B_z'(0) = 0 \Longrightarrow a_1 - a_2 = 0$$
$$(2)\ B_z'(1) = 0 \Longrightarrow b_2 = 0$$
$$(3)\ B_{z^+}(z) = B_{z^-}(z) \Longrightarrow b_1 + b_2 z = a_1 + a_2 z$$

$$(4)\ B'_{z^+}(z) - B'_{z^-}(z) = -1 \Longrightarrow b_2 - a_2 = -1$$

Therefore, we obtain

$$a_1 = 1, a_2 = 1, b_1 = 1 + z, b_2 = 0$$

Then, we get the reproducing kernel function as:

$$B_z(t) = \begin{cases} .1 + t, & t \leq z \\ 1 + t, & t > z \end{cases}$$

Theorem 3.2 *We obtain the reproducing kernel function $B_z(t)$ of the reproducing kernel Hilbert space $V_2^1[0, 1]$ as:*

$$B_z(t) = \begin{cases} .1 + t, & t \leq z \\ 1 + t, & t > z \end{cases}$$

Proof We need to prove that

$$\langle u, B_z \rangle_{V_2^1[0,1]} = u(z)$$

We have

$$\langle u, B_z \rangle_{V_2^1[0,1]} = u(0)B_z(0) + \int_0^1 u'(t)B_z'(t)dt$$

Then, we get

$$\begin{aligned} \langle u, B_z \rangle_{V_2^1[0,1]} &= u(0).1 + \int_0^z u'(t)B_z'(t)dt + \int_z^1 u'(t)B_z'(t)dt \\ &= u(0) + \int_0^z u'(t).1dt + \int_z^1 u'(t).0dt \\ &= u(0) + \int_0^z u'(t)dt \\ &= u(0) + u(z) - u(0) \\ &= u(z). \end{aligned}$$

This completes the proof. □

4 Solutions of the Problem

We take into consideration the solutions of the problem in the reproducing kernel Hilbert space $V_2^3[0, 1]$. We present the bounded linear operator $F : V_2^3[0, 1] \to V_2^1[0, 1]$ as:

$$Fv = v''(t) + Kv(t), \tag{4.1}$$

we have the following problem.

$$Fv = E(t, v), \tag{4.2}$$

with the homogen initial condition

$$v(0) = 0. \tag{4.3}$$

Lemma 4.1 *F is a bounded linear operator.*

Proof We should show that:

$$\|Fv\|^2_{V_2^1[0,1]} \leq J \|v\|^2_{V_2^3[0,1]} .$$

We get

$$\|Fv\|^2_{V_2^1[0,1]} = \langle Fv, Fv\rangle_{V_2^1[0,1]} = [Fv(0)]^2 + \int_0^1 [Fv(t)]'^2 \, dt. \tag{4.4}$$

We obtain

$$v(t) = \langle v(\cdot), A_z(\cdot)\rangle_{V_2^3[0,1]} ,$$

and

$$Fv(t) = \langle v(\cdot), FA_z(\cdot)\rangle_{V_2^3[0,1]}$$

by reproducing property. Therefore, we reach

$$|Fv| \leq \|v\|_{V_2^3[0,1]} \|FA_z\|_{V_2^1[0,1]} = J_1 \|v\|_{V_2^3[0,1]} ,$$

Then, we get

$$[Fv]^2 \leq J_1^2 \|v\|^2_{V_2^3[0,1]} .$$

We know that

$$(Fv)'(t) = \left\langle v(\cdot), (FA_z)'(\cdot)\right\rangle_{V_2^3[0,1]} ,$$

and

$$\left|(Fv)'\right| \leq \|v\|_{V_2^3[0,1]} \left\|(FA_z)'\right\|_{V_2^3[0,1]} = J_2 \|v\|_{V_2^3[0,1]}.$$

Therefore, we obtain

$$[Fv]'^2 \leq J_2^2 \|v\|^2_{V_2^3[0,1]}.$$

Finally, we obtain

$$\begin{aligned}\|Fv\|^2_{V_2^1[0,1]} &= [Fv(0)]^2 + \int_0^1 \left[(Fv)'(t)\right]^2 dt \\ &\leq (J_1^2 + J_2^2) \|v\|^2_{V_2^3[0,1]},\end{aligned}$$

where $J = J_1^2 + J_2^2$. This completes the proof. □

We present $\sigma_i(t) = A_{t_i}(t)$ and $\phi_i(t) = F^*\sigma_i(t)$, where F^* is conjugate operator of F. The orthonormal system $\left\{\widehat{\phi}_i(t)\right\}_{i=1}^{\infty}$ of $V_2^3[0, 1]$ can be obtained by Gram-Schmidt orthogonalization operation of $\{\phi_i(t)\}_{i=1}^{\infty}$,

$$\widehat{\phi}_i(t) = \sum_{k=1}^{i} \gamma_{ik}\phi_k(t), \quad (\beta_{ii} > 0, \quad i = 1, 2, \ldots). \tag{4.5}$$

Theorem 4.1 *Let $\{t_i\}_{i=1}^{\infty}$ be dense in $[0, 1]$ and $\phi_i(t) = F_z A_z(t)|_{z=t_i}$. Then the sequence $\{\phi_i(t)\}_{i=1}^{\infty}$ is a complete system in $V_2^3[0, 1]$.*

Proof We have

$$\begin{aligned}\phi_i(t) &= (F^*\sigma_i)(t) = \left\langle (F^*\sigma_i)(t), A_z(t)\right\rangle \\ &= \langle(\sigma_i)(t), F_z A_z(t)\rangle = F_z A_z(t)|_{z=t_i}.\end{aligned}$$

Assume that $\langle v(t), \phi_i(t)\rangle = 0$, $(i = 1, 2, \ldots)$. Therefore, we obtain

$$\left\langle v(t), (F^*\sigma_i)(t)\right\rangle = \langle Fv(\cdot), \sigma_i(\cdot)\rangle = (Fv)(t_i) = 0.$$

Since $\{t_i\}_{i=1}^{\infty}$ is dense in $[0, 1]$, $(Fv)(t) = 0$. Then, by the property of the inverse operator F^{-1}, we reach $v \equiv 0$. □

Theorem 4.2 *Assume that $v(t)$ is the exact solution of* (4.2). *Thus, we can construct this exact solution by:*

$$v(t) = \sum_{i=1}^{\infty}\sum_{k=1}^{i} \gamma_{ik} E(t_k, v_k)\widehat{\phi}_i(t). \tag{4.6}$$

where $\{t_i\}_{i=1}^{\infty}$ is dense in $[0, 1]$.

Proof We obtain

$$v(t) = \sum_{i=1}^{\infty} \left\langle v(t), \widehat{\phi}_i(t) \right\rangle_{V_2^3[0,1]} \widehat{\phi}_i(t)$$

by the completeness of the system. Then, we get

$$\begin{aligned} v(t) &= \sum_{i=1}^{\infty} \left\langle v(t), \widehat{\phi}_i(t) \right\rangle_{V_2^3[0,1]} \widehat{\phi}_i(t) \\ &= \sum_{i=1}^{\infty} \sum_{k=1}^{i} \gamma_{ik} \left\langle v(t), \phi_k(t) \right\rangle_{V_2^3[0,1]} \widehat{\phi}_i(t) \end{aligned}$$

by Eq. (4.5). We know that $\phi_i(t) = F^* \sigma_i(t)$. Thus, we get

$$\begin{aligned} v(t) &= \sum_{i=1}^{\infty} \left\langle v(t), \widehat{\phi}_i(t) \right\rangle_{V_2^3[0,1]} \widehat{\phi}_i(t) \\ &= \sum_{i=1}^{\infty} \sum_{k=1}^{i} \gamma_{ik} \left\langle v(t), \phi_k(t) \right\rangle_{V_2^3[0,1]} \widehat{\phi}_i(t) \\ &= \sum_{i=1}^{\infty} \sum_{k=1}^{i} \gamma_{ik} \left\langle v(t), F^* \sigma_k(t) \right\rangle_{V_2^3[0,1]} \widehat{\psi}_i(x) \end{aligned}$$

We use the properties of the adjoint operator F^* and obtain

$$\begin{aligned} v(t) &= \sum_{i=1}^{\infty} \left\langle v(t), \widehat{\phi}_i(t) \right\rangle_{V_2^3[0,1]} \widehat{\phi}_i(t) \\ &= \sum_{i=1}^{\infty} \sum_{k=1}^{i} \gamma_{ik} \left\langle v(t), \phi_k(t) \right\rangle_{V_2^3[0,1]} \widehat{\phi}_i(t) \\ &= \sum_{i=1}^{\infty} \sum_{k=1}^{i} \gamma_{ik} \left\langle v(t), F^* \sigma_k(t) \right\rangle_{V_2^3[0,1]} \widehat{\phi}_i(t) \\ &= \sum_{i=1}^{\infty} \sum_{k=1}^{i} \gamma_{ik} \left\langle F v(t), \sigma_k(t) \right\rangle_{V_2^1[0,1]} \widehat{\phi}_i(t) \end{aligned}$$

Then we get

$$
\begin{aligned}
v(t) &= \sum_{i=1}^{\infty} \left\langle v(t), \widehat{\phi}_i(t) \right\rangle_{V_2^3[0,1]} \widehat{\phi}_i(t) \\
&= \sum_{i=1}^{\infty} \sum_{k=1}^{i} \gamma_{ik} \left\langle v(t), \phi_k(t) \right\rangle_{V_2^3[0,1]} \widehat{\phi}_i(t) \\
&= \sum_{i=1}^{\infty} \sum_{k=1}^{i} \gamma_{ik} \left\langle v(t), F^* \sigma_k(t) \right\rangle_{V_2^3[0,1]} \widehat{\phi}_i(t) \\
&= \sum_{i=1}^{\infty} \sum_{k=1}^{i} \gamma_{ik} \left\langle F v(t), \sigma_k(t) \right\rangle_{V_2^1[0,1]} \widehat{\phi}_i(t) \\
&= \sum_{i=1}^{\infty} \sum_{k=1}^{i} \gamma_{ik} \left\langle E(t, v), A_{t_k} \right\rangle_{V_2^1[0,1]} \widehat{\phi}_i(t)
\end{aligned}
$$

by Eq. (4.2). We use the reproducing property and obtained the desired result as:

$$
\begin{aligned}
v(t) &= \sum_{i=1}^{\infty} \left\langle v(t), \widehat{\phi}_i(t) \right\rangle_{V_2^3[0,1]} \widehat{\phi}_i(t) \\
&= \sum_{i=1}^{\infty} \sum_{k=1}^{i} \gamma_{ik} \left\langle v(t), \phi_k(t) \right\rangle_{V_2^3[0,1]} \widehat{\phi}_i(t) \\
&= \sum_{i=1}^{\infty} \sum_{k=1}^{i} \gamma_{ik} \left\langle v(t), F^* \sigma_k(t) \right\rangle_{V_2^3[0,1]} \widehat{\phi}_i(t) \\
&= \sum_{i=1}^{\infty} \sum_{k=1}^{i} \gamma_{ik} \left\langle F v(t), \sigma_k(t) \right\rangle_{V_2^1[0,1]} \widehat{\phi}_i(t) \\
&= \sum_{i=1}^{\infty} \sum_{k=1}^{i} \gamma_{ik} \left\langle E(t, v), A_{t_k} \right\rangle_{V_2^1[0,1]} \widehat{\phi}_i(t) \\
&= \sum_{i=1}^{\infty} \sum_{k=1}^{i} \gamma_{ik} E(t_k, v_k) \widehat{\phi}_i(t).
\end{aligned}
$$

This completes the proof. □

We can obtain the approximate solution $v_n(t)$ as:

$$
v_n(t) = \sum_{i=1}^{n} \sum_{k=1}^{i} \gamma_{ik} E(t_k, v_k) \widehat{\phi}_i(t). \tag{4.7}
$$

Table 1 Approximate solutions (AS), exact solutions (ES), absolute errors (AE) and relative errors (RE) of the problem

x	AS	ES	AE	RE
0.0	1.0000000000	1.0000000000	0.0000000000	0.000000000000
0.1	0.9931500223	0.9950041653	0.0018541430	0.001863452501
0.2	0.9764874848	0.9800665778	0.0035790930	0.003651887618
0.3	0.9502835173	0.9553364891	0.0050529718	0.005289206324
0.4	0.9148977192	0.9210609940	0.0061632748	0.006691494744
0.5	0.8707751817	0.8775825619	0.0068073802	0.007756968399
0.6	0.8184389144	0.8253356149	0.0068967005	0.008356237603
0.7	0.7584668739	0.7648421873	0.0063753134	0.008335462538
0.8	0.6914504934	0.6967067093	0.0052562159	0.007544373880
0.9	0.6179592650	0.6216099683	0.0036507033	0.005872980625
1.0	0.5385516750	0.5403023059	0.0017506309	0.003240095185

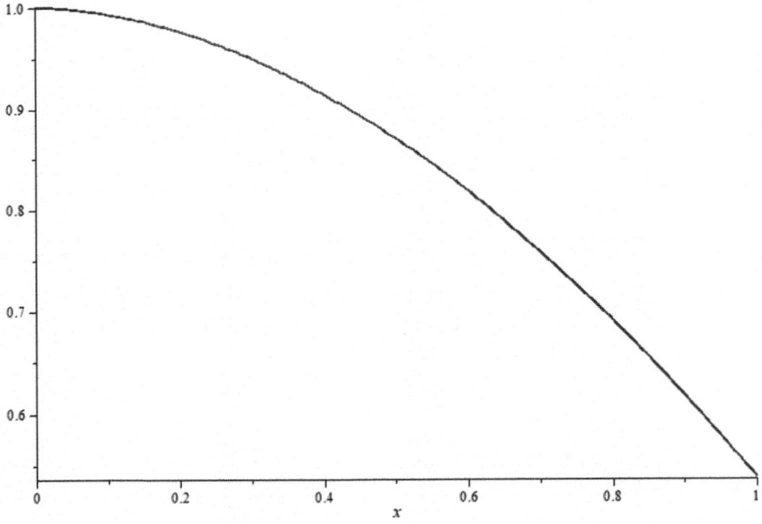

Fig. 1 Approximate solutions of the problem

5 Applications of the Method

We obtain the numerical results of the problem by the reproducing kernel Hilbert space method in this section. We demonstrate our results by the Table 1 and Figs. 1, 2, 3, 4, 5, 6, 7, 8. In Table 1, we give the exact solutions, approximate solutions, absolute errors and relative errors of the problem. In Fig. 1, we demonstrate the approximate solutions of the problem. In Fig. 2, we present the exact solution of the problem. In Figs. 3 and 4, we compare the exact and approximate solutions of the problem. In

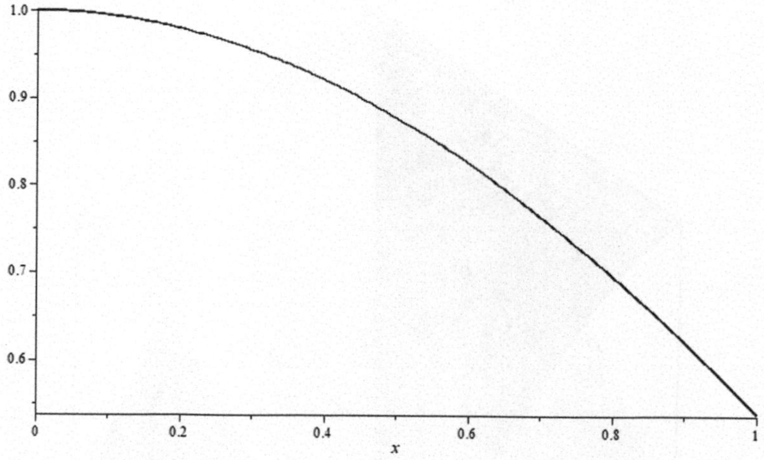

Fig. 2 Exact solutions of the problem

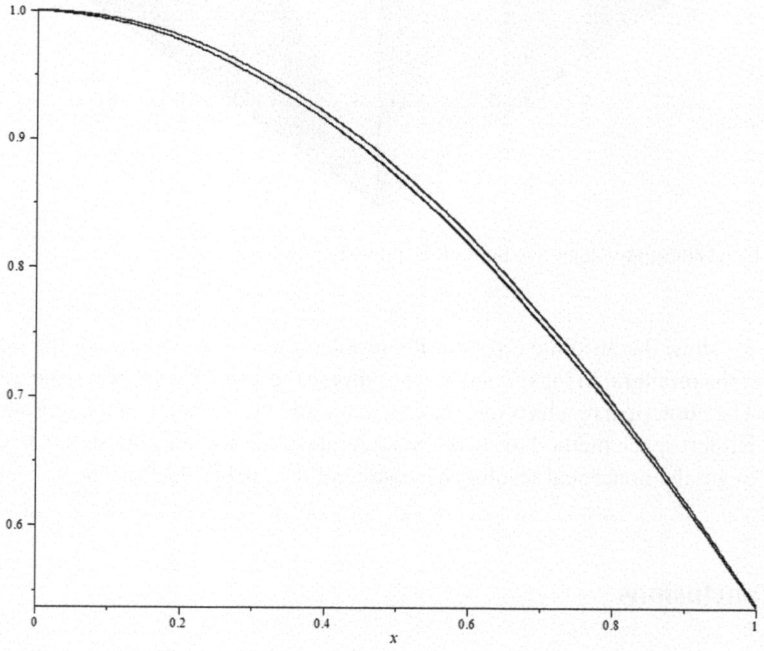

Fig. 3 Exact and approximate solutions of the problem

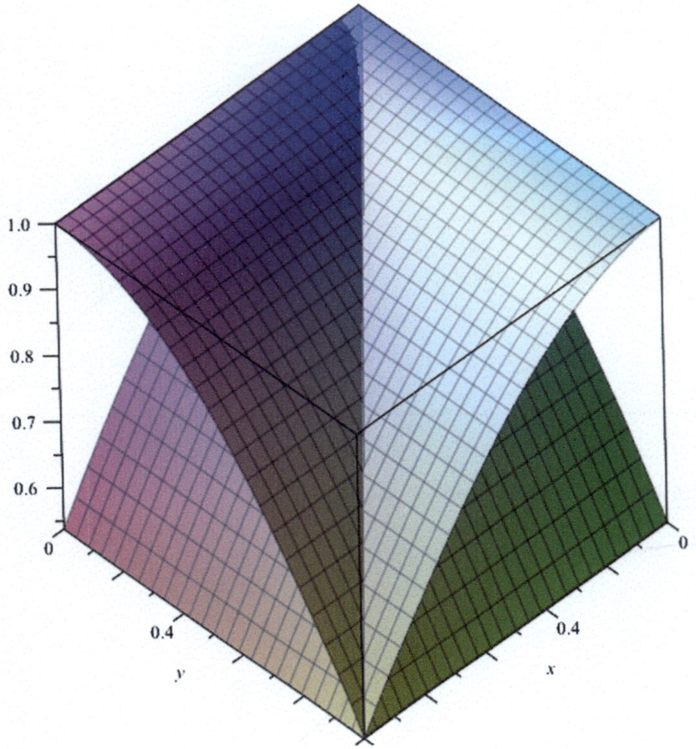

Fig. 4 Exact and approximate solutions of the problem

Fig. 5, we show the absolute errors of the problem. In Fig. 6, we present the relative errors of the problem. In Figs. 7 and 8, we compare the absolute errors and the relative errors. The numerical results prove the efficiency and the accuracy of the reproducing kernel Hilbert space method to solve the integral equations. We choose the 36 dense points to get the numerical results. We obtain all results by MAPLE 18.

6 Conclusions

In this chapter, we searched the integral equations by the reproducing kernel Hilbert space method. We created very useful reproducing kernel Hilbert spaces. We obtained some important reproducing kernel functions in these spaces. We presented the proof of some scientific theorems. We proved the efficiency of the reproducing kernel Hilbert space method by the reproducing kernel Hilbert space method.

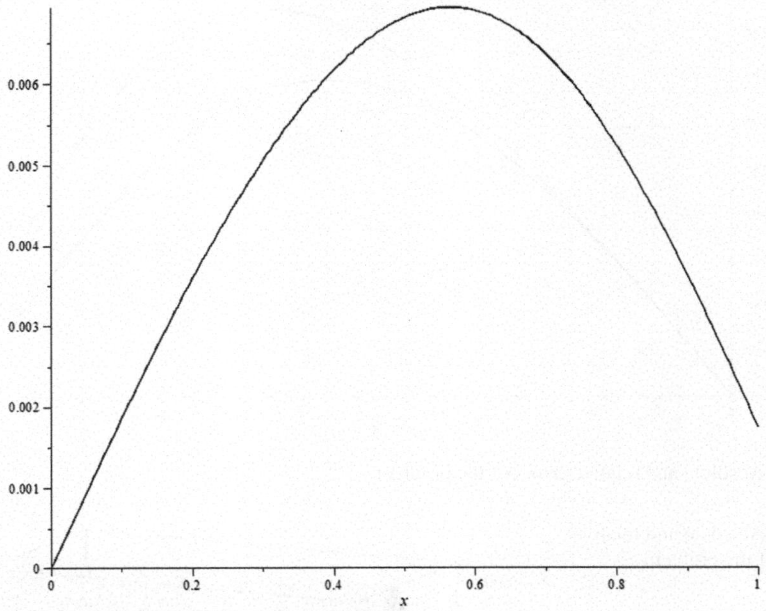

Fig. 5 Absolute errors of the problem

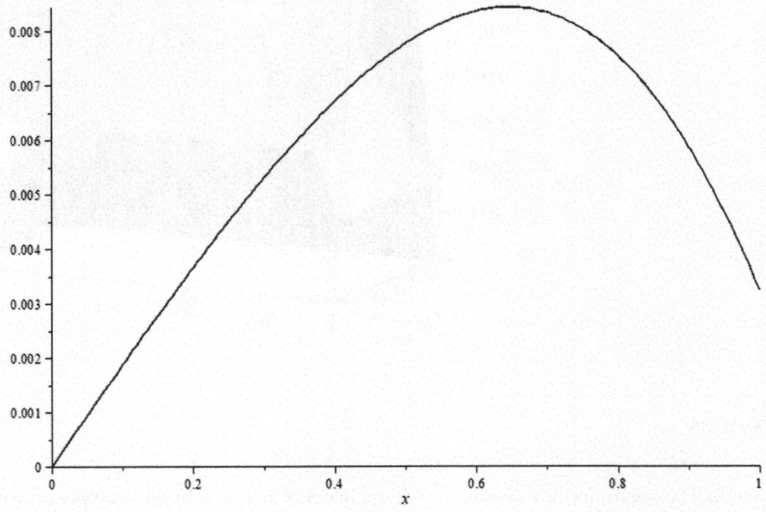

Fig. 6 Relative errors of the problem

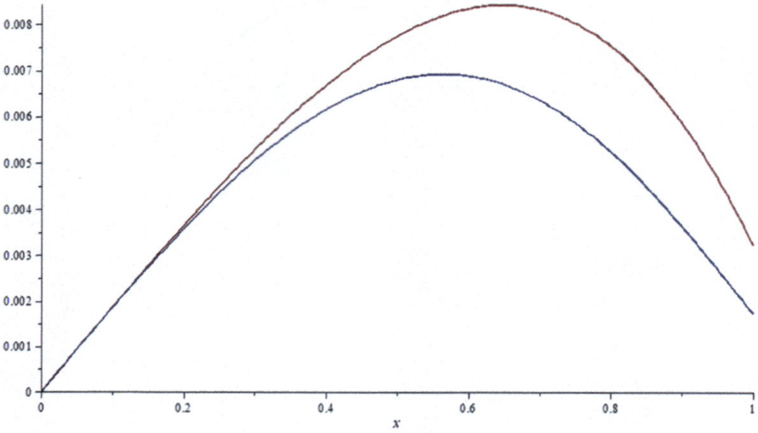

Fig. 7 Absolute and relative errors of the problem

Fig. 8 Absolute and relative errors of the problem

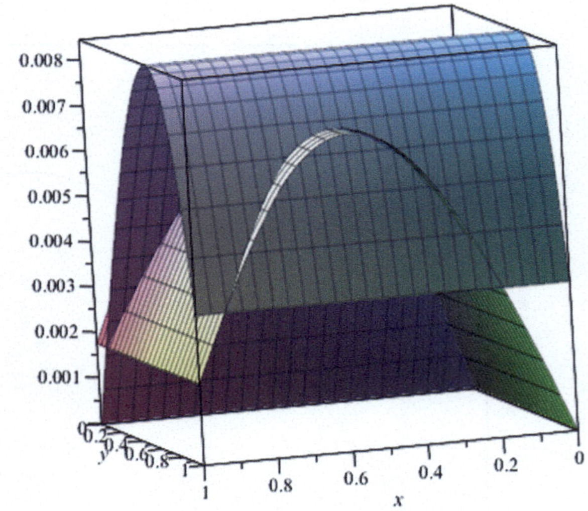

References

1. Costarelli, D.: Approximate solutions of Volterra integral equations by an interpolation method based on ramp functions. Comput. Appl. Math. **38**, 159 (2019)
2. Brunner, H.: Collocation methods for volterra integral and related functional equations. Cambridge Monogr. Appl. Comput. Math. **15** (2004)
3. Brunner, H.: Volterra Integral Equations: An Introduction to Theory and Applications. Cambridge University Press, Cambridge (2017)
4. Calvetti, D., Golub, G.H., Gragg, W.B., Reichel, L.: Computation of Gauss-Kronrod quadrature rules. Math Comput **69**, 1035–1052
5. Cao, Y., Herdman, T., Xu, Y.: A hybrid collocation method for volterra integral equations with weakly singular kernels. SIAM J. Numer. Anal. **41**(1), 364–381 (2003)

6. Cao, F., Chen, Z.: The approximation operators with sigmoidal functions. Comput. Math. Appl. **58**(4), 758–765 (2009)
7. Cao, F., Chen, Z.: The construction and approximation of a class of neural networks operators with ramp functions. J. Comput. Anal. Appl. **14**(1), 101–112 (2012)
8. Coroianu, L., Gal, S.G.: Approximation by truncated max-product operators of Kantorovich-type based on generalized kernels. Math. Methods Appl. Sci. **41**(17), 7971–7984 (2018)
9. Costarelli, D.: Interpolation by neural network operators activated by ramp functions. J. Math. Anal. Appl. **419**(1), 574–582 (2014)
10. Lü, X., Cui, M.G.: Solving a singular system of two nonlinear ODEs. Appl. Math. Comput. **198**, 534–543 (2008)
11. Tahmasbi, A., Fard, O.S.: Numerical solution of linear Volterra integral equations system of the second kind. Appl. Math. Comput. **201**, 547–552 (2008)
12. Wang, Y., Chaolu, T., Jing, P.: New algorithm for second-order boundary value problems of integro-differential equation. J. Comput. Appl. Math. **229**, 1–6 (2009)
13. Wang, Y., Chaolu, T., Chen, Z.: Using reproducing kernel for solving a class of singular weakly nonlinear boundary value problems. Int. J. Comput. Math. **87**, 367–380 (2010)
14. Wang, Y., Su, L., Cao, X., Li, X.: Using reproducing kernel for solving a class of singularly perturbed problems. Comput. Math. Appl. **61**, 421–430 (2011)
15. Yang, L.H., Shen, J.H., Wang, Y.: The reproducing kernel method for solving the system of the linear Volterra integral equations with variable coefficients. J. Comput. Appl. Math. **236**, 2398–2405 (2012)
16. Chen, Z., Jiang, W.: An approximate solution for a mixed linear Volterra-Fredholm integral equation. Appl. Math. Lett. **25**, 1131–1134 (2012)
17. Zhao, Z.H., Lin, Y.Z., Niu, J.: Convergence order of the reproducing kernel method for solving boundary value problems. Math. Model. Anal. **21**, 466–477 (2016)
18. Li, X.Y., Wu, B.Y.: Error estimation for the reproducing kernel method to solve linear boundary value problems. J. Comput. Appl. Math. **243**, 10–15 (2013)
19. Cui, M., Yingzhen, L.: Nonlinear Numerical Analysis in the Reproducing Kernel Space. Nova Science Publishers Inc., New York (2009)
20. Aronszajn, N.: Theory of reproducing kernels. Trans. Am. Math. Soc. **68**, 337–404 (1950)
21. Akgül, A.: Reproducing kernel Hilbert space method based on reproducing kernel functions for investigating boundary layer flow of a Powell-Eyring non-Newtonian fluid. J. Taibah Univ. Sci. **13**(1), 858–863 (2019)
22. Akgül, A.: New reproducing kernel functions. Math. Probl. Eng. **2015**, 158134 (2015)
23. Akgül, A.: On solutions of variable-order fractional differential equations. Int. J. Optim. Control Theor. Appl. (IJOCTA) **7**(1), 112–116 (2017)
24. Akgül, A., Akgül, E.K., Baleanu, D.: Numerical solutions of fractional differential equations of Lane-Emden type by an accurate technique. Adv. Differ. Equ. **2015**(1), 220 (2015)
25. Kumar, S., Kumar, R., Agarwal, R.P., Samet, B.: A study on population dynamics of two interacting species by haar wavelet and Adam's-Bashforth-Moulton methods. Math. Methods Appl. Sci. **43**(8), 5564–5578 (2020)
26. Kumar, S., Ahmadian, A., Kumar, R., Kumar, D., Singh, J., Baleanu, D., Salimi, M.: An efficient numerical method for fractional SIR epidemic model of Infectious disease by using Bernstein wavelets. Mathematics **8**, 558 (2020)
27. Baleanu, D., Jleli, M., Kumar, S., Samet, B.: A fractional derivative with two singular kernels and application to a heat conduction problem. Adv. Differ. Equ. **28**, (2020)
28. Akgül, A.: A novel method for a fractional derivative with non-local and non-singular kernel. Chaos, Solitons Fractals **114**, 478–482 (2018)
29. Akgül, A., Cordero, A., Torregrosa, J.R.: Solutions of fractional gas dynamics equation by a new technique. Math. Methods Appl. Sci. **43**(3), 1349–1358
30. Akgül, A., Cordero, A., Torregrosa, J.R.: A fractional Newton method with 2th-order of convergence and its stability. Appl. Math. Lett. **98**, 344–351
31. Atangana, A., Akgül, A., Owolabi, K.M.: Analysis of fractal fractional differential equations. Alexandria Eng. J. (2020)

32. Atangana, A., Akgül, A.: Can transfer function and Bode diagram be obtained from Sumudu transform. Alexandria Eng. J.(2020)
33. Yusuf, A., Qureshi, S., Shah, S.F.: Mathematical analysis for an autonomous financial dynamical system via classical and modern fractional operators. Chaos, Solitons Fractals **132**, 109552
34. Šremr, J.: Absolutely continuous functions of two variables in the sense of Carathéodory. Electron. J. Differ. Equ. **2010**, 1–11 (2010)
35. Akgül, A., Grow, D.: Existence of unique solutions to the telegraph equation in binary reproducing Kernel Hilbert spaces. Differ. Equ. Dyn. Syst. (2020)

Restricted Global Convergence Domains for Integral Equations of the Fredholm-Hammerstein Type

J. A. Ezquerro and M. A. Hernández-Verón

Abstract Our study focuses on obtaining restricted global convergence domains for Newton's method, that usually are balls, when it is applied to solve nonlinear integral equations of the Fredholm-Hammerstein type. In particular, our work maintains the characteristics of the method of the successive approximations (lineal convergence), based on the Fixed Point Theorem, but it is applied to Newton's method (quadratic convergence). Moreover, this work allows us to address the solution of nonlinear integral equations of the Fredholm-Hammerstein type that have several solutions and obtains results about the location and separation of the different solutions. Throughout the work, we include numerical examples that show the interest of the results obtained.

Keywords Fredholm integral equation · Fixed point theorem · Newton's method · Global convergence · Convergence ball.

2010 Mathematics Subject Classification: 45B05 · 47H30 · 65J15.

1 Introduction

The problems of engineering, mathematics or mathematical physics that are modeled by an integral equation are numerous, understanding that an integral equation is an equation in which the unknown function appears under the integral sign. Basic, simple and usual examples of integral equations are those that are related to boundary

This research was partially supported by Ministerio de Ciencia, Innovación y Universidades under grant PGC2018-095896-B-C21.

J. A. Ezquerro (✉) · M. A. Hernández-Verón
Department of Mathematics and Computation, University of La Rioja, Calle Madre de Dios 53, 26006 Logroño. La Rioja, Spain
e-mail: jezquer@unirioja.es

M. A. Hernández-Verón
e-mail: mahernan@unirioja.es

H. Singh et al. (eds.), *Topics in Integral and Integro-Differential Equations*, Studies in Systems, Decision and Control 340,
https://doi.org/10.1007/978-3-030-65509-9_5

value problems for differential equations, the theory of elasticity, potential theory, electrostatics, radiative heat transfer problems, etc. [3, 5, 6, 15, 19]. In particular, the Fredholm integral equations are among the most interesting and, within these, the Fredholm-Hammerstein integral equations of the type [8, 11, 14, 16]

$$x(s) = u(s) + \lambda \int_a^b \mathfrak{K}(s,t)\mathcal{G}(t,x(t))\,dt, \quad s \in [a,b], \quad \lambda \in \mathbb{R},$$

where $u(s) \in \mathcal{C}[a,b]$, the kernel $\mathfrak{K}(s,t)$ is a known function in $[a,b] \times [a,b]$, $\mathcal{G}$ is a known continuous function in $[a,b] \times [a,b]$ and $x(s)$ is the known function to find. These type of integral equations appear in the theory of radiative transference [7], the theory of transport of neutrons [1] and the kinetic theory of gases [18].

In this work, we focus on a particular case of integral equations of the last type, that are known as integral equations of Fredholm-Hammerstein type with Nemystkii operator [12] and are of the form

$$x(s) = u(s) + \lambda \int_a^b \mathfrak{K}(s,t)\mathcal{H}(x)(t)\,dt, \quad s \in [a,b], \quad \lambda \in \mathbb{R}, \tag{1}$$

where $\mathcal{H}$ is the Nemytskii operator $\mathcal{H} : \mathcal{C}[a,b] \to \mathcal{C}[a,b]$ such that $\mathcal{H}(x)(t) = H(x(t))$, where $H : \mathbb{R} \to \mathbb{R}$.

In general, we use numerical methods to solve integral equations of type (1), since they cannot be solved exactly. Throughout this work we do a qualitative study of this type of equations that consists of locating a solution and separating it from other possible solutions in addition to addressing its numerical solution. For this study, we start considering restricted domains of existence and uniqueness of fixed points, which is based on the well-known Fixed Point Theorem [4]:

> If the operator $\Theta : \mathcal{C}[a,b] \to \mathcal{C}[a,b]$ is a contraction, then Θ has a unique fixed point in $\mathcal{C}[a,b]$ that can be approximated from the method of successive approximations $x_{n+1} = \Theta(x_n)$, $n \geq 0$, with x_0 given in $\mathcal{C}[a,b]$.

Also, we remember that the operator Θ is a contraction if $\|\Theta(x) - \Theta(y)\| < \kappa\|x - y\|$ with $\kappa \in [0,1)$, for all $x, y \in \mathcal{C}[a,b]$. If the operator Θ is derivable, it is enough that the condition $\|\Theta'(x)\| < 1$, for all $x \in \mathcal{C}[a,b]$, is satisfied to see that Θ is a contraction.

Then, if we consider the operator $\Theta : \mathcal{C}[a,b] \to \mathcal{C}[a,b]$ such that

$$[\Theta(x)](s) = u(s) + \lambda \int_a^b \mathfrak{K}(s,t)\mathcal{H}(x)(t)\,dt, \quad s \in [a,b], \quad \lambda \in \mathbb{R}, \tag{2}$$

it is clear that a fixed point of operator (2) is a solution of integral equation (1). In addition, we observe that

$$\|\Theta(x) - \Theta(y)\| \leq |\lambda|\ell\|\mathcal{H}(x) - \mathcal{H}(y)\|, \quad x, y \in \mathcal{C}[a,b],$$

where $\ell = \left\| \int_a^b \mathfrak{K}(s,t)|\,dt \right\|$, so that if the Nemytskii operator $\mathcal{H}$ is Lipschitz continuous with Lipschitz constant μ, we have

$$\|\Theta(x) - \Theta(y)\| \leq |\lambda|\ell\mu\|x - y\|, \quad x, y \in \mathcal{C}[a,b],$$

and, as a consequence, the operator Θ is a contraction if the contractivity factor $\kappa = |\lambda|\ell\mu < 1$. In this case, by the Fixed Point Theorem mentioned above, we obtain that there is a fixed point x^* of the operator Θ and the sequence $x_{n+1} = \Theta(x_n)$, $n \geq 0$, converges linearly to x^* from all $x_0 \in \mathcal{C}[a,b]$. As we can see, this theorem provides global convergence for the method of successive approximations, since the method can be started at any point in the full space, but it does not allow us to locate the fixed point in a domain. Moreover, there is the additional problem: the linear speed of convergence of the method of successive approximations.

Therefore, the aim of this work is threefold. On the one hand, we look for a better location of a fixed point, so that we obtain a restricted domain where the fixed point is instead of in the full space $\mathcal{C}[a,b]$. On the other hand, we look for maintaining the global convergence, so that any point in the domain of existence of the fixed point obtained can be considered as the starting point of the iterative process considered and so obtain its convergence to the fixed point. And thirdly, we look for improving the linear speed of convergence of the method of successive approximations. For this, we use Newton's method that has quadratic convergence.

For the last, we observe that a solution of the equation $F(x) = 0$, where $F : \mathcal{C}[a,b] \to \mathcal{C}[a,b]$ is the operator given by $[F(x)](s) = x(s) - [\Theta(x)](s)$, is a fixed point of the operator Θ. So, if we consider Newton's method,

$$x_{n+1} = N_F(x_n) = x_n - [F'(x_n)]^{-1}F(x_n), \quad n \geq 0, \quad \text{with } x_0 \text{ given in } \mathcal{C}[a,b],$$

where $F'(x_n)$ is the first Fréchet derivative of the operator $F(x)$ at the point x_n and $[F'(x_n)]^{-1}$ is the inverse operator, to approximate a solution x^* of $F(x) = 0$, then x^* is such that $x^* = N_F(x^*)$, so that x^* is a fixed point of the operator $N_F : \mathcal{C}[a,b] \to \mathcal{C}[a,b]$ and $x_{n+1} = N_F(x_n)$, $n \geq 0$, is a sequence with quadratic convergence [2, 9].

Now, in the current situation, it is important to note that there is only one fixed point x^* in the full space $\mathcal{C}[a,b]$, but this does not have to happen in general, since there may be more fixed points. In this case, the best location of the fixed point proposed in this work also looks for separating the fixed points. For this, we look for uniqueness results of the fixed points in the restricted domains obtained.

The study developed in this work allow us to analyze Newton's method from a different point of view than usual, which allows us to obtain domains of global convergence, as well as domains of existence and uniqueness of fixed points for the operator Θ. Moreover, from the domains of global convergence obtained previously, we get what we call balls of convergence, that gives us an idea of the accessibility of the method, since the method converges starting at any point on the ball. Furthermore, from the domains of existence of fixed points, we locate fixed points and, from their domains of uniqueness, they are separated.

This work is organized as follows. In Sect. 2, we present a basic result, based on the Fixed Point Theorem, and we rely on it to obtain the first results in the next section. In Sect. 3, we introduce the Nemystkii operator with bounded second derivative and an auxiliary function that allow us to obtain the first results of restricted global convergence. In Sect. 4, we introduce the Nemystkii operator with Lipschitz continuous first derivative and a technique different from that used in Sect. 3 to obtain restricted global convergence domains. In addition, from a concrete example, we see how these domains are improving.

Throughout the paper, we denote $\overline{B(x,\epsilon)} = \{y \in \mathcal{C}[a,b];\ \|y-x\| \leq \epsilon\}$, $B(x,\epsilon) = \{y \in \mathcal{C}[a,b];\ \|y-x\| < \epsilon\}$ and the set of bounded linear operators from $\mathcal{C}[a,b]$ to $\mathcal{C}[a,b]$ by $\mathcal{L}(\mathcal{C}[a,b],\mathcal{C}[a,b])$, and use the infinity norm in $\mathcal{C}[a,b]$.

2 Motivation

As we have indicated in the introduction, we look for locating fixed points in restricted domains in $\mathcal{C}[a,b]$. For this, we use the following modification of the Fixed Point Theorem, given in the introduction, to establish a result on fixed points in a set of $\mathcal{C}[a,b]$ instead of in the full space $\mathcal{C}[a,b]$:

> If Ω is a convex and compact set of $\mathcal{C}[a,b]$ and the operator $\Upsilon : \Omega \to \Omega$ is a contraction, then the operator Υ has a unique fixed point in Ω and it can be approximated by the method of successive approximations, $x_{n+1} = \Upsilon(x_n)$, $n \geq 0$, from any $x_0 \in \Omega$.

Taking into account the situation brought up in the introduction, if the Nemytskii operator $\mathcal{H}$ is Lipschitz continuous, we have

$$\begin{aligned}\|\Theta(x)\| &\leq \|u\| + |\lambda|\ell\|\mathcal{H}(x)\| \\ &\leq \|u\| + |\lambda|\ell\left(\|\mathcal{H}(0)\| + \mu\|x\|\right),\end{aligned}$$

so that we need the condition

$$\|u\| + |\lambda|\ell\|\mathcal{H}(0)\| + (|\lambda|\ell\mu - 1)R \leq 0$$

to be verified if we want to guarantee that $\Theta : \overline{B(0,R)} \to \overline{B(0,R)}$. The last condition can only be verified if $|\lambda|\ell\mu < 1$ and, when this happens, the operator Θ is a contraction. Therefore, by the modification of the Fixed Point Theorem given above, there is a unique fixed point x^* of Θ in $\overline{B(0,R)}$ and the sequence $x_{n+1} = \Theta(x_n)$, $n \geq 0$, converges linearly to x^* from all $x_0 \in \overline{B(0,R)}$.

As we have just seen, with the simple modification of the Fixed Point Theorem that we just presented, we already obtain a better location of the fixed point x^* in a restricted domain $\overline{B(0,R)} \subset \mathcal{C}[a,b]$, which also has the advantage that it separates the fixed point x^* from any other fixed point that may exist.

Following this idea and considering the operator $N_F : \mathcal{C}[a,b] \to \mathcal{C}[a,b]$ with $N_F(x) = x - [F'(x)]^{-1}F(x)$ and $[F(x)](s) = x(s) - [\Theta(x)](s)$, we obtain the quadratic convergence that Newton's method guarantees. We also locate functions $\widetilde{x} \in \mathcal{C}[a,b]$ that, in general, provide better locations, $\overline{B(\widetilde{x}, R)}$, of the fixed point x^* than those defined from the null function, $\overline{B(0, R)}$, that, sometimes, can be obtained from a previous location, which can be done from considering what condition $\|x^*\|$ must satisfy, where x^* is a fixed point, from Eq. (1).

3 Nemystkii Operator with Bounded Second Derivative

In this section, we seek to obtain results of existence and uniqueness of fixed points for the operator N_F. For this, we use Newton's method and the modification of the Fixed Point Theorem given in Sect. 2, so that the method is convergent if the operator N_F is a contraction in $\mathcal{C}[a,b]$.

Observe first that solving Eq. (1) is equivalent to solving the equation $F(x) = 0$, where $F : \mathcal{C}[a,b] \to \mathcal{C}[a,b]$ and

$$[F(x)](s) = x(s) - u(s) - \lambda \int_a^b \mathfrak{K}(s,t)\mathcal{H}(x(t))\,dt, \quad s \in [a,b]. \tag{3}$$

For operator (3), if $\mathcal{H}$ is sufficiently differentiable, we have

$$[F'(x)y](s) = y(s) - \lambda \int_a^b \mathfrak{K}(s,t)[\mathcal{H}'(x)y](t)\,dt = y(s) - \lambda \int_a^b \mathfrak{K}(s,t)H'(x(t))y(t)\,dt,$$

$$[F''(x)(yz)](s) = -\lambda \int_a^b \mathfrak{K}(s,t)[\mathcal{H}''(x)yz](t)\,dt = -\lambda \int_a^b \mathfrak{K}(s,t)H''(x(t))y(t)z(t)\,dt$$

and $\|F''(x)\| \leq |\lambda|\ell\|\mathcal{H}''(x)\|$, where $\ell = \left\| \int_a^b \mathfrak{K}(s,t)\,dt \right\|$, so that $\|F''(x)\| \leq |\lambda|\ell M = M_\star$ if $\|\mathcal{H}''(x)\| \leq M$.

In addition, from the Banach lemma on invertible operators, it follows that there exists the operator $\Gamma = [F'(x)]^{-1}$, for all $x \in \mathcal{C}[a,b]$, with

$$\|\Gamma\| \leq \frac{1}{1 - |\lambda|\ell\|\mathcal{H}'(x)\|}, \qquad \|\Gamma F(x)\| \leq \frac{\|\widetilde{x} - u\| + |\lambda|\ell\|\mathcal{H}(x)\|}{1 - |\lambda|\ell\|\mathcal{H}'(x)\|},$$

provided that $|\lambda|\ell\|\mathcal{H}'(x)\| < 1$. This condition is very interesting for locating functions $\widetilde{x} \in \mathcal{C}[a,b]$ for which the operator $[F'(\widetilde{x})]^{-1}$ exists and so that $\widetilde{x}$ can act as auxiliary functions to locate fixed points of the operator N_F.

After that, we consider an auxiliary function $\widetilde{x} \in \mathcal{C}[a,b]$ satisfying the following conditions:

(A1) For some $\widetilde{x} \in \mathcal{C}[a,b]$, there exists $\widetilde{\Gamma} = [F'(\widetilde{x})]^{-1} \in \mathcal{L}(\mathcal{C}[a,b], \mathcal{C}[a,b])$ with $\|\widetilde{\Gamma}\| \leq \beta$ and $\|\widetilde{\Gamma} F(\widetilde{x})\| \leq \eta$.

(A2) There exists a constant $M \geq 0$ such that $\|\mathcal{H}''(x)\| \leq M$ for all $x \in \mathcal{C}[a,b]$.

3.1 A First Study of Restricted Global Convergence

The first is to locate a domain that contains a fixed point of the operator N_F. For this, we consider a domain $\Omega = \overline{B(\widetilde{x}, R)}$, $R \in \mathbb{R}_+$, such as we can see in the following lemma.

Lemma 1 *Let $\widetilde{x} \in \mathcal{C}[a,b]$ and suppose conditions (A1)–(A2). If $M_\star \beta \eta \leq \frac{1}{6}$, where $M_\star = |\lambda|\ell M$, then there exists $R > 0$ such that $R \in [R_-, R_+]$, where $R_- = \frac{1-\sqrt{1-6M_\star\beta\eta}}{3M_\star\beta}$ and $R_+ = \frac{1+\sqrt{1-6M_\star\beta\eta}}{3M_\star\beta}$, and $N_F : \overline{B(\widetilde{x}, R)} \to \overline{B(\widetilde{x}, R)}$.*

Proof First, we observe that, under the hypotheses of the lemma, $[R_-, R_+] \neq \emptyset$, so that we can choose R in this interval.

Second, we see that there exists $\Gamma = [F'(x)]^{-1}$, for all $x \in \overline{B(\widetilde{x}, R)}$. Indeed, from

$$\|I - \widetilde{\Gamma} F'(x)\| \leq \|\widetilde{\Gamma}\| \|F'(\widetilde{x}) - F'(x)\| \leq \beta M_\star \|x - \widetilde{x}\| \leq M_\star \beta R < 1,$$

since $R \in [R_-, R_+]$, it follows, by the Banach's lemma on invertible operators, that the operator Γ exists with $\|\Gamma\| \leq \frac{\beta}{1-M_\star\beta R}$ and $\|\Gamma F'(\widetilde{x})\| \leq \frac{1}{1-M_\star\beta R}$. As a consequence, the operator N_F is well defined in the domain indicated.

Third, we see that $N_F(x) \in \overline{B(\widetilde{x}, R)}$ for all $x \in \overline{B(\widetilde{x}, R)}$. Indeed, from

$$\begin{aligned}
\|N_F(x) - \widetilde{x}\| &= \|x - \widetilde{x} - \Gamma F(x)\| \\
&= \left\|-\Gamma \left(F(x) + F'(x)(\widetilde{x} - x)\right)\right\| \\
&= \left\|-\Gamma F(\widetilde{x}) + \Gamma \int_0^1 \left(F'(x + t(\widetilde{x} - x)) - F'(x)\right) dt (\widetilde{x} - x)\right\| \\
&\leq \|\Gamma F(\widetilde{x})\| + \frac{1}{2} M_\star \|\Gamma\| \|\widetilde{x} - x\|^2 \\
&\leq \|\Gamma F'(\widetilde{x})\| \|\widetilde{\Gamma} F(\widetilde{x})\| + \frac{1}{2} M_\star \|\Gamma\| \|\widetilde{x} - x\|^2 \\
&\leq \frac{M_\star \beta R^2 + 2\eta}{2(1 - M_\star \beta R)},
\end{aligned}$$

it follows that $N_F(x) \in \overline{B(\widetilde{x}, R)}$, since

$$\frac{M_\star \beta R^2 + 2\eta}{2(1 - M_\star \beta R)} \leq R,$$

provided that $M_\star \beta \eta \leq \frac{1}{6}$ and $R \in [R_-, R_+]$. ■

After that, we see that the operator N_F is a contraction.

Lemma 2 *Let $\widetilde{x} \in \mathcal{C}[a,b]$ and suppose conditions (A1)–(A2). If $M_\star \beta \eta \leq \frac{1}{6}$, then there exists $R > 0$ such that $R \leq \delta_- = \frac{4-\sqrt{10+6M_\star\beta\eta}}{3M_\star\beta}$ and the operator N_F is a contraction.*

Proof If $x \in \overline{B(\widetilde{x}, R)}$, then

$$F(x) = F(\widetilde{x}) + F'(\widetilde{x})(x - \widetilde{x}) + \int_0^1 \left(F'(\widetilde{x} + t(x - \widetilde{x})) - F'(\widetilde{x})\right)(x - \widetilde{x})\,dt$$

and

$$\widetilde{\Gamma} F(x) = \widetilde{\Gamma} F(\widetilde{x}) + (x - \widetilde{x}) + \widetilde{\Gamma} \int_0^1 \left(F'(\widetilde{x} + t(x - \widetilde{x})) - F'(\widetilde{x})\right)(x - \widetilde{x})\,dt,$$

so that

$$\begin{aligned} \|\widetilde{\Gamma} F(x)\| &\leq \|\widetilde{\Gamma} F(\widetilde{x})\| + \|x - \widetilde{x}\| + \frac{1}{2} M_\star \|\widetilde{\Gamma}\| \|x - \widetilde{x}\|^2 \\ &\leq \frac{1}{2} M_\star \beta R^2 + R + \eta. \end{aligned}$$

Now, if $x, y \in \overline{B(\widetilde{x}, R)}$ and we denote $\Gamma_x = [F'(x)]^{-1}$ and $\Gamma_y = [F'(y)]^{-1}$, then

$$\begin{aligned} N_F(x) - N_F(y) &= x - \Gamma_x F(x) - y + \Gamma_y F(y) \\ &= \Gamma_y \left(F'(y)(x - y) - F'(y)\Gamma_x F(x) + F(y)\right) \\ &= \Gamma_y \left(F(x) - \int_y^x (F'(z) - F'(y))\,dz - F'(y)\Gamma_x F(x)\right) \\ &= \Gamma_y \left(\left(\int_0^1 F''(y + t(x - y))\,dt\,(x - y)\right)\Gamma_x F(x) - \int_y^x (F'(z) - F'(y))\,dz\right) \end{aligned}$$

and

$$\begin{aligned} \|N_F(x) - N_F(y)\| &= \|\Gamma_y\| \left(M_\star \|x - y\| \|\Gamma_x F(x)\| + \frac{1}{2} M_\star \|x - y\|^2\right) \\ &= \|\Gamma_y\| M_\star \left(\|\Gamma_x F(x)\| + \frac{1}{2}\|x - y\|\right) \|x - y\| \\ &\leq \|\Gamma_y\| M_\star \left(\|\Gamma_x F(x)\| + R\right) \|x - y\| \\ &= \alpha \|x - y\|. \end{aligned}$$

Taking now into account that $\|[F'(x)]^{-1}\| \leq \frac{\beta}{1 - M_\star \beta R}$, for all $x \in \overline{B(\widetilde{x}, R)}$, we have $\|\Gamma_y\| \leq \frac{\beta}{1 - M_\star \beta R}$. Moreover,

$$\|\Gamma_x F(x)\| \leq \|\Gamma_x F'(\widetilde{x})\| \|\widetilde{\Gamma} F(x)\| \leq \frac{\frac{1}{2} M_\star \beta R^2 + R + \eta}{1 - M_\star \beta R}.$$

Thus, we can take $\alpha = \dfrac{M_\star\beta(2\eta + 4R - M_\star\beta R^2)}{2(1 - M_\star\beta R)^2}$. Therefore, $\alpha < 1$ if $R \leq \delta_-$ and the operator N_F is then a contraction. ■

Once we have seen that the operator $N_F : \overline{B(\widetilde{x}, R)} \to \overline{B(\widetilde{x}, R)}$ is a contraction, we locate the value of R.

Lemma 3 *If $M_\star\beta\eta \leq \frac{-3+\sqrt{13}}{4} = 0.1513\ldots$, there always exists $R > 0$ such that $R \in \left[R_-, \delta_-\right]$.*

Proof If $M_\star\beta\eta \leq 0.1513\ldots$, we observe that $R_- \leq \delta_-$, since $4(M_\star\beta\eta)^2 + 6(M_\star\beta\eta) - 1 \leq 0$, so that $\left[R_-, \delta_-\right] \neq \emptyset$. Besides, as $R_+ > \delta_-$, the proof is complete. ■

Finally, from the modification of the Fixed Point Theorem given above, we obtain the existence of a unique fixed point of N_F in $\overline{B(\widetilde{x}, R)}$ and its approximation by Newton's method.

Theorem 4 *Let $\widetilde{x} \in \mathcal{C}[a, b]$ and suppose conditions (A1)–(A2). If $M_\star\beta\eta \leq \frac{-3+\sqrt{13}}{4} = 0.1513\ldots$, then there exists $R > 0$ such that $R \in \left[R_-, \delta_-\right]$, the operator $N_F : \overline{B(\widetilde{x}, R)} \to \overline{B(\widetilde{x}, R)}$ has a unique fixed point x^* and $x_{n+1} = N_F(x_n)$, $n \geq 0$, converges quadratically to x^* from any $x_0 \in \overline{B(\widetilde{x}, R)}$.*

Proof As $N_F(x) \in \overline{B(\widetilde{x}, R)}$, for all $x \in \overline{B(\widetilde{x}, R)}$, and the operator $N_F : \overline{B(\widetilde{x}, R)} \to \overline{B(\widetilde{x}, R)}$ is a contraction, then the proof is concluded from the application of the modification of the Fixed Point Theorem, given previously, to the operator N_F in $\overline{B(\widetilde{x}, R)}$. The fact that Newton's method has quadratic convergence under conditions (A1)–(A2) follows easily from the well-known theorem of Newton-Kantorovich (see [9]). ■

Since R_- is the smallest value that R can take and δ_- the largest, we can consider $\overline{B(\widetilde{x}, R_-)}$ as ball of location of the fixed point and $\overline{B(\widetilde{x}, \delta_-)}$ as ball of uniqueness. Besides, we observe that we have global convergence in any ball $\overline{B(\widetilde{x}, R)}$ with $R \in \left[R_-, \delta_-\right]$.

Next, we illustrate the previous result with an example where an integral equation of form (1) is involved.

Example 5 We consider the integral equation of Fredholm-Hammerstein type with Nemystkii operator given by

$$x(s) = \sin(\pi s) + \lambda \int_0^1 \cos(\pi s)\sin(\pi t)\left(x(t)^2 - \cos x(t)\right) dt, \quad s \in [a, b]. \tag{4}$$

Equations of this type has been studied in [13, 16] by different authors.

First, we observe that if we consider the corresponding operator Θ defined in (2), namely

$$\Theta(x) = \sin(\pi s) + \lambda \int_0^1 \cos(\pi s)\sin(\pi t)\left(x(t)^2 - \cos x(t)\right) dt,$$

then

$$\|\Theta(x) - \Theta(y)\| \leq |\lambda|\ell\|x + y\|\|x - y\| = \kappa\|x - y\|,$$

so that the contractivity factor κ depends on x and y and is not bounded in $\mathcal{C}[a, b]$. As a consequence, we cannot apply the Fixed Point Theorem to obtain a fixed point of Θ (namely, a solution of (4)).

Then, we can locate previously a solution $x^*(s)$ of Eq. (4) in a domain $\Omega = \overline{B(0, \varrho)} \subset \mathcal{C}[a, b]$. For this, we note that $x^*(s)$ satisfy

$$\|x^*(s)\| - \|\sin(\pi s)\| - |\lambda|\ell\|H(x^*(s))\| \leq \|x^*(s)\| - 1 - \frac{2}{\pi}|\lambda|\left(\|x^*(s)\|^2 + 1\right) \leq 0, \tag{5}$$

where $\ell = \left\|\int_a^b \cos(\pi s)\sin(\pi t)\,dt\right\| = \frac{2}{\pi}$. In Table 1, we see different values of $|\lambda|$ that satisfy the last inequality. So, according to the values of $|\lambda|$, the solution $x^*(s)$ can be in the domain $\overline{B(0, \varrho)}$. Notice that inequality (5) is not satisfied if $|\lambda| > 0.3253\ldots$, so that we cannot locate previously a solution of integral equation (4) for these values of $|\lambda|$.

Next, we apply the modification of the Fixed Point Theorem given in Sect. 2 to locate a solution $x^*(s)$ of integral equation (4) in a domain $\Omega = \overline{B(0, \varrho)} \subset \mathcal{C}[a, b]$. For this, we have seen in Sect. 2 that

$$\|\Theta(x)\| \leq \|u\| + |\lambda|\ell\left(\|\mathcal{H}(0)\| + \mu\|x\|\right),$$

where μ is the Lipschitz constant of the Nemytskii operator $\mathcal{H}$. As

$$\|\mathcal{H}(x) - \mathcal{H}(y)\| \leq (1 + \|x + y\|)\,\|x - y\| \leq (1 + 2\varrho)\|x - y\|,$$

then $\mu = 1 + 2\varrho$, so that $\|\Theta(x)\| \leq \varrho$ if

$$1 + \frac{2}{\pi}(1 + \varrho + 2\varrho^2)|\lambda| - \varrho \leq 0. \tag{6}$$

In addition, if

$$\frac{2}{\pi}(1 + 2\varrho)|\lambda| < 1, \tag{7}$$

then

$$\|\Theta'(x)\| \leq |\lambda|\ell\|\mathcal{H}'(x)\| \leq \frac{2}{\pi}(1 + 2\varrho)|\lambda| < 1$$

and the operator Θ is a contraction.

Since condition (7) is deduced from condition (6), it follows that $\Theta(x) \in \overline{B(0, \varrho)}$ and $\Theta : \overline{B(0, \varrho)} \to \overline{B(0, \varrho)}$ is a contraction operator if condition (6) holds. Then,

Table 1 $\|x^*\| \leq \varrho$ for different values of $|\lambda|$ in Eq. (4)

$\|\lambda\|$	ϱ
0.05	1.0681...
0.10	1.1474...
0.15	1.2430...
0.20	1.3643...
0.25	1.5333...
0.30	1.8319...

Table 2 $x^* \in \overline{B(0, \varrho)}$ for different values of $|\lambda|$ in Eq. (4)

ϱ	$\|\lambda\| \leq$
1.0681...	0.0246...
1.1474...	0.0484...
1.2430...	0.0715...
1.3643...	0.0940...
1.5333...	0.1157...
1.8319...	0.1369...

from the modification of the Fixed Point Theorem given in Sect. 2, there exists a unique fixed point x^* of Θ in $\overline{B(0, \varrho)}$. So, for the value of ϱ chosen and (6), we have

$$|\lambda| \leq \frac{\pi(\varrho - 1)}{2(1 + \varrho + 2\varrho^2)}, \tag{8}$$

provided that $|\lambda| \leq 0.3253\ldots$, and we then obtain the integral equations given by (4) that have a unique solution in $\overline{B(0, \varrho)}$ for these values of $|\lambda|$. Thus, by this procedure, we can locate a unique fixed point of Θ (namely, a solution of (4)) in the ball $\overline{B(0, \varrho)}$ for some integral equations of form (4).

In Table 1, we see what domains can contain fixed points. In addition, we can choose $\Omega = \overline{B(0, \varrho)}$ as the domain where we apply the last, so that the operator Θ has a unique fixed point in $\overline{B(0, \varrho)}$ if ϱ satisfies condition (6) for the corresponding value of $|\lambda|$. Then, from Table 2, we see when integral equation (4) has a unique solution in $\overline{B(0, \varrho)}$, for those values of $|\lambda|$ that are given by (8) in the second column of the table, so that we can then locate the solution in the corresponding ball given these values of $|\lambda|$.

Notice that, in the previous study, we can only consider values of λ such that $|\lambda| \leq 0.1369\ldots$.

Now, we use integral equation (4) to illustrate Theorem 4 and see what conclusions we can deduce. Observe that $\|\mathcal{H}''(x)\| \leq 3 = M$ and $M_\star = |\lambda|\ell M = \frac{6}{\pi}|\lambda|$. If we choose $\widetilde{x}(s) = u(s) = \sin \pi s$, then $\beta = \frac{\pi}{\pi - 2(2+\sin 1)|\lambda|}$ and $\eta = \frac{2|\lambda|}{\pi - 2(2+\sin 1)|\lambda|}$, since $\|\mathcal{H}(\widetilde{x})\| = 1$ and $\|\mathcal{H}'(\widetilde{x})\| = 3$, provided that $|\lambda| < \frac{\pi}{2(2+\sin 1)} = 0.5528\ldots$

Table 3 Radii of the balls $\overline{B(\sin \pi s, R)}$ where the existence of a unique solution of (4) is guaranteed from Theorem 4

$\lvert\lambda\rvert$	$R \in [R_-, \delta_-]$
0.0246...	[0.0164..., 5.6718...]
0.0484...	[0.0339..., 2.7421...]
0.0715...	[0.0530..., 1.7615...]
0.0940...	[0.0738..., 1.2679...]
0.1157...	[0.0972..., 0.9689...]
0.1369...	[0.1238..., 0.7669...]

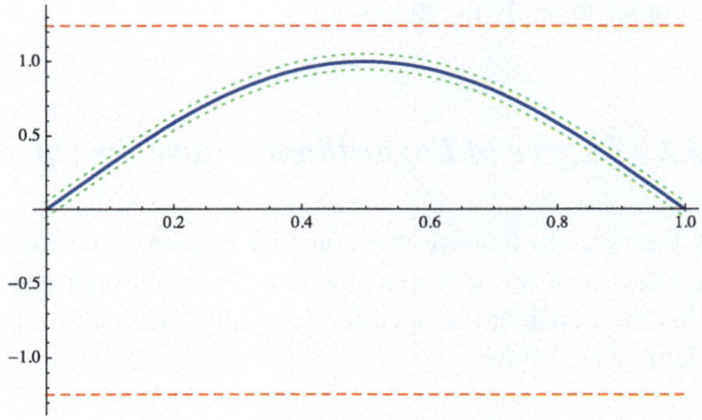

Fig. 1 Location of a solution of (4) with $\lambda = 0.0715\ldots$: $\overline{B(0, 1.2430\ldots)}$ and $\overline{B(\sin \pi s, 0.0530\ldots)}$

Then, $M_\star \beta\eta \leq 0.1513\ldots$ if

$$(0.7204\ldots)|\lambda|^2 + (0.5477\ldots)|\lambda| - (0.1513\ldots) \leq 0,$$

which is satisfied provided that $|\lambda| \leq 0.2153\ldots$ Some values of $|\lambda|$ that satisfies the last inequality are given in Table 3, along with the corresponding radii of the balls $\overline{B(\sin \pi s, R)}$ where the existence of a unique fixed point of the operator N_F is guaranteed by Theorem 4, and then a solution of integral equation (4). Remember that we consider R_- as the radius of the ball of localization of the fixed point because it is the smallest value that R can take and δ_- as the radius of the ball of uniqueness because it is the largest value that R can take. In addition, we use the biggest ball, $B(\sin \pi s, \delta_-)$, to separate the fixed point from other possible fixed points.

Observe that the location of the solution of (4) for those values of $|\lambda|$ improves significantly those obtained previously, as we can see in Fig. 1, where the domains $\overline{B(0, \varrho)}$, with $\varrho = 1.2430\ldots$, and $\overline{B(\sin \pi s, R)}$, with $R = 0.0530\ldots$, containing a solution of (4) with $\lambda = 0.0715\ldots$, are drawn. Note that the first domain is traced by the dashed line and the second one by the dotted line.

Moreover, it is important to highlight that, from Theorem 4, we can extend the previous study to values of $|\lambda| \leq 0.2153\ldots$ Some additional values of $|\lambda|$ are given in Table 4.

Table 4 Radii of the balls $\overline{B(\sin \pi s, R)}$ where the existence of a unique solution of (4) is guaranteed from Theorem 4

$\|\lambda\|$	$R \in [R_-, \delta_-]$
0.14	[0.1281..., 0.7423...]
0.16	[0.1598..., 0.6044...]
0.18	[0.2007..., 0.4947...]
0.20	[0.2603..., 0.4045...]

Finally, it is important to note that we do not need to locate the solution previously in order to apply the theorem.

3.2 Degree of Logarithmic Convexity Operator

Other variant that we can consider to prove that the operator N_F is a contraction comes from the fact that N_F is a derivable operator such that $\|N_F'(x)\| < 1$. For this, we consider the operator L_F, called degree of logarithmic convexity [9], that is defined as follows

$$L_F(x) : \mathcal{C}[a,b] \xrightarrow{F''(x)[F'(x)]^{-1}F(x)} \mathcal{C}[a,b] \xrightarrow{[F'(x)]^{-1}} \mathcal{C}[a,b].$$

Therefore,

$$L_F(x) = [F'(x)]^{-1}F''(x)\left([F'(x)]^{-1}F(x)\right) \in \mathcal{L}(\mathcal{C}[a,b], \mathcal{C}[a,b]),$$

with $[F'(x)]^{-1} \in \mathcal{L}(\mathcal{C}[a,b], \mathcal{C}[a,b])$. This operator satisfies $L_F(x) = N_F'(x)$, as we can see in the following [9].

Lemma 6 *Consider two operators $V : \mathcal{C}[a,b] \to \mathcal{C}[a,b]$ and $W : \mathcal{C}[a,b] \to \mathcal{L}(\mathcal{C}[a,b], \mathcal{C}[a,b])$, that are Fréchet differentiable at the point $x_0 \in \mathcal{C}[a,b]$, and, given $x \in \mathcal{C}[a,b]$, define the operator*

$$\begin{array}{rcccc} N_F(x) : \mathcal{C}[a,b] & \xrightarrow{V} & \mathcal{C}[a,b] & \xrightarrow{W(x)} & \mathcal{C}[a,b] \\ y & \longrightarrow & V(y) & \longrightarrow & W(x)V(y). \end{array}$$

Then, N_F is differentiable at x_0 and $N_F'(x_0) : \mathcal{C}[a,b] \to \mathcal{C}[a,b]$ is a lineal operator given by

$$N_F'(x_0)[-] = \left(W'(x_0)[-]V\right)(x_0) + W(x_0)V'(x_0)[-].$$

From the previous lemma, we prove the following result.

Theorem 7 *If $N_F'(x_0)$ is defined as in Lemma 6, then $N_F'(x_0) = L_F(x_0)$.*

Proof Let $T(x) = [\Gamma(x)F](x)$, where the operator $\Gamma : \mathcal{C}[a,b] \to \mathcal{L}(\mathcal{C}[a,b], \mathcal{C}[a,b])$ is such that $\Gamma(x) = [F'(x)]^{-1} \in \mathcal{L}(\mathcal{C}[a,b], \mathcal{C}[a,b])$, for $x \in \mathcal{C}[a,b]$. In addition, $N'_F(x_0) = I - T'(x_0)$. Moreover, since

$$T'(x_0)(z) = [\Gamma'(x_0)(z)F](x_0) + [\Gamma(x_0)F'(x_0)](z) = [\Gamma'(x_0)(z)F](x_0) + z,$$

for $z \in \mathcal{C}[a,b]$, it follows that $N'_F(x_0)(z) = -[\Gamma'(x_0)(z)F](x_0)$.

On the other hand, we can write $\Gamma = PF'$, where

$$\Gamma : \mathcal{C}[a,b] \xrightarrow{F'} \mathcal{L}(\mathcal{C}[a,b], \mathcal{C}[a,b]) \xrightarrow{P} \mathcal{L}(\mathcal{C}[a,b], \mathcal{C}[a,b]),$$

so that $P : \mathcal{L}(\mathcal{C}[a,b], \mathcal{C}[a,b]) \to \mathcal{L}(\mathcal{C}[a,b], \mathcal{C}[a,b])$ is the inverse operator $P(F) = F^{-1}$. If we now apply the chain rule for differentiating compositions of operators, we have $\Gamma'(x_0) \in \mathcal{L}(\mathcal{C}[a,b], \mathcal{L}(\mathcal{C}[a,b], \mathcal{C}[a,b])$ and

$$\Gamma'(x_0)[-] = P'(F'(x_0))F''(x_0)[-].$$

In addition, for $z \in \mathcal{C}[a,b]$,

$$\Gamma'(x_0)(z) = P'(F'(x_0))(F''(x_0)(z)) = -\Gamma(x_0)F''(x_0)(z)\Gamma(x_0) \in \mathcal{L}(\mathcal{C}[a,b], \mathcal{C}[a,b]).$$

Taking now into account the symmetry of the operator $F''(x_0)$, for $z \in \mathcal{C}[a,b]$, it follows that

$$\Gamma(x_0)F''(x_0)(z)\Gamma(x_0)F(x_0) = \Gamma(x_0)F''(x_0)\Gamma(x_0)F(x_0)(z) = L_F(x_0)(z).$$

The proof is complete. ■

Now we are ready to prove what we are looking for.

Theorem 8 *Let $\widetilde{x} \in \mathcal{C}[a,b]$ and suppose conditions (A1)–(A2). If $M_\star \beta\eta \leq \frac{1}{6}$ and $\|L_F(x)\| \leq K < 1$, for all $x \in B(\widetilde{x}, R)$, then the operator $N_F : \overline{B(\widetilde{x}, R)} \to \overline{B(\widetilde{x}, R)}$ has a unique fixed point x^* for all $R \in \left[R_-, R_+\right]$ and $x_{n+1} = N_F(x_n)$, $n \geq 0$, converges quadratically to x^* from any $x_0 \in \overline{B(\widetilde{x}, R)}$.*

Proof Under the conditions required, we have seen in Lemma 1 that $N_F(x) \in \overline{B(\widetilde{x}, R)}$ for all $x \in \overline{B(\widetilde{x}, R)}$, where $R \in \left[R_-, R_+\right]$. Moreover, as

$$\|N'_F(x)\| \leq \|L_F(x)\| \leq K < 1,$$

for all $x \in \overline{B(\widetilde{x}, R)}$, then the operator N_F is a contraction. Therefore, from the application of the modification of the Fixed Point Theorem, given previously, to the operator N_F in $\overline{B(\widetilde{x}, R)}$, the proof is complete. ■

In this case, we have $\overline{B(\widetilde{x}, R_-)}$ as ball of location of the fixed point and $\overline{B(\widetilde{x}, R_+)}$ as ball of uniqueness. We note that the most favorable ball of global convergence in this case is the last.

The existence condition of the operator L_F in a certain set seems very strong, since it implies the existence of the inverse operator $[F'(x)]^{-1}$ at each point in the set. However, by requiring conditions (A1)–(A2), it is guaranteed the existence of L_F in a domain, as we can see in the following result.

Theorem 9 *Let $F : C[a,b] \to C[a,b]$ given by (3). Suppose that condition (A1)–(A2) are satisfied. Then, the operator $L_F(x)$ exists, for all $x \in B\left(\widetilde{x}, \frac{1}{M_\star\beta}\right)$ and*

$$\|L_F(x)\| \leq \frac{M_\star\beta\,\|\widetilde{\Gamma}F(x)\|}{(1-M_\star\beta\|x-\widetilde{x}\|)^2}.$$

Proof Observe that

$$\widetilde{\Gamma}\int_{\widetilde{x}}^{x} F''(v)\,dv = \widetilde{\Gamma}F'(x) - I$$

and

$$\|\widetilde{\Gamma}F'(x) - I\| = \left\|\widetilde{\Gamma}\int_{\widetilde{x}}^{x} F''(v)\,dv\right\| \leq M_\star\beta\|x-\widetilde{x}\| < 1.$$

Then, from the Banach lemma on invertible operators, there exists $[\widetilde{\Gamma}F'(x)]^{-1}$, for all $x \in B\left(\widetilde{x}, \frac{1}{M_\star\beta}\right)$, and

$$\|[\widetilde{\Gamma}F'(x)]^{-1}\| \leq \frac{1}{1-M_\star\beta\|x-\widetilde{x}\|}.$$

Now, from

$$L_F(x) = [\widetilde{\Gamma}F'(x)]^{-1}\widetilde{\Gamma}F''(x)[\widetilde{\Gamma}F'(x)]^{-1}\widetilde{\Gamma}F(x), \tag{9}$$

the proof is complete. ■

Moreover, we observe that $B(\widetilde{x}, R) \subset B\left(\widetilde{x}, \frac{1}{M_\star\beta}\right)$ if $R \in [R_-, R_+]$, since $R_+ < \frac{1}{M_\star\beta}$. As a consequence, we can establish the following result.

Theorem 10 *Let $F : C[a,b] \to C[a,b]$ given by (3). Suppose that condition (A1)–(A2) are satisfied. Then, the operator $L_F(x)$ exists, for all $x \in B(\widetilde{x}, R)$, with $R \in [R_-, R_+]$, and*

$$\|L_F(x)\| \leq \frac{(M_\star\beta R)^2 + 2(M_\star\beta R) + 2M_\star\beta\eta}{2\,(1-M_\star\beta R)^2}.$$

Proof We have seen above that $\|\Gamma F'(\widetilde{x})\| \leq \frac{1}{1-M_\star\beta R}$ and $\|\widetilde{\Gamma}F(x)\| \leq \frac{1}{2}M_\star\beta R^2 + R + \eta$. Now, from (9), it follows that

$$\|L_F(x)\| \leq \|\Gamma F'(\widetilde{x})\|\|\widetilde{\Gamma}\|\|F''(x)\|\|\Gamma F'(\widetilde{x})\|\|\widetilde{\Gamma}F(x)\|$$
$$\leq \frac{M_\star\beta\left(\frac{1}{2}M_\star\beta R^2 + R + \eta\right)}{2\left(1 - M_\star\beta R\right)^2},$$

for all $x \in B(\widetilde{x}, R)$, and the proof is complete. ■

In addition, as $N_F'(x) = L_F(x)$, for $x \in B(\widetilde{x}, R)$, with $R \in [R_-, R_+]$, we can give the next result.

Theorem 11 *Let $\widetilde{x} \in \mathcal{C}[a, b]$ and suppose conditions (A1)–(A2). If $M_\star\beta\eta \leq \frac{-5+4\sqrt{2}}{4} = 0.1642\ldots$, then there exists $R > 0$ such that $R \in [R_-, \sigma_-]$, where $\sigma_- = \frac{3-\sqrt{7+2M_\star\beta\eta}}{M_\star\beta}$, the operator $N_F : \overline{B(\widetilde{x}, R)} \to \overline{B(\widetilde{x}, R)}$ has a unique fixed point x^* and $x_{n+1} = N_F(x_n)$, $n \geq 0$, converges quadratically to x^* from any $x_0 \in \overline{B(\widetilde{x}, R)}$.*

Proof First, from Lemma 1, we know that $N_F(x) \in \overline{B(\widetilde{x}, R)}$, for all $x \in \overline{B(\widetilde{x}, R)}$, and $N_F : \overline{B(\widetilde{x}, R)} \to \overline{B(\widetilde{x}, R)}$.

Then, from Theorem 10, we have

$$\|N_F'(x)\| \leq \frac{(M_\star\beta R)^2 + 2(M_\star\beta R) + 2M_\star\beta\eta}{2\left(1 - M_\star\beta R\right)^2},$$

since $N_F'(x) = L_F(x)$. Then, $N_F(x)$ is a contraction if

$$\frac{(M_\star\beta R)^2 + 2(M_\star\beta R) + 2M_\star\beta\eta}{2\left(1 - M_\star\beta R\right)^2} < 1.$$

As $M_\star\beta\eta \leq \frac{1}{6}$, the last is true if $R \leq \sigma_-$.

After that, we observe that $R_- \leq \sigma_-$ if $M_\star\beta\eta \leq \frac{1}{6}$, so that $[R_-, \sigma_-] \neq \emptyset$. Also, if $M_\star\beta\eta \leq 0.1642\ldots$, we have $16(M_\star\beta\eta)^2 + 40(M_\star\beta\eta) - 7 \leq 0$ and, as a consequence, $R_+ > \sigma_-$.

Finally, as a consequence of the above, the operator N_F is a contraction. Therefore, from the application of the modification of the Fixed Point Theorem, given previously, to the operator N_F in $\overline{B(\widetilde{x}, R)}$, the proof is complete. ■

Now, the ball of location of the fixed point is $\overline{B(\widetilde{x}, R_-)}$ and the ball of uniqueness is $\overline{B(\widetilde{x}, \sigma_-)}$, which in turn is the most favorable ball of global convergence.

Example 12 To illustrate Theorem 11, we use again integral equation (4) and $\widetilde{x}(s) = u(s) = \sin \pi s$. For Theorem 11, we have that $M_\star\beta\eta \leq 0.1642\ldots$ if

$$(0.6785\ldots)|\lambda|^2 + (0.5941\ldots)|\lambda| - (0.1642\ldots) \leq 0,$$

which is satisfied provided that $|\lambda| \leq 0.2207\ldots$ Some values of $|\lambda|$ that satisfies the previous inequality are given in Table 5, along with the corresponding radii of the

Table 5 Radii of the balls $\overline{B(\sin \pi s, R)}$ where the existence of a unique solution of (4) is guaranteed from Theorem 11

$\|\lambda\|$	$R \in [R_-, \sigma_-]$
0.0246...	[0.0164..., 7.1957...]
0.0484...	[0.0339..., 3.4794...]
0.0715...	[0.0530..., 2.2359...]
0.0940...	[0.0738..., 1.6101...]
0.1157...	[0.0972..., 1.2312...]
0.1369...	[0.1238..., 0.9754...]

Table 6 Radii of the balls $\overline{B(\sin \pi s, R)}$ where the existence of a unique solution of (4) is guaranteed from Theorem 11

$\|\lambda\|$	$R \in [R_-, \sigma_-]$
0.14	[0.1281..., 0.9443...]
0.16	[0.1598..., 0.7698...]
0.18	[0.2007..., 0.6311...]
0.20	[0.2603..., 0.5171...]
0.22	[0.4008..., 0.4206...]

balls $\overline{B(\sin \pi s, R)}$ where the existence of a unique fixed point of the operator N_F is guaranteed by Theorem 11, and then a solution of integral equation (4).

Moreover, from Theorem 11, we can also extend the previous study to values of $|\lambda| \leq 0.2207\ldots$ Some additional values of $|\lambda|$ are given in Table 6. In addition, for Theorem 11, we can consider values of $|\lambda|$ greater than for Theorem 4. For the first, $|\lambda| \leq 0.2207\ldots$ and, for the second, $|\lambda| \leq 0.2153\ldots$ We have included in Table 6 the value $|\lambda| = 0.22$ that cannot be considered for Theorem 4.

Observe that the interval in which the radius R moves has been extended with respect to that obtained from Theorem 4, so that the location of the fixed point remains the same, but the separation from other fixed points has been improved, since the balls of uniqueness that we can consider, $\overline{B(\sin \pi s, \sigma_-)}$, are bigger than the balls $\overline{B(\sin \pi s, \delta_-)}$ obtained from Theorem 4.

4 Nemystkii Operator with Lipschitz Continuous First Derivative

On the other hand, we can also prove the restricted global convergence to a ball of Newton's method from other point of view. So, in this section, we suppose the following conditions:

(B1) For some $\widetilde{x} \in \mathcal{C}[a,b]$, there exists $\widetilde{\Gamma} = [F'(\widetilde{x})]^{-1} \in \mathcal{L}(\mathcal{C}[a,b], \mathcal{C}[a,b])$ with $\|\widetilde{\Gamma}\| \leq \beta$ and $\|\widetilde{\Gamma}F(\widetilde{x})\| \leq \eta$.

(B2) There exists a constant $L \geq 0$ such that $\|\mathcal{H}'(x) - \mathcal{H}'(y)\| \leq L\|x-y\|$ for all $x, y \in \mathcal{C}[a,b]$.

4.1 Restricted Global Convergence

Now, we study the first approximation x_1 of Newton's method. If $x_0 \in B(\widetilde{x}, R)$, then

$$\|I - \widetilde{\Gamma}F'(x_0)\| \leq \|\widetilde{\Gamma}\|\|F'(\widetilde{x}) - F'(x_0)\| \leq L_\star \beta R,$$

where $L_\star = |\lambda|\ell L$. Thus, $\|I - \widetilde{\Gamma}F'(x_0)\| < 1$ if

$$L_\star \beta R < 1. \tag{10}$$

Hence, by the Banach lemma on invertible operators, the operator $\Gamma_0 = [F'(x_0)]^{-1}$ exists,

$$\|\Gamma_0\| \leq \frac{\beta}{1 - L_\star\beta R} = a \quad \text{and} \quad \|\Gamma_0 F'(\widetilde{x})\| \leq \frac{1}{1 - L_\star\beta R},$$

provided that condition (10) is satisfied.

Next, from

$$\|x_1 - x_0\| = \|\Gamma_0 F(x_0)\| \leq \|\Gamma_0 F'(\widetilde{x})\|\|\widetilde{\Gamma}F(x_0)\|$$

and taking into account

$$\widetilde{\Gamma}F(x_0) = \widetilde{\Gamma}F(\widetilde{x}) + (x_0 - \widetilde{x}) + \int_0^1 \widetilde{\Gamma}\left(F'(\widetilde{x} + t(x_0 - \widetilde{x})) - F'(\widetilde{x})\right)(x_0 - \widetilde{x})\,dt,$$

it follows $\|\widetilde{\Gamma}F(x_0)\| \leq \frac{1}{2}L_\star\beta R^2 + R + \eta$ and

$$\|x_1 - x_0\| \leq \frac{L_\star\beta R^2 + 2R + 2\eta}{2(1 - L_\star\beta R)} = b.$$

Besides, as

$$\begin{aligned}
\|x_1 - \widetilde{x}\| &= \|x_0 - \Gamma_0 F(x_0) - \widetilde{x}\| \\
&= \left\| -\Gamma_0 \left(F(x_0) + F'(x_0)(\widetilde{x} - x_0) \right) \right\| \\
&= \left\| -\Gamma_0 \left(F(\widetilde{x}) - \int_0^1 \left(F'(x_0 + t(\widetilde{x} - x_0)) - F'(x_0) \right) (\widetilde{x} - x_0) \right) \right\| \\
&\leq \|\Gamma_0 F(\widetilde{x})\| + \frac{1}{2} L_\star \|\Gamma_0\| \|\widetilde{x} - x_0\|^2 \\
&\leq \|\Gamma_0 F'(\widetilde{x})\| \|\widetilde{\Gamma} F(\widetilde{x})\| + \frac{1}{2} L_\star \beta \frac{\|\widetilde{x} - x_0\|^2}{1 - L_\star \beta R} \\
&< \frac{L_\star \beta R^2 + 2\eta}{2(1 - L_\star \beta R)},
\end{aligned}$$

then $x_1 \in B(\widetilde{x}, R)$, provided that

$$\frac{L_\star \beta R^2 + 2\eta}{2(1 - L_\star \beta R)} \leq R. \tag{11}$$

Note that inequality (11) is equivalent to $3L_\star \beta R^2 - 2R + 2\eta \leq 0$ if condition (10) holds, so that (11) is satisfied if

$$L_\star \beta \eta \leq \frac{1}{6} \quad \text{and} \quad R \in [\rho_-, \rho_+],$$

where $\rho_- = \frac{1 - \sqrt{1 - 6L_\star \beta \eta}}{3L_\star \beta}$ and $\rho_+ = \frac{1 + \sqrt{1 - 6L_\star \beta \eta}}{3L_\star \beta}$.

If we now suppose (11) and

$$\begin{aligned}
\|x_n - x_{n-1}\| &< d^{\,2^{n-2}} \|x_{n-1} - x_{n-2}\|, \quad n \geq 2, \\
\|x_n - \widetilde{x}\| &< \frac{L_\star \beta R^2 + 2\eta}{2(1 - L_\star \beta R)} \leq R, \quad n \geq 0,
\end{aligned} \tag{12}$$

where $d = \frac{1}{2} L_\star ab$, we can prove by mathematical induction on n that

$$\|x_{n+1} - x_n\| < d^{\,2^{n-1}} \|x_n - x_{n-1}\| \quad \text{and} \quad \|x_{n+1} - \widetilde{x}\| < \frac{L_\star \beta R^2 + 2\eta}{2(1 - L_\star \beta R)} \leq R.$$

Next, if

$$L_\star \beta \eta \leq \frac{1}{6} \quad \text{and} \quad R < \frac{5 - \sqrt{13 + 6L_\star \beta \eta}}{3L_\star \beta} = \gamma,$$

then

$$3(L_\star \beta R)^2 - 10(L_\star \beta R) + 2(2 - L_\star \beta \eta) > 0 \tag{13}$$

and, as a consequence, $d < 1$ and condition (10) is satisfied.

From the above-mentioned, we can then give the next result.

Lemma 13 *There always exists $R > 0$, such that conditions* (10), (11) *and* (13) *are satisfied, if*

(a) $L_\star \beta\eta \le \frac{-3+2\sqrt{2}}{3} = 0.1547\ldots$ *and* $R \in [\rho_-, \gamma)$,
(b) $L_\star \beta\eta \in \left[0.1547\ldots, \frac{1}{6}\right]$ *and* $R \in [\rho_-, \rho_+]$,

where $L_\star = |\lambda|\ell L$, $\rho_- = \frac{1-\sqrt{1-6L_\star\beta\eta}}{3L_\star\beta}$, $\rho_+ = \frac{1+\sqrt{1-6L_\star\beta\eta}}{3L_\star\beta}$ *and* $\gamma = \frac{5-\sqrt{13+6L_\star\beta\eta}}{3L_\star\beta}$.

Proof On the one hand, if $L_\star\beta\eta \le 0.1547\ldots$, then $\rho_- < \gamma$ and $[\rho_-, \gamma) \neq \emptyset$. In addition, $3(L_\star\beta\eta)^2 + 6(L_\star\beta\eta) - 1 \le 0$, so that $\rho_+ > \gamma$ and $R \in [\rho_-, \gamma) \subset [\rho_-, \rho_+]$. Therefore, (11) and (13) are satisfied.

On the other hand, if $L_\star\beta\eta \in \left[0.1547\ldots, \frac{1}{6}\right]$, then $3(L_\star\beta\eta)^2 + 6(L_\star\beta\eta) - 1 \ge 0$ and $\rho_+ < \gamma$. As a consequence, (11) and (13) are satisfied.

Finally, we observe that $R < \gamma$ in item (a) and item (b), so that $L_\star\beta R < 1$ in both cases. ■

After that, we can establish the following result that guarantees the global convergence of Newton's method to a solution of $F(x) = 0$ under conditions (B1)–(B2).

Theorem 14 *Let $\widetilde{x} \in \mathcal{C}[a, b]$ and suppose conditions (B1)-(B2). Then, Newtons's method is well-defined and quadratically convergent to a solution x^* of $F(x) = 0$ in $\overline{B(\widetilde{x}, R)}$, where R is according to the condition given in item (a) or item (b) of Lemma 13, from any point $x_0 \in B(\widetilde{x}, R)$.*

Proof From the conditions required, it follows that $d < 1$. Taking now into account (12), we obtain $\|x_{n+1} - x_n\| < \|x_n - x_{n-1}\|$, for all $n \ge 1$. Thus, $\{\|x_{n+1} - x_n\|\}$ is a strictly decreasing sequence, for all for all $n \ge 0$. Therefore, $\{x_n\}$ converges and $\lim_{n\to\infty} x_n = x^*$. Then, by the continuity of F and $\|F(x_n)\| \to 0$, when $n \to \infty$, it follows $F(x^*) = 0$. The fact that Newton's method has quadratic convergence under conditions (B1)-(B2) follows easily from the Ortega's version of the Newton-Kantorovich theorem (see [9]). ■

Remark 15 From the last theorem, we observe that $B(\widetilde{x}, \rho_-)$ is the best ball of location of the solution x^*. In addition, the biggest ball of convergence and uniqueness is $B(\widetilde{x}, \rho_+)$ if $L_\star\beta\eta \le 0.1547\ldots$ or $B\left(\widetilde{x}, \frac{5-\sqrt{13+6L_\star\beta\eta}}{3L_\star\beta}\right)$ if $L_\star\beta\eta \in \left[0.1547\ldots, \frac{1}{6}\right]$.

Example 16 We return to integral equation (4) to illustrate Theorem 14 and newly we choose $\widetilde{x}(s) = u(s) = \sin \pi s$. For Theorem 14, we take into account that

$$\|\mathcal{H}'(x) - \mathcal{H}'(y)\| \le 2\|x - y\| + \|\sin x - \sin y\| \le 3\|x - y\|,$$

so that $L = 3$ and $L_\star = |\lambda|\ell L = \frac{6}{\pi}|\lambda|$.

First, we analyse Theorem 14 under condition of item (a) of Lemma 13. Then, $L_\star\beta\eta \le 0.1547\ldots$ if

$$(0.6785\ldots)|\lambda|^2 + (0.5941\ldots)|\lambda| - (0.1642\ldots) \le 0,$$

Table 7 Radii of the balls $\overline{B(\sin \pi s, R)}$ where the existence of a unique solution of (4) is guaranteed from Theorem 14

$\|\lambda\|$	$R \in [\rho_-, \gamma)$
0.0246...	[0.0164..., 9.44523...)
0.0484...	[0.0339..., 4.57285...)
0.0715...	[0.0530..., 2.94521...)
0.0940...	[0.0738..., 2.12849...)
0.1157...	[0.0972..., 1.63592...)
0.1369...	[0.1238..., 1.30521...)

Table 8 Radii of the balls $\overline{B(\sin \pi s, R)}$ where the existence of a unique solution of (4) is guaranteed from Theorem 14

$\|\lambda\|$	$R \in [\rho_-, \gamma)$
0.14	[0.1281..., 1.2652...)
0.16	[0.1598..., 1.0413...)
0.18	[0.2007..., 0.8651...)
0.20	[0.2603..., 0.7220...)

which is satisfied provided that $|\lambda| \leq 0.2168\ldots$ Some values of $|\lambda|$ that satisfies the last inequality are given in Table 7, along with the corresponding radii of the balls $\overline{B(\sin \pi s, R)}$ where the existence of a unique solution of integral equation (4) is guaranteed by Theorem 14.

Moreover, from Theorem 14, we can also extend the previous study to values of $|\lambda| \leq 0.2168\ldots$ Some additional values of $|\lambda|$ are given in Table 8.

For integral equation (4), we notice that $L_\star = M_\star$, so that $R_- = \rho_-$ and we can then compare the results obtained from Theorems 11 and 14. So, we observe that the interval in which the radius R moves has been extended from Theorem 14 with respect to those obtained from Theorems 4 and 11, so that the location of the solution remains the same, but the separation from other possible solutions has been improved, since the balls of uniqueness that we can consider, $\overline{B(\sin \pi s, \gamma - \epsilon)}$, with ϵ small enough, are bigger than the balls $\overline{B(\sin \pi s, \delta_-)}$ and $\overline{B(\sin \pi s, \sigma_-)}$ obtained from Theorems 4 and 11, respectively.

Second, we analyse Theorem 14 under condition of item (b) of Lemma 13. Then, $L_\star \beta\eta \in \left[0.1547\ldots, \frac{1}{6}\right]$ if

$$(0.6785\ldots)|\lambda|^2 + (0.5941\ldots)|\lambda| - (0.1642\ldots) \geq 0$$

and

$$(4.0228\ldots)|\lambda|^2 + (3.6178\ldots)|\lambda| - 1 \leq 0$$

Table 9 Radii of the balls $\overline{B(\sin \pi s, R)}$ where the existence of a unique solution of (4) is guaranteed from Theorem 14

$\lvert\lambda\rvert$	$R \in [\rho_-, \rho_+]$
0.217	[0.3601 ..., 0.6169 ...]
0.219	[0.3844 ..., 0.5780 ...]
0.221	[0.4242 ..., 0.5237 ...]

which is satisfied provided that $|\lambda| \in [0.2168\ldots, 0.2217\ldots]$ Some values of $|\lambda| \in [0.2168\ldots, 0.2217\ldots]$ are given in Table 9, along with the corresponding radii of the balls $\overline{B(\sin \pi s, R)}$ where the existence of a unique solution integral equation (4) is guaranteed by Theorem 14.

4.2 Uniqueness of Solution

Now, we establish the uniqueness of solution in the following result.

Theorem 17 *Suppose the hypotheses of the last theorem. Then, the solution x^* is unique in $B\left(\widetilde{x}, \frac{2}{L_\star\beta} - R\right)$.*

Proof Let z^* be a solution of $F(x) = 0$ in $B\left(\widetilde{x}, \frac{2}{L_\star\beta} - R\right)$ such that $z^* \neq x^*$. From

$$0 = F(x^*) - F(z^*) = \int_0^1 F'(z^* + t(x^* - z^*))dt(x^* - z^*),$$

if the operator $\Phi = \int_0^1 F'(z^* + t(x^* - z^*))dt$ is invertible, then $z^* = x^*$, since $\Phi(x^* - z^*) = F(x^*) - F(z^*)$. For this, we do

$$\begin{aligned}
\|I - \widetilde{\Gamma}\Phi\| &\leq \|\widetilde{\Gamma}\| \int_0^1 \|F'(\widetilde{x}) - F'(z^* + t(x^* - z^*))\|dt \\
&\leq L_\star\beta \int_0^1 \|\widetilde{x} - (z^* + t(x^* - z^*))\|dt \\
&< L_\star\beta \int_0^1 ((1-t)r + tR)dt \\
&= 1,
\end{aligned}$$

where $r = \frac{2}{L_\star\beta} - R$, and, by the Banach lemma on invertible operators, it follows that the operator Φ is invertible. ■

Table 10 Numerical solution $x^*(s)$ of integral equation (4) with $\lambda = \frac{1}{5}$

n	$x_n(s)$
0	$\sin(\pi s)$
1	$-(0.0027498509174498764180549285 06\ldots)\cos(\pi s) + \sin(\pi s)$
2	$-(0.002749400271813126774908794455\ldots)\cos(\pi s) + \sin(\pi s)$
3	$-(0.002749400271801023943483858466\ldots)\cos(\pi s) + \sin(\pi s)$
4	$-(0.002749400271801023943483858457\ldots)\cos(\pi s) + \sin(\pi s)$

Table 11 Sequence $\{\|F(x_n)\|\}$ for integral equation (4) with $\lambda = \frac{1}{5}$

n	$\|F(x_n)\|$
0	2.7498×10^{-3}
1	4.5079×10^{-7}
2	1.2106×10^{-14}
3	8.7323×10^{-30}

Remark 18 From Theorem 17, we can also deduce that, under the conditions of the theorem, the solution x^* of $F(x) = 0$ is unique in $\overline{B(\widetilde{x}, R)}$. Moreover, under the conditions of Theorem 14, we have $L_\star\beta R < 1$, so that $\frac{2}{L_\star\beta} - R > 0$ and, since $\frac{2}{L_\star\beta} - R \geq \gamma$ and $\frac{2}{L_\star\beta} - R \geq \rho_+$, we obtain the uniqueness of the solution x^* at least in the ball of global convergence obtained from Theorem 14.

Example 19 According to integral equation (4), $x_0(s) = \sin(\pi s)$ is a reasonable choice of starting point for Newton's method, as we can see in [3, 13, 17]. From the initial approximation $x_0(s)$ and $\lambda = \frac{1}{5}$, we apply Newton's method and obtain the approximation

$$x^*(s) = -(0.002749400271801023\ldots)\cos(\pi s) + \sin(\pi s),$$

given in Table 10, after four iterations with the stopping criterion $\|x_n(s) - x_{n-1}(s)\| < 10^{-32}$.

From the sequence $\{\|F(x_n)\|\}$ given in Table 11, we observe that $x^*(s)$ is a good approximation of a solution of (4) with $\lambda = \frac{1}{5}$. In Table 12, we show the errors $\|x^*(s) - x_n(s)\|$.

Finally, for integral equation (4) with $\lambda = \frac{1}{5}$, we have $L_\star\beta\eta = 0.1194\ldots$, so that we are in the situation described in item (a) of Lemma 13 and then $R \in [0.2603\ldots, 0.7220\ldots)$. Thus, we observe that the domain of uniqueness of solution of integral equation (4) with $\lambda = \frac{1}{5}$, obtained from Theorem 17, is $B(\sin \pi s, r)$, where $r \in (2.6196\ldots, 3.0813\ldots]$, which is much greater than the ball of global convergence obtained from Theorem 14, that is, $\overline{B(\sin \pi s, R)}$ with $R \in [0.2603\ldots, 0.7220\ldots)$.

Table 12 Absolute errors for integral equation (4) with $\lambda = \frac{1}{5}$

n	$\|x^* - x_n\|$
0	2.7494×10^{-3}
1	4.5064×10^{-7}
2	1.2102×10^{-14}
3	8.7295×10^{-30}

5 Conclusions

The convergence of Newton's method is usually studied as that of a one-point iterative process by using the technique of majorizing sequences [9] or the construction of recurrence relations [10]. Both techniques allow us to guarantee the convergence of the method for a certain starting point x_0 that verifies certain conditions. In this work, we have also seen that, considering the Newtonian operator N_F, we can apply fixed point techniques to this operator and thus obtain results of global convergence restricted to certain balls. This new approach, carried out in this work for Newton's method and applied to nonlinear integral equations of the Fredholm-Hammerstein type, opens the doors to obtain global convergence results for any operator. This will be our next aim. It may also be interesting to establish restricted global convergence results for high-order one-point iterative processes and even to review this approach for multipoint iterative processes.

References

1. Argyros, I.K.: On a class of nonlinear integral equations arising in neutron transport. Aequationes Math. **36**(1), 99–111 (1988)
2. Argyros, I.K.: On the Newton-Kantorovich hypothesis for solving equations. J. Comput. Appl. Math. **169**(2), 315–332 (2004)
3. Awawdeh, F., Adawi, A., Al-Shara, S.: A numerical method for solving nonlinear integral equations. Int. Math. Forum **4**, 805–817 (2009)
4. Berinde, V.: Iterative Approximation of Fixed Point. Springer, New York (2005)
5. Brunner, H.: Collocation Methods for Volterra Integral and Related Functional Differential Equations. Cambridge University Press, Cambridge (2004)
6. Bruns, D.D., Bailey, J.E.: Nonlinear feedback control for operating a nonisothermal CSTR near an unstable steady state. Chem. Eng. Sci. **32**, 257–264 (1977)
7. Chandrasekhar, S.: Radiative Transfer. Dover, New York (1960)
8. Davis, H.T.: Introduction to Nonlinear Differential and Integral Equations. Dover, New York (1962)
9. Ezquerro, J.A., Hernández-Verón, M.A.: Newton's Method: An Updated Approach of Kantorovich's Theory. Frontiers in Mathematics. Birkhäuser/Springer, Cham (2017)
10. Ezquerro, J.A., Hernández-Verón, M.A.: Mild Differentiability Conditions for Newton's Method in Banach Spaces. Frontiers in Mathematics. Birkhäuser/Springer, Cham (2020)
11. Ganesh, M., Joshi, M.C.: Numerical solvability of Hammerstein integral equations of mixed type. IMA J. Numer. Anal. **11**(1), 21–31 (1991)

12. Matkowski, J.: Functional equations and Nemytskii operators. Funkcial. Ekvac. **25**, 127–132 (1982)
13. Nadir, M., Khirani, A.: Adapted Newton-Kantorovich method for nonlinear integral equations. J. Math. Stat. **12**(3), 176–181 (2016)
14. Porter, D., Stirling, D.S.G.: Integral Equations. Cambridge University Press, Cambridge (1990)
15. Rashidinia, J., Zarebnia, M.: New approach for numerical solution of Hammerstein integral equations. Appl. Math. Comput. **185**, 147–154 (2007)
16. Rashidinia, J., Parsa, A.: Analytical-numerical solution for nonlinear integral equations of Hammerstein type. Int. J. Math. Model. Comput. **2**(1), 61–69 (2012)
17. Saberi-Nadja, J., Heidari, M.: Solving nonlinear integral equations in the Urysohn form by Newton-Kantorovich-quadrature method. Comput. Math. Appl. **60**, 2018–2065 (2010)
18. Shizgal, B.: Integral equations in the kinetic theory of gases and related topics. In: Spectral Methods in Chemistry and Physics. Springer, Dordrecht (2015)
19. Singh, H., Baleanu, D., Srivastava, H.M., Dutta, H., Jha, N.K.: Solution of multi-dimensional Fredholm equations using Legendre scaling functions. Appl. Numer. Math. **150**, 313–324 (2020)

Boundary Integral Equation Formulation for Fractional Order Theory of Thermo-Viscoelasticity

M. A. Elhagary

Abstract This work presents a formulation of the boundary integral equation method for fractional order theory of thermo-viscoelasticity. Fundamental solutions of corresponding differential equations are obtained in the Laplace transform domain. A reciprocity theorem is established. The implementation of the boundary element method is discussed for the solution of the above boundary equations Special attention is given to the representation of primary fields, namely temperature and displacements. The initial, mixed boundary value problem is considered as an example illustrating the BIE formulation.

Keywords Boundary integral equations · Fractional calculus · Fundamental solution · Reciprocity theorem · Thermoelasticity · Thermoviscoelasticity

1 Introduction

The boundary integral equation method (BIEM) has been playing a very important role in many branches of applied mathematics. Although the implementation of (BIEM) is not simple, particularly the calculation of the Hadamard finite part for improper integrals, it is better than other numerical techniques like the finite element method in the requationuires fewer grid points on the boundary, not the whole region. The (BIEM) method is readily applicable, especially, to solve the elasticity and thermoelasticity problems in an infinite region. Sladek and Sladek [1, 2] set up the BIEM formulation for coupled thermoelasticity, while Anwer and Sherief [3, 4] completed the formulation for generalized thermoelasticity for one and two relaxation times. Some BIE formulations for the generalized thermoviscoelasticity were introduced by Elkaramany and Ezzat [5, 6]. Elhagary [7] represented the formulation for thermoviscoelasticity with one relaxation time. Tiwari and Mukhopadhyay [8] introduced the BIEM formulation for fractional order thermoelasticity.

M. A. Elhagary (✉)
Department of Mathematics, Faculty of Science, Damiatta University, New Damiatta, Egypt
e-mail: melhagry@gmail.com

H. Singh et al. (eds.), *Topics in Integral and Integro-Differential Equations*,
Studies in Systems, Decision and Control 340,
https://doi.org/10.1007/978-3-030-65509-9_6

During the second half of the 20th century, in the years after World War Two, thermal problems of the theory of elasticity became increasingly important. The reasons for this continuing interest are the engineering problems that appeared in widely diverse fields. In the aeronautics field, the high heat producing during the work of modern aircraft reduces the strength of the aircraft structure. Likewise, in the technology of modern propulsive systems, like jet and rocket engines, the high temperatures that result from combustion processes are the source of unwanted thermal stresses. This phenomenon is also found in spacecraft and missile technologies and large steam turbine mechanics. In the nuclear field, the extremely high temperatures arising inside nuclear reactors affect their design and operations.

Danilovskaya [9] was the first to solve an actual problem in the theory of elasticity with non-uniform heat. The problem was based on what became known as the theory of uncoupled thermoelasticity. In this theory, the temperature is governed by a parabolic partial differential equation, which does not contain any elastic terms. It was not much later that many attempts were made to remedy the shortcomings of this theory.

Biot [10] eliminated the paradox inherent that elastic changes do not affect the temperature when he investigated the theory of coupled thermoelasticity. The heat equation for this theory is of the diffusion type predicting infinite speeds of propagation for heat waves contrary to physical observations.

Physically, the theory of generalized thermoelasticity with one relaxation time gives more realistic results than the coupled or uncoupled theories of thermoelasticity. This theory was introduced by Lord and Shulman [11] for the special case of an isotropic body. They used a modified Fourier law of heat conduction including both the heat flux and its time derivative. The heat equation associated with this theory is hyperbolic and hence eliminates the paradox of infinite speeds of propagation inherent in both the uncoupled and the coupled theories of thermoelasticity. In 1980, Dhaliwal and Sherief [12] extended this theory to include the anisotropic case. In the context of this theory, two crack problems [13, 14] have been solved by Sherief and El-Maghraby. Sherief et al. [15] solved a dynamical problem for an infinitely long hollow cylinder for a short time. Saleh [16] solved a one-dimensional problem for a half-space under the action of a body force. Elhagary [17] solved a two-dimensional problem for two media in the generalized theory of thermoelasticity.

The linear theory of thermo-viscoelasticity remains a very crucial area of research, this is due to the extensive use of polymers in many industrial applications. The viscoelasticity and thermoviscoelasticity theories were developed by many researchers like Gurtin and Sternberg [18], Pobedrya [19], Medri [20], Baranoglu and Mengi [21], Lychev [22], and Kartashov [23, 24].

Sherief et al. [25] introduce the generalized theory of thermoviscoelasticity with one relaxation time. This theory is an extension of the generalization theory of thermoelasticity with one relaxation time. In this paper, Sherief et al. solved a one-dimensional thermoviscoelastic problem for a half space for short times using asymptotic expansions. Elhagary [26] solved a thermo-mechanical shock problem for the generalized theory of thermoviscoelasticity. A two-dimensional problem for a half-space in the generalized theory of thermo-viscoelasticity was solved by Sherief et al.

[27]. Sherief et al. [28] studied the wave propagation study for axi-symmetric 2D problems of a generalized thermo-visco-elastic half space.

On the other hand, Fractional calculus becomes a powerful tool that simulates physical processes in science and engineering. Many scientific areas are currently paying attention to the fractional calculus concepts and we can refer its adoption in viscoelasticity and damping, diffusion and wave propagation, electromagnetism, chaos and fractals, heat transfer, biology, electronics, signal processing, robotics, system identification, traffic systems, genetic algorithms, percolation, modeling and identification, telecommunications, chemistry, irreversibility, physics, control systems as well as the economy, and finance. Some applications of fractional calculus were illustrated in [29]. By using fractional derivatives for the description of viscoelastic materials, Caputo and Mainardi [30, 31] and Caputo [32] had obtained powerful experimental results. They were able to obtain the relationship between the theory of linear viscoelasticity and fractional order derivatives. Adolfsson et al. [33, 34] and Povstenko [35–37] constructed some models of fractional order viscoelasticity and thermoelasticity, respectively. Sene [38–40] used the fractional derivatives to get the solutions for some problems in fluid dynamics. Recently, the fractional order theory of thermoelasticity was derived by Sherief et al. [41]. In the context of this theory, some problems were solved by Sherief and Abd El-Latief [42, 43] and Raslan [44, 45]. Ezzat and El Karamany [46–48] solved some problems of fractional order in electro-magneto thermoelasticity. Recently, Sherief and Elhagary [49] introduced the fractional order theory of thermo-viscoelasticity and solved a half space problem as an application. The motivation behind the introduction of this theory is that it expects retarded response to the physical effects, as it exists in nature, as opposed to instantaneous response predicted by the generalized theory of thermoelasticity. This late response stems from the fact that fractional derivatives are actually, integrals over time. This is materially the result of Van der Walles' weak forces.

In this work, the author will give a formulation of the boundary integral equation method for thermo-viscoelasticity, which could be helpful for the solution of problems in the present theory by using the boundary element method.

2 The Mathematical Model

We shall consider a homogeneous isotropic viscoelastic solid occupying the region **V** and bound by a smooth surface **S**. The initial state of the medium is assumed to be quiescent. In this work, the summation convention is used. Also, dots denote time derivatives and comma denotes material derivatives.

In the following, we shall use the Caputo definition of fractional derivatives of an order $\beta \in (0, 1]$ of the absolutely continuous function f(t) given by [49]

$$\frac{d^{\beta}}{dt^{\beta}}f(t) = \frac{1}{\Gamma(1-\alpha)}\int_{0}^{t}(t-s)^{-\beta}f(s)\,ds$$

where f(t) is a Lebesgue integrable function, β > 0.

The system of governing equations describing the fractional order theory of thermoviscoelasticity with one relaxation time consists of [49]

(i) The equation of motion:

$$\rho\ddot{u}_i = \rho F_i + \int_0^t G_2(t-\tau)\frac{\partial u_{i,jj}(x,\tau)}{\partial\tau}d\tau + \int_0^t (G_1(t-\tau)+G_2(t-\tau))\frac{\partial u_{j,ij}(x,\tau)}{\partial\tau}d\tau - \alpha_t\int_0^t G(t-\tau)\frac{\partial T_{,i}(x,\tau)}{\partial\tau}d\tau \tag{1}$$

where ρ is the density, α_t is the coefficient of linear thermal expansion and G_1, G_2, and G are the relaxation functions. T is the absolute temperature, u_i is the displacement component in the x_i-direction. F_i is a component of externally body forces per unit mass in the x_i-direction.

(ii) The equation of heat conduction:

$$kT_{,ii} = \rho c_E\left(\frac{\partial}{\partial t} + \tau_0\frac{\partial^{\beta+1}}{\partial t^{\beta+1}}\right)T - \alpha_t T_0\left[\int_0^t\left(\frac{\partial}{\partial t} + \tau_0\frac{\partial^{\beta+1}}{\partial t^{\beta+1}}\right)e_{ij}(x,t-\tau)\frac{\partial G(\tau)}{\partial\tau}d\tau - G(0)\left(\frac{\partial}{\partial t} + \tau_0\frac{\partial^{\beta+1}}{\partial t^{\beta+1}}\right)e_{ij}\right] - \rho\left(1+\tau_0\frac{\partial^{\beta}}{\partial t^{\beta}}\right)Q \tag{2}$$

where k is the thermal conductivity, c_E the specific heat at constant strain, τ_0 is constant with the dimensions of time that acts as relaxation time and T_0 is a reference temperature chosen so that $|(T - T_0)/T_0| << 1$. Q is the strength of the applied heat source per unit mass. The cubical dilatation e has the form $e = e_{ii} = u_{i,i}$

(iii) the constitutive equations giving the stresses σ_{ij}:

$$\sigma_{ij} = 2\int_0^t G_2(t-\tau)\frac{\partial e_{ij}(\mathrm{x},\tau)}{\partial\tau}d\tau + \delta_{ij}\left[\int_0^t G_1(t-\tau)\frac{\partial e(\mathrm{x},\tau)}{\partial\tau}d\tau - \alpha_t\int_0^t G(t-\tau)\frac{\partial T(\mathrm{x},\tau)}{\partial\tau}d\tau\right] \quad (3)$$

where δ_{ij} denotes the Kronecker delta and e_{ij} are the strain components given by

$$e_{ij} = \frac{1}{2}(u_{i,j} + u_{j,i}) \quad (4)$$

Let us introduce the non-dimensional variables

$$x_i^* = c_1\xi\ x_i,\ u_i^* = c_1\xi\ u_i,\ t^* = c_1^2\xi\ t,\ \tau_0^* = c_1^{2\beta}\xi^\beta\tau_0,$$
$$G_i^* = \frac{G_i}{\lambda+2\mu},\ i=1,2,\ \ \theta = \alpha_t(T-T_0),\ \sigma_{xx}^* = \frac{\sigma_{xx}}{(\lambda+2\mu)},$$
$$Q^* = \frac{\rho\alpha_t Q}{k\ c_1^2\xi^2},\ F_i^* = \frac{\rho F_i}{(\lambda+2\mu)c_1\xi}$$

where λ, μ are Lame's constants, $c_1 = \sqrt{(\lambda+2\mu)/\rho}$ is the velocity of longitudinal waves and $\xi = \rho c_E/k$ is the thermal viscosity.

Under these non-dimensional variables, Eqs. (1)–(3), respectively, take the following form (dropping the asterisks for convenience):

$$\ddot{u}_i = F_i + \int_0^t G_2(t-\tau)\frac{\partial u_{i,jj}(\mathrm{x},\tau)}{\partial\tau}d\tau + \int_0^t (G_1(t-\tau)+G_2(t-\tau))\frac{\partial u_{j,ij}(\mathrm{x},\tau)}{\partial\tau}d\tau - \int_0^t G(t-\tau)\frac{\partial\theta_{,i}(\mathrm{x},\tau)}{\partial\tau}d\tau \quad (5)$$

$$\theta_{,ii} = \left(\frac{\partial}{\partial t} + \tau_0\frac{\partial^{\beta+1}}{\partial t^{\beta+1}}\right)\theta - \varepsilon\left[\int_0^t\left(\frac{\partial}{\partial t} + \tau_0\frac{\partial^{\beta+1}}{\partial t^{\beta+1}}\right)e_{ij}(x,t-\tau)\frac{\partial G(\tau)}{\partial\tau}d\tau - G(0)\left(\frac{\partial}{\partial t} + \tau_0\frac{\partial^{\beta+1}}{\partial t^{\beta+1}}\right)e_{ij}\right] - \left(1+\tau_0\frac{\partial^\beta}{\partial t^\beta}\right)Q \quad (6)$$

$$\sigma_{ij} = 2\int_0^t G_2(t-\tau)\frac{\partial e_{ij}(\mathrm{x},\tau)}{\partial\tau}d\tau + \delta_{ij}\left[\int_0^t G_1(t-\tau)\frac{\partial e(\mathrm{x},\tau)}{\partial\tau}d\tau - \alpha_t\int_0^t G(t-\tau)\frac{\partial\theta(\mathrm{x},\tau)}{\partial\tau}d\tau\right] \quad (7)$$

where $\varepsilon = \frac{\alpha_t^2 T_0(\lambda+2\mu)}{\rho c_E}$.

The above equations are supplemented with the following boundary conditions:

$$\sigma_{ij}n_j = p_{i0}(\mathrm{x},t) \quad \text{on} \quad S_1 \subset S \quad (8)$$

$$u_i = u_{i0}(\mathrm{x},t) \quad \text{on} \quad S_2 \subset S \quad (9)$$

$$\theta = \theta_0(\mathrm{x},t) \quad \text{on} \quad S_3 \subset S \quad (10)$$

$$\theta_{,n} = \theta_{0,n}(\mathrm{x},\ t) \quad \text{on} \quad S_4 \subset S \quad (11)$$

where S_1, S_2, S_3, and S_4 are different partitions of the bounded surface S such that,

$$S_1 \cup S_2 = S_3 \cup S_4 = S$$
$$S_1 \cap S_2 = S_3 \cap S_4 = \phi$$

The Initial conditions will be assumed homogeneous. Using Eqs. (4) and (7), the component of the traction vector on the surface S has the form

$$\begin{aligned} p_i(r,t) &= \left[2\int_0^t G_2(t-\tau)\frac{\partial e_{ij}(\mathrm{x},t-\tau)}{\partial\tau}d\tau + \delta_{ij}\left[\int_0^t G_1(t-\tau)\frac{\partial e(\mathrm{x},t-\tau)}{\partial\tau}d\tau \right.\right. \\ &\quad \left.\left. -\alpha_t\int_0^t G(t-\tau)\frac{\partial\theta(\mathrm{x},t-\tau)}{\partial\tau}d\tau\right]\right]n_j \\ &= \int_0^t G_2(t-\tau)n_k(r)\frac{\partial u_{i,k}(\mathrm{x},t-\tau)}{\partial\tau}d\tau + \int_0^t G_2(t-\tau)n_j(r)\frac{\partial u_{j,i}(\mathrm{x},t-\tau)}{\partial\tau}d\tau \\ &\quad + \int_0^t G_1(t-\tau)n_i(r)\frac{\partial u_{j,j}(\mathrm{x},t-\tau)}{\partial\tau}d\tau - \int_0^t G(t-\tau)\,n_i(r)\frac{\partial\theta(\mathrm{x},t-\tau)}{\partial\tau}d\tau \end{aligned} \quad (12)$$

3 Formulation in the Laplace Transform Domain

Applying the Laplace transform defined by the relation

$$\bar{f}(r,s) = \int_0^\infty e^{-st} f(r,t)\,dt$$

to both sides of Eqs. (5–7), we arrive at

$$s^2\bar{u}_i = \bar{F}_i + \bar{\gamma}_4\bar{u}_{i,jj} + \bar{\gamma}_3\bar{u}_{j,ij} - \bar{\gamma}_2\bar{\theta}_{,i} \tag{13}$$

$$\bar{\theta}_{,ii} = s\left(1+\tau_0 s^\beta\right)\left[\bar{\theta} + \bar{\alpha}\,\bar{u}_{k,k}\right] - \bar{Q}\left(1+\tau_0 s^\beta\right) \tag{14}$$

$$\bar{\sigma}_{ij} = 2\bar{\gamma}_4\bar{e}_{ij} + \delta_{ij}\left[\bar{\gamma}_5\bar{u}_{k,k} - \bar{\gamma}_2\bar{\theta}\right] \tag{15}$$

where

$$\bar{\gamma}_1 = s(\bar{G}_1 + 2\bar{G}_2),\quad \bar{\gamma}_2 = s\,(3\bar{G}_1 + 2\bar{G}_2),\quad \bar{\gamma}_3 = s\,(\bar{G}_1 + \bar{G}_2),$$
$$\bar{\gamma}_4 = s\bar{G}_2,\quad \bar{\gamma}_5 = s\bar{G},\quad and\ \bar{\alpha} = \varepsilon(2G(0) - \bar{\gamma}_2)$$

Using Helmholtz's theorem [50], the displacement and body forces can be expressed as

$$u_i = \varphi_{,i} + \varepsilon_{ijk}\psi_{j,k} \tag{16}$$

$$F_i = X_{,i} + \varepsilon_{ijk}Y_{j,k} \tag{17}$$

Substitution from (16) and (17) into (13), (14) and (15), we get

$$\left(\nabla^2 - \frac{s^2}{\bar{\gamma}_1}\right)\bar{\phi} + \frac{\bar{X}}{\bar{\gamma}_1} = \bar{m}\bar{\theta} \tag{18}$$

$$\left(\nabla^2 - \frac{s^2}{\bar{\gamma}_4}\right)\bar{\psi}_j = -\frac{\bar{Y}_j}{\bar{\gamma}_4} \tag{19}$$

$$\left(\nabla^2 - s\left(1+\tau_0 s^\beta\right)\right)\bar{\theta} - \bar{\alpha}s\left(1+\tau_0 s^\beta\right)\nabla^2\bar{\phi} = \left(1+\tau_0 s^\beta\right)\bar{Q} \tag{20}$$

where $\bar{m} = \frac{\bar{\gamma}_2}{\bar{\gamma}_1}$.

4 Fundamental Solutions in the Laplace Transform Domain

The following two cases to describe the action of body force and the heat source of very large magnitude that act for very short period of time.

Case I

Instantaneous concentrated heat source located at the point $\mathbf{x} = \mathbf{y}$ is acting upon a viscoelastic body in the absence of externally applied body forces. The fundamental solutions corresponding to this case will be denoted by primes. This case can be stated as: $Q = \delta(\mathrm{x} - \mathrm{y})\delta(t), \;\; F_i = 0$.

Taking the Laplace transforms of Q and $\mathrm{F_i}$ we get, $\bar{Q} = \delta(\mathrm{x} - \mathrm{y}), \; \bar{F_i} = 0$.

The governing Eqs. (18–20) under the above assumptions reduce to

$$\left(\nabla^2 - \frac{s^2}{\bar{\gamma}_1}\right)\bar{\phi}' = \bar{m}\bar{\theta}' \tag{21}$$

$$\left(\nabla^2 - \frac{s^2}{\bar{\gamma}_4}\right)\bar{\psi}_j = 0 \tag{22}$$

$$\left(\nabla^2 - s\left(1 + \tau_0 s^\beta\right)\right)\bar{\theta}' - \bar{\alpha}s\left(1 + \tau_0 s^\beta\right)\nabla^2\bar{\phi}' = \left(1 + \tau_0 s^\beta\right)\delta(\mathrm{x} - \mathrm{y}) \tag{23}$$

It follows from Eq. (22) that

$$\bar{\psi}_j' = 0 \tag{24}$$

Eliminating $\bar{\theta}'$ from Eqs. (21) and (23) we get

$$\left(\nabla^2 - k_1^2\right)\left(\nabla^2 - k_2^2\right)\bar{\varphi}' = \bar{m}(1 + \tau_0 s)\delta(x - y) \tag{25}$$

where k_1^2 and k_2^2 are the roots with positive real parts of the characteristic equation

$$k^4 - k^2\left[\frac{s^2}{\bar{\gamma}_1} + s\left(1 + \tau_0 s^\beta\right)(1 + \bar{m}\bar{\alpha})\right] + \frac{s^3\left(1 + \tau_0 s^\beta\right)}{\bar{\gamma}_1} = 0$$

The roots of this equation satisfy the relations

$$k_1^2 + k_2^2 = \frac{s^2}{\bar{\gamma}_1} + s\left(1 + \tau_0 s^\beta\right)(1 + \bar{m}\bar{\alpha}) \tag{26a}$$

$$k_1^2 k_2^2 = \frac{s^3\left(1 + \tau_0 s^\beta\right)}{\bar{\gamma}_1} \tag{26b}$$

The solution of Eq. (25) takes the form

$$\bar{\varphi}'(r,s)=\frac{\bar{m}(1+\tau_0 s)}{4\pi\left(k_1^2-k_2^2\right)}\left[e^{-k_1 r}-e^{k_2 r}\right] \tag{27}$$

where $r=\sqrt{(x_i-y_i)(x_i-y_i)}$

In obtaining the above expressions, we have used Helmholtz equation in the space [51], namely,

$$\frac{1}{\nabla^2-k^2}[\delta(\mathrm{r})]=\frac{-1}{4\pi\, r}e^{-k\,r}$$

Substituting from Eqs. (24) and (27) into Eq. (16), we obtain

$$\bar{u}_i'(r,s)=\frac{-r_{,i}}{r^2}f_1 \tag{28}$$

where $r_{,i}=\frac{(x_i-y_i)}{r}$.

and

$$f_1=\frac{\bar{m}\left(1+\tau_0 s^{\beta}\right)}{4\pi\left(k_1^2-k_2^2\right)}\left[e^{-k_1 r}(1+k_1 r)-e^{k_2 r}(1+k_2 r)\right]$$

It easy to show that

$$\bar{u}_{i,j}'=\frac{1}{r^3}\left[r_{,i}r_{,j}f_2+\left(2r_{,i}r_{,j}-\delta_{ij}\right)f_1\right] \tag{29}$$

where $f_2=\frac{\bar{m}\left(1+\tau_0 s^{\beta}\right)}{4\pi\left(k_1^2-k_2^2\right)}\left[e^{-k_1 r}\left(1+k_1 r+k_1^2 r^2\right)-e^{k_2 r}\left(1+k_2 r+k_2^2 r^2\right)\right]$.

From Eqs. (21) and (27), we obtain

$$\bar{\theta}'=\frac{1}{r}f_3 \tag{30}$$

where $f_3=\frac{\left(1+\tau_0 s^{\beta}\right)}{4\pi\left(k_1^2-k_2^2\right)}\left[e^{-k_1 r}\left(k_1^2+\frac{s^2}{\bar{\gamma}_1}\right)-e^{k_2 r}\left(k_2^2+\frac{s^2}{\bar{\gamma}_1}\right)\right]$.

The Laplace transform of the traction vector, in this case, is obtained from Eq. (12)

$$\bar{p}_i(\mathrm{r},s)=\bar{\gamma}_4\left[\bar{u}_{l,k}'+\bar{u}_{k,l}'\right]n_k+\left(\bar{\gamma}_5\bar{u}_{k,k}'-\bar{\gamma}_2\bar{\theta}'\right)n_l \tag{31}$$

Using the above expression, the traction vector takes the form

$$\bar{p}_l'(\mathrm{r},s)=\frac{1}{r^3}\left\{\left[2\bar{\gamma}_4(f_2+2f_1)r_{,l}r_{,k}-\delta_{lk}f_1\right]n_k+\left[\bar{\gamma}_5(f_2-f_1)-\bar{\gamma}_2 r^2 f_3\right]n_l\right\} \tag{32}$$

From Eq. (29), we get

$$\frac{\partial \bar{\theta}'}{\partial n} = -\bar{\theta}'_{,i} n_i = \frac{r_{,i}}{r^2} n_i f_4 \tag{33}$$

where $f_4 = \frac{(1+\tau_0 s^\beta)}{4\pi(k_1^2 - k_2^2)}\left[(k_1^2 - s^2)(1 + k_1 r)e^{-k_1 r} - (k_2^2 - s^2)(1 + k_2 r)e^{k_2 r}\right]$.

Case II

Instantaneous concentrated body force located at the point **x** = **y**, is acting upon a viscoelastic body with no heat source acting inside the medium. The fundamental solutions, in this case, will be denoted by a superscript (*j*). This case can be stated as:

$$Q = 0,\, F_i = F_i^{(j)} = \delta_{ij}\delta(\mathrm{x} - \mathrm{y})\delta(t)$$

Taking the Laplace transforms of $\mathrm{F_i}$ and Q we get, $\bar{Q} = 0, \bar{F}_i = \bar{F}_i^{(j)} = \delta_{ij}\delta(\mathrm{x} - \mathrm{y})$.

In this case, Eqs. (18–20) take the form

$$\left(\nabla^2 - \frac{s^2}{\bar{\gamma}_1}\right)\bar{\phi}^{(j)} + \frac{\bar{X}^{(j)}}{\bar{\gamma}_1} = \bar{m}\bar{\theta}^{(j)} \tag{34}$$

$$\left(\nabla^2 - \frac{s^2}{\bar{\gamma}_4}\right)\bar{\psi}_i^{(j)} = \frac{-\bar{Y}_i^{(j)}}{\bar{\gamma}_4} \tag{35}$$

$$\left(\nabla^2 - s\left(1 + \tau_0 s^\beta\right)\right)\bar{\theta}^{(j)} - \bar{\alpha} s\left(1 + \tau_0 s^\beta\right)\nabla^2\bar{\varphi}^{(j)} = 0 \tag{36}$$

The scalar and vector potentials in Eq. (17) have the forms

$$X = X^{(j)} = \frac{-1}{4\pi}\left(\frac{1}{r}\right)_{,j} \qquad Y_i = Y_i^{(j)} = \frac{-1}{4\pi}\varepsilon_{ilj}\left(\frac{1}{r}\right)_{,j} \tag{37}$$

Eliminating $\bar{\theta}^{(j)}$ from Eqs. (34) and (36) and substituting in the resulting equations for $X^{(j)}$ given in Eq. (37), we get

$$\left(\nabla^2 - k_1^2\right)\left(\nabla^2 - k_2^2\right)\bar{\phi}^{(j)} = \frac{1}{4\pi\bar{\gamma}_1}\left[\nabla^2 - s\left(1 + \tau_0 s^\beta\right)\right]\left(\frac{1}{r}\right)_{,j} \tag{38}$$

To solve this equation, we make the change of variables $\bar{\phi}^{(j)} = g_{,j}$ to arrive at

$$g = \frac{1}{4\pi\bar{\gamma}_1\left(k_1^2 - k_2^2\right)}\left[\nabla^2 - s\left(1 + \tau_0 s^\beta\right)\right](\Gamma_1 - \Gamma_2)$$

where Γ_i is the solution of the equation

$$\left(\nabla^2 - k_i^2\right)\Gamma_i = \frac{1}{r} \qquad i = 1, 2$$

The solution of these equations bounded both at the origin and infinity is given by

$$\Gamma_i = \frac{1}{k_1^2 r}\left[e^{-k_i r} - 1\right] \qquad i = 1, 2$$

Collecting the previous results and using Eqs. (26a), we arrive at

$$\bar{\phi}^{(j)} = \frac{r_{,j}}{4\pi s^2 r^2}\left[1 + \frac{1}{\left(k_1^2 - k_2^2\right)}\left\{(1 + k_1 r)\left(k_2^2 - \frac{s^2}{\bar{\gamma}_1}\right)e^{-k_1 r} - (1 + k_2 r)\left(k_1^2 - \frac{s^2}{\bar{\gamma}_1}\right)e^{-k_2 r}\right\}\right] \tag{39}$$

Substituting from the second equation of (37) into Eq. (35) and using the substitution

$$\bar{\psi}_l^{(j)} = \varepsilon_{ilj} R_{,l}$$

where $R = \frac{1}{4\pi s^2 r}\left(e^{-\frac{s}{\sqrt{\bar{\gamma}_4}} r} - 1\right)$,

we get as before

$$\bar{\psi}_l^{(j)} = \frac{1}{4\pi s^2 r^2}\varepsilon_{lmj} r_{,m}\left[\left(1 + \frac{sr}{\sqrt{\bar{\gamma}_4}}\right)e^{-\frac{sr}{\sqrt{\bar{\gamma}_4}}} - 1\right] \tag{40}$$

Substituting from Eqs. (39) and (40) into Eq. (16) and using the relation

$$r_{,ij} = \frac{\delta_{ij} - r_{,i} r_{,j}}{r}$$

we have

$$\bar{u}_i^{(j)} = \frac{1}{r}\left[U_1 \delta_{ij} - U_2 r_{,i} r_{,j}\right] \tag{41}$$

where

$$U_1 = \frac{1}{4\pi s^2 r^2}\left\{\left(1 + \frac{sr}{\sqrt{\bar{\gamma}_4}}\right)e^{-\frac{sr}{\sqrt{\bar{\gamma}_4}}} + s^2 \sum_{n=1}^{2} (-1)^{n-1} A_n (1 + k_n r) e^{-k_n r}\right\}$$

where $A_n = \frac{k_n^2 - s\left(1 + \tau_0 s^\beta\right)}{\bar{\gamma}_1 k_n^2 \left(k_1^2 - k_2^2\right)}$.

and

$$U_2 = \frac{1}{4\pi\, s^2 r^2}$$
$$\left\{ \left(3 + \frac{3s\, r}{\sqrt{\bar{\gamma}_4}} + \frac{s^2 r^2}{\bar{\gamma}_4} \right) e^{-\frac{s\, r}{\sqrt{\bar{\gamma}_4}}} + s^2 \sum_{n=1}^{2} (-1)^{n-1} A_n \left(3 + 3k_n r + k_n^2 r^2 \right) e^{-k_n r} \right\}$$

Eliminate $\bar{\varphi}^{(j)}$ from Eqs. (34) and (36) and substituting in the resulting equations for $X^{(j)}$ given in Eq. (37), we get

$$\left(\nabla^2 - k_1^2 \right) \left(\nabla^2 - k_2^2 \right) \bar{\theta}^{(j)} = \frac{-s\bar{\alpha}\left(1 + \tau_0 s^\beta \right)}{4\pi\, \bar{\gamma}_1} \nabla^2 \left(\frac{1}{r} \right)_{,j} \tag{42}$$

As the previous procedure, it can be checked that

$$\bar{\theta}^{(j)} = \frac{s\bar{\alpha}}{\bar{\gamma}_2} \bar{u}_i' \tag{43}$$

This equation together with Eq. (28) gives

$$\bar{\theta}^{(j)} = \frac{-s\bar{\alpha}}{\bar{\gamma}_2\, r^2} f_1\, r_{,j} \tag{44}$$

Equation (44) leads to

$$\frac{\partial \bar{\theta}^{(j)}}{\partial n} = \bar{\theta}_{,i}^{(j)} n_i = \frac{\bar{\alpha}}{\bar{\gamma}_2 r^3} \left[\delta_{ij} f_1 - r_{,i} r_{,j} f_5 \right] \tag{45}$$

where

$$f_5 = \frac{\bar{m}\left(1 + \tau_0 s^\beta \right)}{4\pi \left(k_1^2 - k_2^2 \right)} \left[e^{-k_1 r} \left(3 + 3k_1 r + k_1^2 r^2 \right) - e^{k_2 r} \left(3 + 3k_2 r + k_2^2 r^2 \right) \right]$$

The Laplace transform of the traction vector for this case can be obtained from Eq. (12)

$$\bar{p}_i^{(j)}(\mathrm{r}, s) = \bar{\gamma}_4 \left[\bar{u}_{l,k}^{(j)} + \bar{u}_{k,l}^{(j)} \right] n_k + \left(\bar{\gamma}_5 \bar{u}_{k,k}^{(j)} - \bar{\gamma}_2 \bar{\theta}^{(j)} \right) n_l \tag{46}$$

From Eq. (41) we obtain

$$\bar{u}_{k,l}^{(j)} = \frac{1}{r^2} \left[U_3 \delta_{kj} r_{,l} + U_2 \left(\delta_{kl} r_{,j} - \delta_{jl} r_{,k} \right) + U_4 r_{,j} r_{,k} r_{,l} \right] \tag{47}$$

and

$$\bar{u}_{k,k}^{(j)} = \frac{1}{r^2}[U_3 + 4U_2 + U_4]r_{,j} \tag{48}$$

where $U_3 = r\frac{\partial U_1}{\partial r} - U_1$ and $U_4 = r\frac{\partial U_2}{\partial r} - 3U_2$.

Substituting from Eqs. (44), (47) and (48) into Eq. (46) we get,

$$\begin{aligned}\bar{p}_i^{(j)}(\mathrm{r}, s) = \frac{1}{r^2}\{&\bar{\gamma}_4(U_2 + U_3)r_{,l}n_j + \left[(U_2 + U_3)\delta_{jl} + 2U_4 r_{,j} r_{,l}\right]r_{,k}n_k \\ &+\left[\bar{\gamma}_5(U_3 + 4U_2 + U_4) + \bar{\alpha} f_1\right]r_{,j}n_l\}\end{aligned} \tag{49}$$

5 Reciprocity Theorem

In this section, a reciprocity theorem of the generalized thermoviscoelasticity with one relaxation time is established. Two systems of causes (Q, F_i) and (Q', F'_i) which result in two systems of effects (u_i, θ) and (u'_i, θ') are assumed.

Multiplying Eq. (15) by e'_{ij} and integrating over the volume V, we obtain

$$\int_V \bar{\sigma}_{ij}\bar{e}'_{ij}dV = \int_V \left[2\bar{\gamma}_4\bar{e}_{ij}\bar{e}'_{ij} + \bar{\gamma}_5\bar{e}\bar{e}' - \bar{\gamma}_2\bar{e}'\bar{\theta}\right]dV \tag{50}$$

The analogous equation for the above equation is

$$\int_V \bar{\sigma}'_{ij}\bar{e}_{ij}dV = \int_V \left[2\bar{\gamma}_4\bar{e}_{ij}\bar{e}'_{ij} + \bar{\gamma}_5\bar{e}\bar{e}' - \bar{\gamma}_2\bar{e}\bar{\theta}'\right]dV \tag{51}$$

Subtracting the above equations, we get

$$\int_V \left[\bar{\sigma}_{ij}\bar{e}'_{ij} - \bar{\sigma}'_{ij}\bar{e}_{ij}\right]dV = \bar{\gamma}_2\int_V \left[\bar{e}\bar{\theta}' - \bar{e}'\bar{\theta}\right]dV \tag{52}$$

Substituting by the non-dimensional transformed equations of motion in the form

$$\bar{\sigma}_{ji,j} = s^2\bar{u}_i + \bar{F}_i \quad \text{and} \quad \bar{\sigma}'_{ji,j} = s^2\bar{u}'_i + \bar{F}'_i$$

and the transformed expressions form of Eq. (4) in Eq. (52) and integrating by parts, we have

$$\int_S \left[\bar{\sigma}_{ij}\bar{u}'_i - \bar{\sigma}'_{ij}\bar{u}_i\right]n_j dS + \int_V \left[\bar{F}_i\bar{u}'_i - \bar{F}'_i\bar{u}_i\right]dV = \bar{\gamma}_2\int_V \left[\bar{e}\bar{\theta}' - \bar{e}'\bar{\theta}\right]dV \tag{53}$$

Multiplying Eq. (14) by $\bar{\theta}'$ and following the same procedure as above, we find

$$\int_S \left[\bar{\theta}'\bar{\theta}_{,n} - \bar{\theta}\bar{\theta}'_{,n}\right]dS = \left(s + \tau_0 s^{1+\beta}\right)\bar{\alpha}\int_V \left[\bar{e}\bar{\theta}' - \bar{e}'\bar{\theta}\right]dV - \left(1 + \tau_0 s^{\beta}\right)\int_V \left[\bar{Q}\bar{\theta}' - \bar{Q}'\bar{\theta}\right]dV \tag{54}$$

Eliminating the term $\int_V \left[\bar{e}\bar{\theta}' - \bar{e}'\bar{\theta}\right]dV$ between (53) and (54), we get

$$\left(s + \tau_0 s^{1+\beta}\right)\bar{\alpha}\int_S \left[\bar{\sigma}_{ij}\bar{u}'_i - \bar{\sigma}'_{ij}\bar{u}_i\right]n_j dS - \bar{\gamma}_2\int_S \left[\bar{\theta}'\bar{\theta}_{,n} - \bar{\theta}\bar{\theta}'_{,n}\right]dS$$
$$= \bar{\gamma}_2\left(1 + \tau_0 s^{\beta}\right)\int_V \left[\bar{Q}\bar{\theta}' - \bar{Q}'\bar{\theta}\right]dV - \left(s + \tau_0 s^{1+\beta}\right)\bar{\alpha}\int_V \left[\bar{F}_i\bar{u}'_i - \bar{F}'_i\bar{u}_i\right]dV \tag{55}$$

6 Boundary Integral Equations

To obtain the integral representations for the transformed displacement and temperature distributions, we consider the following cases.

First, we take $\bar{F}'_i = 0,\ \bar{Q}' = \delta(\mathrm{x} - \mathrm{y})$ where $\mathrm{y} \in \mathrm{V} \cup \mathrm{S}$. Thus Eq. (55) leads to

$$\left(1 + \tau_0 s^{\beta}\right)\Delta(\mathrm{x})\bar{\theta}(\mathrm{x}, s) = \int_S \left[\bar{\theta}'\bar{\theta}_{,n} - \bar{\theta}\bar{\theta}'_{,n}\right]dS - \left(s + \tau_0 s^{1+\beta}\right)\bar{\beta}_1\int_S \left[\bar{\sigma}_{ij}\bar{u}'_i - \bar{\sigma}'_{ij}\bar{u}_i\right]n_j dS$$
$$- \left(s + \tau_0 s^{1+\beta}\right)\bar{\beta}_1\int_V \bar{F}_i\bar{u}'_i dV + \left(1 + \tau_0 s^{\beta}\right)\int_V \bar{Q}\bar{\theta}' dV \tag{56}$$

where $\bar{\beta}_1 = \frac{\bar{\alpha}}{\bar{\gamma}_2}$ and

$$\Delta(\mathrm{x}) = \begin{cases} 1 & if\ \mathrm{x} \in \mathrm{V} \\ 0 & if\ \mathrm{x} \notin \mathrm{V} \cup \mathrm{S} \\ \frac{1}{2} & if\ \mathrm{x} \in \mathrm{S} \end{cases}$$

The corresponding transformed displacement $\bar{u}'_i$ and temperature $\bar{\theta}'$ are the fundamental solutions as obtained previously in case I.

Secondly, we take $\bar{F}_i^{(j)} = \delta_{ij}\delta(\mathrm{x} - \mathrm{y}),\ \bar{Q}^{(j)} = 0$, the corresponding transformed displacement $\bar{u}_i^{(j)}$ and temperature $\bar{\theta}^{(j)}$ are the fundamental solutions as obtained previously in case II. As before, we get

$$
\begin{aligned}
(1+\tau_0 s^\beta)\Delta(\mathrm{x})\bar{u}_j(\mathrm{x},s) =&(1+\tau_0 s^\beta)\int\limits_V \bar{F}_i\,\bar{u}_i^{(j)} dV + (1+\tau_0 s^\beta) \\
&\int\limits_S \left[\bar{\sigma}_{ij}\bar{u}_i^{(j)} - \bar{\sigma}_{ij}^{(j)}\bar{u}_i\right] n_j dS \\
&- \bar{\beta}_2 \int\limits_S [\bar{\theta}\,\bar{\theta}_{,n}^{(j)} - \bar{\theta}^{(j)}\bar{\theta}_{,n}] dS - \bar{\beta}_2(1+\tau_0 s^\beta)\int\limits_V \bar{Q}\bar{\theta}^{(j)} dV
\end{aligned}
\tag{57}
$$

where $\bar{\beta}_2 = \frac{\bar{\gamma}_2}{s\bar{\alpha}}$.

Using the convolution theorem of the Laplace transforms [52], namely

$$
L^{-1}[\bar{f}_1(s)\bar{f}_2(s)] = \int\limits_0^t f_1(\tau) f_2(t-\tau) d\tau
$$

We get

$$
\Delta(\mathrm{x})\left[\theta(\mathrm{x},t) + \tau_0 \frac{\partial^\beta \theta(\mathrm{x},t)}{\partial t^\beta}\right] = W_1(\mathrm{x},t) \tag{58}
$$

$$
\Delta(\mathrm{x})\left[u_j(\mathrm{x},t) + \tau_0 \frac{\partial^\beta u_j(\mathrm{x},t)}{\partial t^\beta}\right] = W_2^{(j)}(\mathrm{x},t) \tag{59}
$$

where

$$
\begin{aligned}
W_1(\mathrm{x},t) =& \int\limits_S\int\limits_0^t \left[\theta'(\mathrm{x}-\eta,t-\tau)\frac{\partial\theta(\eta,\tau)}{\partial n(\eta)} - \theta(\eta,\tau)\frac{\partial\theta'(\mathrm{x}-\eta,t-\tau)}{\partial n(\eta)}\right] d\tau\, dS_\eta \\
&+ \int\limits_S\int\limits_0^t \left[\sigma_{ij}(\eta,\tau)\left[\frac{\partial}{\partial\tau} + \tau_0\frac{\partial^{1+\beta}}{\partial\tau^{1+\beta}}\right]\int\limits_0^{t-\tau} \beta_1(\zeta) u_i'(\mathrm{x}-\eta,t-\tau-\zeta) d\zeta\right] d\tau\, dS_\eta \\
&- \int\limits_S\int\limits_0^t \left[u_i(\eta,\tau)\left[\frac{\partial}{\partial\tau} + \tau_0\frac{\partial^{1+\beta}}{\partial\tau^{1+\beta}}\right]\int\limits_0^{t-\tau} \beta_1(\zeta) \sigma_{ij}'(\mathrm{x}-\eta,t-\tau-\zeta) d\zeta\right] d\tau\, dS_\eta \\
&- \int\limits_V\int\limits_0^t \left[F_i(\mathrm{y},\tau)\left[\frac{\partial}{\partial\tau} + \tau_0\frac{\partial^{1+\beta}}{\partial\tau^{1+\beta}}\right]\int\limits_0^{t-\tau} \beta_1(\zeta) u_i'(\mathrm{x}-\mathrm{y},t-\tau-\zeta) d\zeta\right] d\tau\, dV_{\mathrm{y}}
\end{aligned}
$$

$$+\int_V\int_0^t\left[Q(\mathrm{y},\tau)\left[1+\tau_0\frac{\partial^\beta}{\partial\tau^\beta}\right]\theta'(\mathrm{x}-\mathrm{y},t-\tau)\right]d\tau\,dV_{\mathrm{y}}$$

and

$$W_2^{(j)}(\mathrm{x},t)=\int_V\int_0^t\left[F_i(\mathrm{y},\tau)\left[\frac{\partial}{\partial\tau}+\tau_0\frac{\partial^{1+\beta}}{\partial\tau^{1+\beta}}\right]u_i^{(j)}(\mathrm{x}-\mathrm{y},t-\tau)\right]d\tau\,dV_{\mathrm{y}}$$

$$+\int_S\int_0^t\left[\sigma_{ij}(\eta,\tau)\left[1+\tau_0\frac{\partial^\beta}{\partial\tau^\beta}\right]u_i^{(j)}(\mathrm{x}-\eta,t-\tau)\right.$$

$$\left.-u_i(\eta,\tau)\left[1+\tau_0\frac{\partial^\beta}{\partial\tau^\beta}\right]\sigma_{ij}^{(j)}(\mathrm{x}-\eta,t-\tau)\right]d\tau\,dS_\eta$$

$$+\int_S\int_0^t\left[\theta(\eta,\tau)\frac{\partial u_i^{(j)}(\mathrm{x}-\eta,t-\tau)}{\partial n(\eta)}-u_i^{(j)}(\mathrm{x}-\eta,t-\tau)\frac{\partial\theta(\eta,\tau)}{\partial n(\eta)}\right]d\tau\,dS_\eta$$

$$-\int_V\int_0^t\left[Q(\mathrm{y},\tau)\left[1+\tau_0\frac{\partial^\beta}{\partial\tau^\beta}\right]\int_0^{t-\tau}\beta_2(\zeta)\theta^{(j)}(\mathrm{x}-\mathrm{y},t-\tau-\zeta)d\zeta\right]d\tau\,dV_{\mathrm{y}}$$

the solution of Eqs. (58) and (59) can be written as

$$\Delta(\mathrm{x})\theta(\mathrm{x},t)=\frac{1}{\tau_0}\int_0^t E_{\beta,\beta}\left(-\tau^\beta/\tau_0\right)W_1(\mathrm{x},\mathrm{t}-\tau)d\tau \tag{60}$$

$$\Delta(\mathrm{x})u_j(\mathrm{x},t)=\frac{1}{\tau_0}\int_0^t E_{\beta,\beta}\left(-\tau^\beta/\tau_0\right)W_2^{(j)}(\mathrm{x},\mathrm{t}-\tau)d\tau \tag{61}$$

where $E_{\beta,\gamma}(z)$ is known as "Mittag-Leffler" function which is defined as

$$E_{\beta,\gamma}(z)=\sum_{k=0}^{\infty}\frac{z^k}{\Gamma(\beta k+\gamma)}$$

Letting $\mathrm{x}\to\xi\in\mathrm{S}$ in Eqs. (58) and (59), we arrive at

$$\theta(\xi,t)=\frac{2}{\tau_0}\int_0^t E_{\beta,\beta}\left(-\tau^\beta/\tau_0\right)W_1(\mathrm{x},\mathrm{t}-\tau)d\tau \tag{62}$$

$$u_j(\xi,t) = \frac{2}{\tau_0}\int_0^t E_{\beta,\beta}\left(-\tau^\beta/\tau_0\right)W_2^{(j)}(\mathrm{x},\mathrm{t}-\tau)d\tau \tag{63}$$

Equations (62) and (63) with the boundary conditions (8)–(11) and the limiting behavior of the solutions as $r \to 0$, namely

$$\bar{\theta}'(r,s) \approx \frac{1+\tau_0 s^\beta}{4\pi k r} + O(1), \qquad \bar{u}_i\prime(r,s) \approx O(1)$$
$$\bar{\sigma}_{ij}(r,s) \approx O\left(\frac{1}{r}\right), \qquad \bar{u}_i^{(j)}(r,s) \approx \bar{u}_i^{(j)}(r) + O(1) \tag{64}$$

can be used to set up the system of linear equations of the boundary integral method. This completes our formulation. Also, it is clear that all the above functions trend to zero when $r \to \infty$.

7 Example

Now, we shall consider an example to explain the current formulation. We consider a formulation in which we determine the primary variables $u(x,t)$ and $\theta(x,t)$ subjected to homogenous initial conditions and the boundary conditions (8)–(11).

Taking $\Delta(x) = 1$ Eqs. (56) and (57) became

$$\begin{aligned}\bar{\theta}(\mathrm{x},s) = {} & \frac{1}{\left(1+\tau_0 s^\beta\right)}\left\{\int_{S_3}\left[\bar{\theta}_0'\bar{\theta}_{,n} - \bar{\theta}_0\bar{\theta}'_{,n}\right]dS + \int_{S_4}\left[\bar{\theta}'\bar{\theta}_{0n} - \bar{\theta}\bar{\theta}_{0n}'\right]dS\right\} \\ & - s\bar{\beta}_1\left\{\int_{S_1}\left[\bar{p}_{i0}\bar{u}_i' - \bar{p}_{i0}'\bar{u}_i\right]dS + \int_{S}\left[\bar{\sigma}_{ij}\bar{u}_{i0}' - \bar{\sigma}_{ij}'\bar{u}_{i0}\right]n_j dS\right\} \\ & - s\bar{\beta}_1\int_V \bar{F}_i\bar{u}_i' dV + \int_V \bar{Q}\bar{\theta}' dV\end{aligned} \tag{65}$$

$$\begin{aligned}\bar{u}_j(\mathrm{x},s) = {} & \left\{\int_{S_1}\left[\bar{p}_{i0}\bar{u}_i^{(j)} - \bar{p}_{i0}^{(j)}\bar{u}_i\right]n_j dS + \int_{S_2}\left[\bar{\sigma}_{ij}\bar{u}_{i0}^{(j)} - \bar{\sigma}_{ij}^{(j)}\bar{u}_{i0}\right]n_j dS\right\} \\ & - \frac{\bar{\beta}_2}{\left(1+\tau_0 s^\beta\right)}\left\{\int_{S_3}\left[\bar{\theta}_0\bar{\theta}_{,n}^{(j)} - \bar{\theta}_0^{(j)}\bar{\theta}_{,n}\right]dS + \int_{S_4}\left[\bar{\theta}\bar{\theta}_{0n}^{(j)} - \bar{\theta}^{(j)}\bar{\theta}_{0n}\right]dS\right\} \\ & + \int_V \bar{F}_i\bar{u}_i^{(j)}dV - \bar{\beta}_2\int_V \bar{Q}\bar{\theta}^{(j)}dV\end{aligned} \tag{66}$$

the function $u_i(x, t)$ is unknown on the part S_1, while the function $\theta(x, t)$ is unknown on the S_4. Also, we assume that the fundamental solutions satisfy the following conditions, $\theta'(x, t) = \theta^{(j)}(x, t) = 0$ on the parts S_1 and S_4.and $u'_i(x, t) = u_i^{(j)}(x, t) = 0$ on the parts S_2 and S_3.

Therefore by using these conditions and taking $x \to \xi$, we obtain the Eqs. (65) and (66) as the system of two Fredholm's integral equations:

$$\begin{aligned}0 = &\frac{-1}{(1 + \tau_0 s^\beta)}\left\{\int_{S_3} \bar{\theta}_0(y, s)\frac{\partial\bar{\theta}'(\xi, y, s)}{\partial n'(\xi)}dS + \int_{S_4} \bar{\theta}(y, s)\frac{\partial\bar{\theta}'_0(\xi, y, s)}{\partial n'(\xi)}dS\right\}\\
&+ s\bar{\beta}_1\left\{\int_{S_1} \bar{p}'_{i0}(y, s)\frac{\partial\bar{u}_i(\xi, y, s)}{\partial n'(\xi)}dS + \int_{S_2} \bar{\sigma}'_{ij}(y, s)\frac{\partial\bar{u}_{i0}(\xi, y, s)}{\partial n'(\xi)}n_j dS\right\}\\
&- s\bar{\beta}_1\int_V \bar{F}_i(y, s)\frac{\partial\bar{u}'_i(\xi, y, s)}{\partial n'(\xi)}dV + \int_V \bar{Q}(y, s)\frac{\partial\bar{\theta}'(\xi, y, s)}{\partial n'(\xi)}dV\end{aligned} \tag{67}$$

$$\begin{aligned}0 = &-\left\{\int_{S_1} \bar{p}_{i0}^{(j)}(y, s)\bar{u}_i(\xi, y, s)dS + \int_{S_2} \bar{\sigma}_{ij}^{(j)}(\xi, y, s)\bar{u}_{i0}(y, s)n_j dS\right\}\\
&- \frac{\bar{\beta}_2}{(1 + \tau_0 s^\beta)}\left\{\int_{S_3} \bar{\theta}_0(\xi, y, s)\bar{\theta}_{,n}^{(j)}(y, s)\,dS + \int_{S_4} \bar{\theta}(y, s)\bar{\theta}_{0n}^{(j)}(\xi, y, s)dS\right\}\\
&+ \int_V \bar{F}_i(y, s)\bar{u}_i^{(j)}(\xi, y, s)dV - \bar{\beta}_2\int_V \bar{Q}(y, s)\bar{\theta}^{(j)}(\xi, y, s)dV\end{aligned} \tag{68}$$

where $n'(\xi)$ is the outer normal to the surface S_4.

Equations (67) and (68) do not seem to have analytical solutions for general boundary shapes. Therefore, it is necessary to use appropriate numerical techniques depending on the nature of the problem to be solved. The integrals have to be discretized and the problem reduces to finding the solution of a system of linear algebraic equations The final solution can be gotten by using a suitable numerical method of Laplace transformed inversion. Readers can find the materialize of the behavior of the solutions of the fractional order theory of thermo-viscoelasticity in the physical domain in [49].

References

1. Sladek, V., Sladek, J.: Boundary integral equation method in thermoelasticity. Part I. General analysis. Appl. Math. Model. **7**, 241–253 (1983)
2. Sladek, V., Sladek, J.: Boundary integral equation method in thermoelasticity. Part III. Uncoupled thermoelasticity. Appl. Math. Model. **8**, 413–418 (1984)
3. Anwar, M., Sherief, H.: Boundary integral equation formulation of generalized thermoelasticity in a laplace transform domain. Appl. Math. Model. **12**, 161–166 (1988)
4. Anwar, M., Sherief, H.: Boundary integral equation formulation for thermoelasticity with two relaxation times. J. Therm. Stress. **17**, 257–270 (1994)
5. Elkaramany, A., Ezzat, M.: On the boundary integral formulation of thermo-viscoelasticity theory. Int. J. Eng. Sci. **40**(17), 1943–1956 (2002)
6. Elkaramany, A., Ezzat, M.: Boundary integral equation formulation for the generalized thermoviscoelasticity with two relaxation times. Appl. Math. Comput. **151**, 347–362 (2004)
7. Elhagary, M.: Boundary integral equation formulation for the generalized thermoviscoelasticity with one relaxation time. Eng. Anal. Boundary Elem. **104**, 209–214 (2019)
8. Tiwari, R., Mukhopadhyay, S.: Boundary integral equations formulation for fractional order thermoelasticity. CMST **20**, 49–58 (2014)
9. Danilovskaya, V.: Thermal stresses in an elastic half-space due to a sudden heating of its boundary (in Russian). Prikl. Mat. Mekh. **14**, 316–321 (1950)
10. Biot, M.: Thermoelasticity and irreversible thermo-dynamics. J. Appl. Phys. **27**, 240–253 (1956)
11. Lord, H., Shulman, Y.: A generalized dynamical theory of thermoelasticity. J. Mech. Phys. Solids **15**, 299–309 (1967)
12. Dhaliwal, R., Sherief, H.: Generalized thermoelasticity for anisotropic media. Quart. Appl. Math. **33**, 1–8 (1980)
13. Sherief, H., El-Maghraby, N.: An internal penny-shaped crack in an infinite thermoelastic solid. J. Therm. Stress. **26**, 333–352 (2003)
14. Sherief, H., El-Maghraby, N.: A mode-I crack problem for an infinite space in generalized thermoelasticity. J. Therm. Stress. **28**, 465–484 (2005)
15. Sherief, H., Elmisiery, A., Elhagary, M.: Generalized thermoelastic problem for an infinitely long hollow cylinder for short times. J. Therm. Stress. **27**, 885–902 (2004)
16. Saleh, H.: Problem in generalized thermoelasticity for a half-space under the action of a body force. J. Therm. Stress. **28**, 253–266 (2005)
17. Elhagary, M.: A two-dimensional problem for two media in the generalized theory. J. Therm. Stress. **33**, 1–15 (2010)
18. Gurtin, E., Sternberg, E.: On the linear theory of viscoelasticity. Arch. Ration. Mech. Anal. **11**, 182–191 (1962)
19. Pobedrya, B.: Coupled problems in thermoviscoelasticity. Mech. Compos. Mater. **5**, 353–358 (1969)
20. Medri, M.: Coupled thermoviscoelasticity: a way to the stress-strain analysis of polymeric industrial components. Meccanica **23**, 226–231 (1988)
21. Baranoglu, B., Mengi, Y.: The use of dual reciprocity boundary element method in coupled thermoviscoelasticity. Comput. Methods Appl. Mech. Engrg. **196**, 379–392 (2006)
22. Lychev, S.: Coupled dynamic thermoviscoelasticity problem. Mech. Solid. **43**, 769–784 (2008)
23. Kartashov, E.: New model ideas in dynamic thermoviscoelasticity in the problem of thermal shock. Doklady Mathematics. **86**, 704–706 (2012)
24. Kartashov, E.: Dynamic thermoviscoelasticity in a thermal shock problem. Thermal. Eng **59**, 1051–1061 (2012)
25. Sherief, H., Allam, M., Elhagary, M.: Generalized theory of thermoviscoelasticity and a half-space problem. Int. J. Thermophys. **32**, 1271–1295 (2011)
26. Elhagary, M.: Thermo-mechanical shock problem for generalized theory of thermoviscoelasticity. Int. J. Thermophys. **34**, 170–188 (2013)

27. Sherief, H., Hamza, F., Abd El-Latief, A.: 2D problem for a half-space in the generalized theory of thermo-viscoelasticity. Mech Time-Depend Mater. **19**, 557–568 (2015)
28. Sherief, H., Hamza, F., Abd El-Latief, A.: Wave propagation study for axi-symmetric 2D problems of a generalized thermo-visco-elastic half space. J. Therm. Stress. **42**, 1–14 (2019)
29. Machado, J., Silva, M., Barbosa, R., Jesus, I., Reis, C., Marcos, M., Galhano, A.: Some applications of fractional calculus in engineering. Math. Prob. Eng. (2010). https://doi.org/10.1155/2010/639801
30. Caputo, M., Mainardi, F.: A new dissipation model based on memory mechanism. Pure. appl. Geophys. **91**, 134–147 (1971)
31. Caputo, M., Mainardi, F.: Linear model of dissipation in an elastic solids. Riv. Nuovo Cimento. **1**, 161–198 (1971)
32. Caputo, M.: Vibrations on an infinite viscoelastic layer with a dissipative memory. J. Acoust. Soc. Am. **56**, 897–904 (1974)
33. Adolfsson, K., Enelund, M.: Fractional derivative viscoelasticity at large deformations. Nonlinear Dyn. **33**, 301–321 (2003)
34. Adolfsson, K., Enelund, M., Larsson, S.: Adaptive discretization of fractional order viscoelasticity using sparse time history. Comput. Methods Appl. Mech. Eng. **193**, 4567–4590 (2004)
35. Povstenko, Y.: Fractional heat conduction and associated thermal stress. J. Therm. Stress. **28**, 83–102 (2005)
36. Povstenko, Y.: Thermoelasticity that uses fractional heat conduction equation. J. Math. Sci. **162**, 296–305 (2009)
37. Povstenko, Y.: Fractional Cattaneo-type equations and generalized thermoelasticity. J. Thermal Stress. **34**, 97–114 (2011)
38. Sene, N.: Stokes' first problem for heated flat plate with Atangana-Baleanu fractional derivative. Chaos, Solitons Fractals **117**, 68–75 (2018)
39. Sene, N.: Integral balance methods for stokes' first, equation described by the left generalized fractional derivative. Physics **1**, 154–166 (2019)
40. Sene, N.: Second-grade fluid model with Caputo-Liouville generalized fractional derivative. Chaos, Solitons Fractals **133**, 109631 (2020)
41. Sherief, H., El-Sayed, A., Abd El-Latief, A.: Fractional order theory of thermoelasticity. Int. J. Solids. Struct. **47**, 269–275 (2010)
42. Sherief, H., AbdEl-Latief, A.: Application of fractional order theory of thermoelasticity to a1D problem for a half-space. ZAMM **94**, 509–515 (2014)
43. Sherief, H., AbdEl-Latief, A.: Effect of variable thermal conductivity on a half-space under the fractional order theory of thermoelasticity. Int. J. Mech. Sci. **74**, 185–189 (2013)
44. Raslan, W.: Application of fractional order theory of thermoelasticity in a thick plate under axisymmetric temperature distribution. J. Therm. Stress. **38**, 733–743 (2015)
45. Raslan, W.: Application of fractional order theory of thermoelasticity to a 1d problem for a spherical shell. JTAM **54**, 295–304 (2016)
46. Ezzat, M., El Karamany, A.: Theory of fractional order in electro-thermoelasticity. Eur. J. Mech. A. **30**, 491–500 (2011)
47. Ezzat, M., El Karamany, A.: A fractional order heat conduction law in magneto-thermoelasticity involving two temperatures. ZAMP **62**, 937–952 (2011)
48. El Karamany, A., Ezzat, M.: On fractional thermoelasticity. Math. Mech. Solids. **16**(3), 334–346
49. Sherief, H., Elhagary, M.: Fractional order theory of thermo-viscoelasticity and application. Mech. Time-Depend. Mater. (2019). https://doi.org/10.1007/s11043-019-09415-2
50. Fung, Y.C.: Foundation of Solid Mechanics. Prentice-Hall, New Delhi (1968)
51. Morse, P., Feshbach, H.: Methods of Theoretical Physics. McGraw-Hill, New York (1953)
52. Churchill, R.: Operational Mathematics, 3rd edn. McGraw-Hill, New York (1972)

Spectral Methods for Solving Integro-differential Equations and Bibiliometric Analysis

Sedigheh Sabermahani, Yadollah Ordokhani, and Parisa Rahimkhani

Abstract The use of a bibliography of scientific mapping is appropriate when the emphasis on empirical collaborations generates extensive, fragmented and controversial research streams. In this chapter, first, the integro-differential equations are studied at the bibliometric viewpoint. Also, we discuss the application of this class of equations with arbitrary order, in science. Next, we present some numerical methods based on the spectral methods for finding numerical solution of some kind of integro-differential equations, such as fractional integro-differential equations and variable-order fractional partial integro-differential equations. Moreover, in each method, the estimation errors are presented. Some numerical examples are included to demonstrate the validity and applicability of the numerical techniques.

Keywords Spectral method · Fractional integro-differential equations · Variable-order fractional partial integro-differential equations · Bibiliometric analysis

1 Introduction

Spectral methods are well-known in applied mathematics. These methods have become important tools to achieve an approximate solution of different kinds of equations, such as integral and integro-differential equations [1, 2], differential equations [3–5], optimal control problems [6, 7], partial differential equations [8], etc. Fourier analysis is the basic idea of spectral methods. Many interesting results about these methods can be found in Refs. [9, 10]. Moreover, Ref. [11] is proposed a discus-

---

S. Sabermahani · Y. Ordokhani (✉) · P. Rahimkhani
Department of Mathematics, Faculty of Mathematical Sciences, Alzahra University, Tehran, Iran
e-mail: ordokhani@alzahra.ac.ir

S. Sabermahani
e-mail: s.saber@alzahra.ac.ir

P. Rahimkhani
e-mail: p.rahimkhani@alzahra.ac.ir

H. Singh et al. (eds.), *Topics in Integral and Integro-Differential Equations*,
Studies in Systems, Decision and Control 340,
https://doi.org/10.1007/978-3-030-65509-9_7

sion for solving boundary value problems using spectral methods based on Hermite, rational Chebyshev, Sinc, Laguerre functions and so on.

Integro-differential equations are one of the most important tools in mathematics. This class of equations occurs naturally in many fields of science and engineering [12]. The research work resulted in a specific topic, where the unknown functions appear with their derivatives and either the unknown functions, or their derivatives, or both, appear under the sign of integration. The origin of the study of integral and integro-differential equations can be found in the works by Lakshmikantham [13]. Several researchers are studied on existence and uniqueness of the solution of these kinds of equations. For instance, Guo [14] investigated the existence of solutions of nth-order integro-differential equations. The existence of solutions for a class of fractional integro-differential equations is studied in Refs. [15, 16]. Also, Matar in [17] proved the existence and uniqueness of solution for a class of Volterra-Fredholm integro-differential equations.

On the other hand, several research works have been done to investigate the stability of different classes of integro-differential equations. For example, stability of a class of integro-differential equations using the fixed-point theory is proved in Ref. [18]. Stability of the zero solution is studied for integro-differential equation with operator coefficients in Ref. [19]. Agarwal et al. studied the stability of partial functional integro-differential equations [20]. Also, stability of a class of integro-differential equations with a delay is proved in Ref. [21].

Since finding the analytical solution of many integro-differential equations with arbitrary-order is complex, numerous numerical methods have been proposed by researchers to solve various types of these equations. As example, adomian decomposition method is applied for solving fractional integro-differential equations [22]. Abazari and Klcman used differential transform method for solving nonlinear integro-differential equations with proportional delay [23]. Continuous Runge-Kutta method is presented in [24] for finding approximate solution of neutral Volterra integro-differential equations. AlHabees et al. proposed reproducing kernel Hilbert space method to solve fractional proportional delay first order integro-differential equations [25]. A class of variable-order fractional integro-differential equations is solved by applying finite difference method in Ref. [26]. Variable-order fractional integro-differential equations with weakly singular kernels are solved by a numerical method based on piecewise cubic spline interpolation and finite difference method [27].

Moreover, due to effectiveness of spectral methods to solve the different kinds of equations, these methods have been of interest, such as Chebyshev pseudo-spectral method [28], CAS wavelets method [29], Legendre collocation method [30], fractional alternative Legendre functions method [31], hp-spectral collocation method [32], block-pulse functions method [33], Müntz-Legendre Petrov-Galerkin method [34], collocation method [35], Legendre spectral method [36], Laguerre method [37], Sinc method [38], spectral method [39, 40], pseudo-operational matrix method [41], fractional order operational matrix method [42], Tau method [43] and so on.

2 Bibiliometric Analysis

Bibliometric is a statistical attempt to quantitative assesses the academic quality of journals, authors, etc. The aim of this section is to evaluate the importance and impact of the research works that have been published with the subject of "Spectral method and Integro-differential equations" during the period 1970–2020.

The needed data in this section are extracted from the SCOPUS database, on 3rd March 2020. The "integro-differential equation*" AND ("spectral method" OR "collocation method" OR "Petrov-Galerkin method" OR "tau method" OR "tau-collocation method" OR "least squares method" OR "pseudo-spectral method" OR "spectral element method") topic search term used to derived data. We have obtained 4249 documents from SCOPUS in the subject area of Mathematics. Table 1 shows the different types of documents.

Moreover, Fig. 1 displays the map of countries' collaboration in this study. In this figure, we can see that the United States is a country with the greatest number of international collaborations. Also, as reported in this figure, countries such as Chine, Iran, United States and Turkey have the highest number of documents. Figure 2 shows the trend of documents during the period 1970–2019. As demonstrated by this figure, the number of publications is increasing. To investigate the temporal trend of producing documents in the considered keyword, we design Fig. 3. This figure demonstrates the plot of top-authors production over time. The data collected from SCOPUS were analyzed by the Bibliometrix R package for comprehensive science mapping analysis [44]. A Three-Fields plot on the considered keyword is shown in Fig. 4. This figure displays the relationship between top journals, top countries and top keywords. Also, Fig. 5 shows the evolution of the Keywords index in the considered span. From the data in the graph, it appears that the trend of the keywords is increasing. The words numerate as Keywords index are words or phrases that are appear frequently in the whole of articles and refer to the process used to make papers findable in databases.

Table 1 Type of documents collected bibliometric data

Document type	Documents
Article	4008
Conference paper	97
Book	48
Review	44
Book chapter	42
Conference review	4
Editorial	2
Note	1
Undefined	3

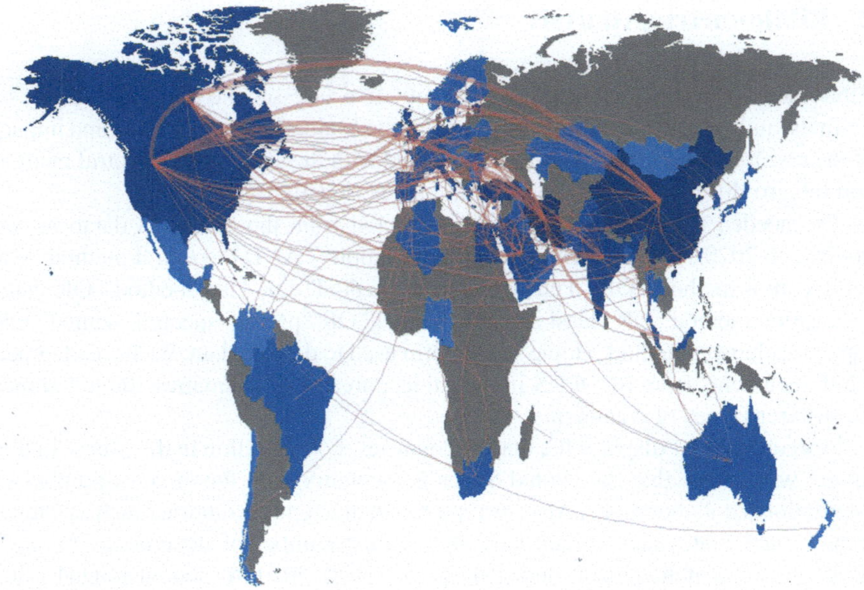

Fig. 1 Countries' collaboration map of considered keyword

Fig. 2 Number of documents per year

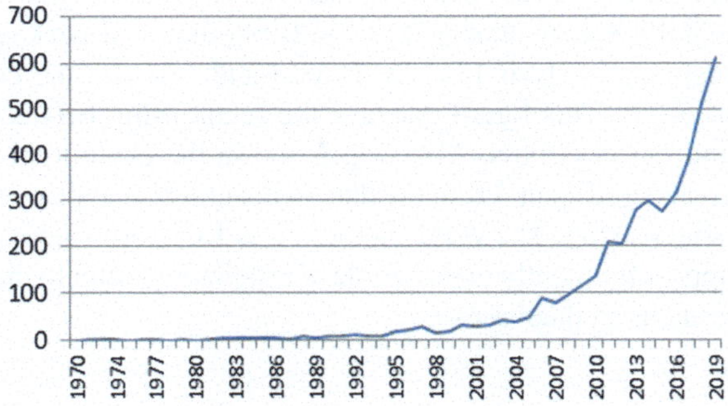

3 Applications

Given the great applications of linear or nonlinear integro-differential equations of arbitrary order to model of various phenomena in different sciences, we recall some of these applications.

- The effects of gene-attenuation on viral-tumor characteristics
  The use of viruses as a cancer treatment has been of interest to researchers. The following delay integro-differential system describes the dynamics of a virus with a constant population of tumor cells in vitro, which may be heterogeneous in the virus and cell populations [45].

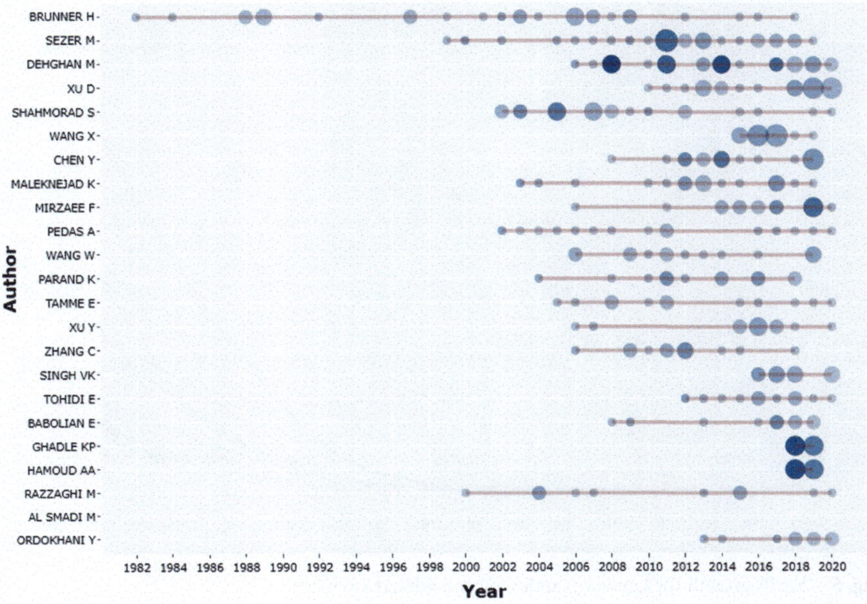

Fig. 3 Top-authors production over the time

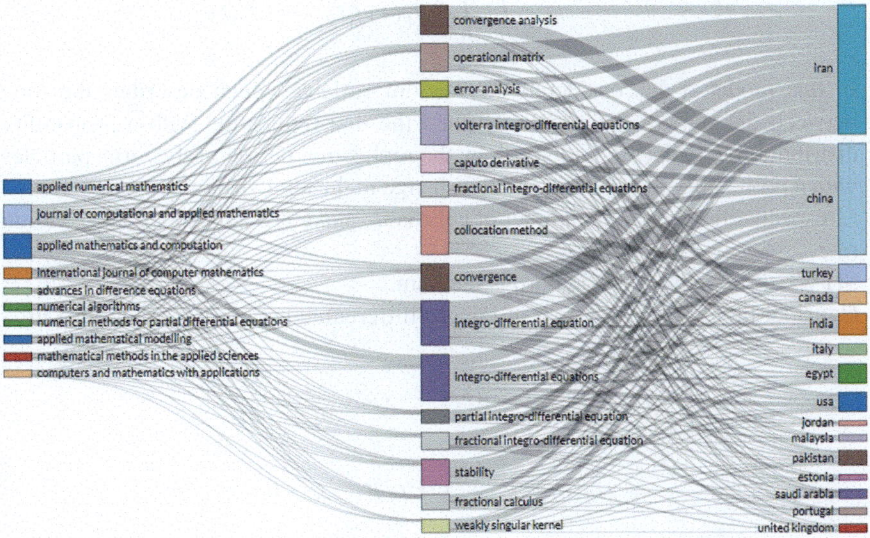

Fig. 4 Three-fields plot of top journals, top countries and top keywords on considered keyword

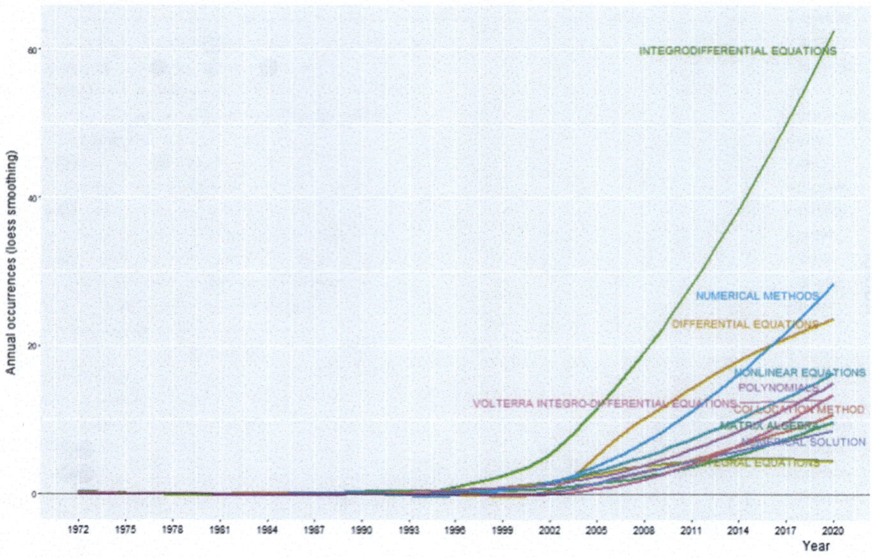

Fig. 5 Words growth for keywords index on considered keyword

$$\frac{dV_T}{dt} = kC_0 \int_0^t \int_t^{\infty} f_\tau(x) f_l(y-x) dy dx$$

In this model, τ is considered the random variable which describes the time between the start of the experiment and the first replication with a probability distribution f_τ. Also, f_l and V_T show the replication period of the virus particles with a probability distribution and the change in the total virus at any point in time t, respectively. The initial number of cells is C_0 and k is the mean viral replication rate.

- SH waves in a visco-elastic porous medium
  Consider the following system of integro-differential hyperbolic equations [46]

$$\rho_s \frac{\partial^2 U}{\partial t^2} = \frac{\partial}{\partial z}\Theta(z,t) - \chi \rho_l^2 \left(\frac{\partial U}{\partial t} - \frac{\partial V}{\partial t}\right),$$

$$\frac{\partial V}{\partial t} = \chi \rho_l (U - V), \quad x, z \in R, \quad x > 0, \quad z > 0,$$

subject to

$$U|_{t \le 0} \equiv 0, \quad V|_{t \le 0} \equiv 0, \quad \Theta(+0, t) = \delta^{'}(t),$$

in which $U(z,t)$ is the velocity of the elastic porous body with a constant partial density ρ_s, $V(z,t)$ is the velocity of the fluid with a constant density ρ_l and $\delta(t)$ is Dirac delta function,

$$\Theta(z,t) = \mu(z)\frac{\partial U}{\partial z} + \int_0^t k(t-\tau)\mu(z)\frac{\partial U}{\partial z}(z,\tau)d\tau,$$

where, χ is a constant and positive coefficient denoting intercomponent friction coefficient, $\mu(z) > 0$ is the Lame coefficient and $k(t)$ is the function characterizing viscosity of the medium.

- Transport model for a nuclear reactor
  The model is defined as [47]

$$t^2\frac{\partial^\alpha U}{\partial x^\alpha} = U + \int_0^1 U(x,\tau)d\tau,$$

subject to

$$(U - t\frac{\partial U}{\partial x})|_{x=0} = -r(0,t),$$

$$(U + t\frac{\partial U}{\partial x})|_{x=1} = -r(1,t),$$

where, $U(x,t)$ is the neutron density that migrates in a direction.

- Population balance model
  For a homogeneous and well-mixed system with particle size, a generic population balance equation is as follows [48]:

$$\begin{aligned}\frac{d(u(x,t))}{dt} &= \int_L^\infty \beta(L,\lambda)g(\lambda)u(t,\lambda)d\lambda \\ &+ \frac{t^2}{2}\int_0^L \frac{F\left((L^3-\lambda^3)^{\frac{1}{3}},\lambda\right)u\left(t,(L^3-\lambda^3)^{\frac{1}{3}}\right)u(t,\lambda)}{(L^3-\lambda^3)^{\frac{2}{3}}}d\lambda \\ &- g(t,L)u(t,L) - u(t,L)\int_0^\infty F(L,\lambda)u(t,\lambda)d\lambda \\ &+ \frac{\partial(G(t,L)u(t,L))}{\partial L} + B_0(t,L).\end{aligned}$$

Each term of the left hand side order expresses as the variation of particle-size distribution with time, birth by breakage, birth by agglomeration, death by breakage, death by agglomeration, growth and primary nucleation [48].
Moreover, population biology [49], transfer of drug resistance in solid tumors [50], the problem of biological species living together [51], tumor growth [52], modeling intrinsic heterogeneity and growth of cancer cells [53], microscopic description of DNA thermal denaturation [54], the model of fluid flow inside the porous formation [55], the scattering of water waves by a floating flexible porous plate [56], direct interaction along light cones at the quantum level [57], a scattering of radio waves on meteoric ionization [58], conduction-radiation problem [59] and linear thermoviscoelastic materials with memory [60], viscoelasticity [61],

a risk model with stochastic return on investments [62], the behavior of a reinforced granular layer overlying soft soil [63], anti-plane cracks in inhomogeneous piezoelectric plane [64] and nano-transistors [65] are samples of applications of inegro-differential equations with arbitrary order in sciences.

4 Spectral Methods and Integro-differential Equations

Given the efficient and important role of spectral methods, three numerical methods are proposed to obtain the approximate solution of some kinds of integro-differential equations.

We consider the following classes of integro-differential equations with arbitrary order as follows:
Problem 1:

Fractional integro-differential equation [31]:

$$D^{\nu}u(x) = F(x, u(x), \int_0^x \kappa_1(x,s)G_1(s,u(s))ds, \int_0^1 \kappa_2(x,s)G_2(s,u(s))ds), \quad (1)$$

with the initial conditions

$$u^{(k)}(0) = \tau_k, \qquad k = 0, 1, \ldots, n-1, \quad (2)$$

where, $n - 1 < \nu \leq n$.
Problem 2:

Variable order fractional integro-differential equations and variable order fractional partial integro-differential equations.
Problem 2.1:

Consider the following variable order fractional nonlinear Fredholm-Volterra integro-differential equation [41]:

$$D^{\nu(x)}u(x) + \sum_{k=0}^{n} p_k(x)u^{(k)}(x) = h(x) + \lambda_f \int_0^1 k_f(x,t)\psi_f(t,u(t))dt$$
$$+ \lambda_v \int_0^x k_v(x,t)\psi_v(t,u(t))dt, \quad (3)$$

with the initial conditions

$$u^{(i)}(0) = \delta_i, \quad i = 0, 1, 2, \ldots, n-1, \quad (4)$$

where, $q - 1 < \nu(x) \leq q$, $u(x)$ is an unknown function, the functions $p_k(x)$, $h(x)$, $k_f(x,t)$, $k_v(x,t)$, $\psi_f(x,u(x))$, $\psi_v(x,u(x))$ and the parameters λ_f, λ_v, δ_i are known and belong to interval $x, t \in [0, 1]$ and also $D^{\nu(x)}$ is the Caputo fractional derivative of order $\nu(x)$.

Problem 2.2:

Consider variable-order fractional nonlinear Volterra partial integro-differential equation [41]:

$$D_t^{\nu(x,t)}u(x,t) + \sum_{k=0}^{2} Q_k(x,t)\frac{\partial^k u(x,t)}{\partial x^k} = h(x,t) + \lambda \int_0^t k(x,t,\eta)\psi(x,\eta,u(x,\eta))d\eta, \quad (5)$$

with the initial condition

$$u(x, 0) = f(x), \qquad x \in [0, 1],$$

and the boundary conditions

$$u(0, t) = \varphi_0(t), \qquad u(1, t) = \varphi_1(t), \qquad t \in [0, 1],$$

where, $0 < \nu(x, t) \leq 1$, $u(x, t)$ is an unknown function, the functions $Q_k(x, t)$, $h(x, t)$, $k(x, t, \eta)$, $f(x)$, $\varphi_1(t)$, $\varphi_2(t)$ and the parameter λ, are known and belong to interval $(x, t) \in \Omega$ and also $D^{\nu(x,t)}$ is the Caputo fractional derivative of order $\nu(x, t)$. The highest order of derivative relative to x and t is equal to 2 and 1, respectively.

4.1 Preliminaries

Here, the necessary definitions and notations needed for the next subsections are given.

4.1.1 Fractional Calculus

In recent decades, fractional calculus has evolved as an important and interesting tool in various branches of sciences. The much interest in this field owes to its extensive applications in the mathematical modeling of many phenomena in economics, medicine, earthquake, solid mechanics, physics and so on [66]. Also, recently, the concept of variable-order fractional derivative and integral have been defined. The field of mathematics is used to modeling of several problems such as alcoholism, glass transition from amorphous networks to shape-memory behavior, viscoelastic, etc. [5].

We recall some basic definitions and properties of fractional calculus theory.

Definition 1 For function u, the Riemann-Liouville fractional integral operator of order $\nu \geq 0$ is defined as follows [66]:

$$I^{\nu}u(x) = \begin{cases} \frac{1}{\Gamma(\nu)} \int_0^x \frac{u(s)}{(x-s)^{1-\nu}} ds, & \nu > 0, \ x > 0, \\ \\ u(x), & \nu = 0. \end{cases} \tag{6}$$

Definition 2 Caputo's fractional derivative of order ν is defined as [6, 66]:

$$D^{\nu}u(x) = \frac{1}{\Gamma(n-\nu)} \int_0^x \frac{u^{(n)}(s)}{(x-s)^{\nu+1-n}} ds, \qquad n-1 < \nu \leq n, \ n \in N. \tag{7}$$

For the Riemann-Liouville fractional integral and the Caputo derivative, we have [6, 66]:

1. $I^{\nu}(\lambda_1 u(x) + \lambda_2 w(x)) = \lambda_1 I^{\nu} u(x) + \lambda_2 I^{\nu} w(x)$,
2. $I^{\nu} x^{\beta} = \frac{\Gamma(\beta+1)}{\Gamma(\beta+\nu+1)} x^{\nu+\beta}, \quad \beta > -1$,
3. $D^{\nu} I^{\nu} u(x) = u(x)$,
4. $I^{\nu} D^{\nu} u(x) = u(x) - \sum_{i=0}^{n-1} u^{(i)}(0) \frac{x^i}{i!}$,
5. $D^{\nu}(\lambda_1 u(x) + \lambda_2 w(x)) = \lambda_1 D^{\nu} u(x) + \lambda_2 D^{\nu} w(x)$,
6. $D^{\nu} c = 0$,

where λ_1, λ_2 and c are real constants.

Definition 3 The Riemann-Liouville variable-order fractional integral operator is defined as [5, 67]:

$$I^{\nu(x)}u(x) = \frac{1}{\Gamma(\nu(x))} \int_0^x (x-s)^{\nu(x)-1} u(s) ds, \quad \nu(x) > 0,$$

where $x > 0$ and $\Gamma(.)$ is Gamma function. For variable-order fractional integration, we have [5, 67]:

$$I^{\nu(x)} x^{\alpha} = \begin{cases} \frac{\Gamma(\alpha+1)}{\Gamma(\alpha+\nu(x)+1)} x^{\alpha+\nu(x)}, & \alpha > -1, \\ 0, & otherwise. \end{cases}$$

Definition 4 The variable-order of Caputo's derivative of $u(x)$ is defined by [5, 67]:

$$D^{\nu(x)}u(x) = \frac{1}{\Gamma(q-\nu(x))} \int_0^x (x-s)^{q-\nu(x)-1} \frac{\partial^q u(s)}{\partial s^q} ds,$$

for $q-1 < \nu(x) \leq q$, $x > 0$ and $q \in N$.

Due to the above definition, variable-order fractional derivative has the following useful property [5, 67]:

$$D^{\nu(x)} x^{m} = \begin{cases} \frac{\Gamma(m+1)}{\Gamma(m-\nu(x)+1)} x^{m-\nu(x)}, & q \leq m \in N, \\ 0, & otherwise. \end{cases}$$

4.1.2 Fractional-Order Alternative Legendre Functions and Their Properties

Alternative Legendre polynomials are introduced by Chelyshov in [68]. Here, we recall definition of the alternative Legendre and fractional-order alternative Legendre functions.

Recently, the set of the alternative Legendre polynomials, $P_m = \{p_{m,i}(t)\}_{i=0}^{m}$, were introduced as follows [31, 70]:

$$p_{m,i}(t) = \sum_{r=0}^{m-i}(-1)^r \binom{m-i}{r}\binom{m+i+r+1}{m-i} t^{i+r}$$
$$= \sum_{r=i}^{m}(-1)^{r-i} \binom{m-i}{r-i}\binom{m+r+1}{m-i} t^{r}, \quad i = 0, 1, \ldots, m, \tag{8}$$

where, m is a fixed non-negative integer. These polynomials are a set of generalized orthogonal polynomials, which satisfy the following relationship [31, 70]:

$$\int_0^1 p_{m,k}(t)p_{m,l}(t)dt = \frac{1}{k+l+1}\delta_{k,l}, \qquad k, l = 0, 1, \ldots, m, \tag{9}$$

in which $\delta_{k,l}$ denotes the Kronecker delta. Also, in contrast to common sets of orthogonal polynomials, every $p_{m,i}(t)$ has degree m [31]. To see this for $m = 3$, these polynomials are given as

$$p_{3,0}(t) = 4 - 30t + 60t^2 - 35t^3, \quad p_{3,1}(t) = 10t - 30t^2 + 21t^3,$$
$$p_{3,2}(t) = 6t^2 - 7t^3, \quad p_{3,3}(t) = t^3.$$

Moreover, Rodrigues's type representation of the alternative Legendre polynomials is as follows [31, 70]:

$$p_{m,i}(t) = \frac{1}{(m-i)!}\frac{1}{t^{i+1}}\frac{d^{m-i}}{dt^{m-i}}(t^{m+i+1}(1-t)^{m-i}), \qquad i = 0, 1, \ldots, m. \tag{10}$$

According to Eq. (9), the following equation is derived [31]:

$$\int_0^1 p_{m,i}(t)dt = \int_0^1 t^m dt = \frac{1}{m+1}, \qquad i = 0, 1, 2, \ldots, m. \tag{11}$$

Given the properties of the alternative Legendre polynomials, some researchers have proposed numerical methods based on these polynomials to solve some equations numerically such as Volterra-Hammerstein integral equations [69], nonlinear

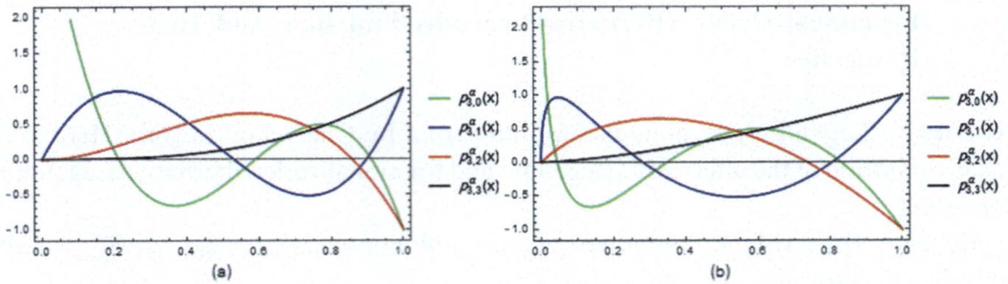

Fig. 6 Graphs of the fractional-order alternative Legendre functions in **a** $\alpha = 1$, **b** $\alpha = \frac{1}{2}$ [31]

Volterra-Fredholm-Hammerstein integral equations [70] and nonlinear fractional differential equations [71].

Next, Rahimkhani and Ordokhani in Ref. [31] introduced a new set of fractional functions based on the alternative Legendre polynomials. They defined the fractional-order alternative Legendre functions (FALFs), $p^{\alpha}_{m,i}(x)$, by a change of variable t to $x^{\alpha}(\alpha > 0)$, on the alternative Legendre polynomials. Then, FALFs are defined as [31]:

$$\begin{aligned} p^{\alpha}_{m,i}(x) &= \sum_{r=0}^{m-i} (-1)^r \binom{m-i}{r} \binom{m+i+r+1}{m-i} x^{(i+r)\alpha} \\ &= \sum_{r=i}^{m} (-1)^{r-i} \binom{m-i}{r-i} \binom{m+r+1}{m-i} x^{r\alpha}, \quad i = 0, 1, \ldots, m. \end{aligned} \tag{12}$$

The following equation is proposed the orthogonality property of FALFs with the weight function $w(x) = x^{\alpha-1}$ [31]:

$$\int_0^1 p^{\alpha}_{m,k}(x) p^{\alpha}_{m,l}(x) x^{\alpha-1} dx = \frac{1}{(k+l+1)\alpha} \delta_{kl}, \qquad k, l = 0, 1, \ldots, m. \tag{13}$$

In the following, for $m = 3$, the FALFs are presented and the graphs of these functions for $\alpha = 1, \frac{1}{2}$, are shown in Fig. 6 [31].

$$p^{\alpha}_{3,0}(x) = 4 - 30x^{\alpha} + 60x^{2\alpha} - 35x^{3\alpha}, \quad p^{\alpha}_{3,1}(x) = 10x^{\alpha} - 30x^{2\alpha} + 21x^{3\alpha},$$

$$p^{\alpha}_{3,2}(x) = 6x^{2\alpha} - 7x^{3\alpha}, \quad p^{\alpha}_{3,3}(x) = x^{3\alpha}.$$

FALFs operational matrix of the fractional integration

Considering the following $(m+1) \times 1$ vector as [31]:

$$\Phi^{\alpha}(x) = [p^{\alpha}_{m,0}(x), p^{\alpha}_{m,1}(x), p^{\alpha}_{m,2}(x), \ldots, p^{\alpha}_{m,m}(x)]^T, \tag{14}$$

and by taking the Riemann-Liouville fractional integration on the above vector, we have

$$I^{\nu}\Phi^{\alpha}(x) = F^{(\nu,\alpha)}\Phi^{\alpha}(x), \tag{15}$$

where, $F^{(\nu,\alpha)}$ is the $(m+1)\times(m+1)$ fractional integration operational matrix of order ν.

To obtain $F^{(\nu,\alpha)}$, using Eq. (12) and properties of Riemann-Liouville fractional integration, the following relation is achieved [31]:

$$\begin{aligned} I^{\nu} p^{\alpha}_{m,i}(x) &= \sum_{r=i}^{m}(-1)^{r-i}\binom{m-i}{r-i}\binom{m+r+1}{m-i} I^{\nu}(x^{r\alpha}) \\ &= \sum_{r=i}^{m}(-1)^{r-i}\binom{m-i}{r-i}\binom{m+r+1}{m-i}\frac{\Gamma(r\alpha+1)}{\Gamma(r\alpha+\nu+1)}x^{r\alpha+\nu} \\ &= \sum_{r=i}^{m}\gamma^{(\nu,\alpha)}_{mi,r}x^{r\alpha+\nu}, \quad i=0,1,\ldots,m, \end{aligned} \tag{16}$$

subject to

$$\gamma^{(\nu,\alpha)}_{mi,r} = (-1)^{r-i}\binom{m-i}{r-i}\binom{m+r+1}{m-i}\frac{\Gamma(r\alpha+1)}{\Gamma(r\alpha+\nu+1)}.$$

Next, $x^{r\alpha+\nu}$ is expanded in terms of the FALFs as

$$x^{r\alpha+\nu} \simeq \sum_{j=0}^{m}\delta^{(\nu,\alpha)}_{r,j}\, p^{\alpha}_{m,j}(x). \tag{17}$$

Using Eq. (17), Eq. (16) can be rewritten as follows [31]:

$$I^{\nu} p^{\alpha}_{m,i}(x) \simeq \sum_{r=i}^{m}\gamma^{(\nu,\alpha)}_{mi,r}\sum_{j=0}^{m}\delta^{(\nu,\alpha)}_{r,j}\, p^{\alpha}_{m,j}(x) = \sum_{j=0}^{m}\left(\sum_{r=i}^{m}\omega^{(\nu,\alpha)}_{mi,j,r}\right)p^{\alpha}_{m,j}(x), \tag{18}$$

where

$$\omega^{(\nu,\alpha)}_{mi,j,r} = \gamma^{(\nu,\alpha)}_{mi,r}\,\delta^{(\nu,\alpha)}_{r,j}.$$

Then, the following relation is obtained [31]:

$$I^{\nu} p^{\alpha}_{m,i}(x) \simeq [\sum_{r=i}^{m}\omega^{(\nu,\alpha)}_{mi,0,r}, \sum_{r=i}^{m}\omega^{(\nu,\alpha)}_{mi,1,r}, \ldots, \sum_{r=i}^{m}\omega^{(\nu,\alpha)}_{mi,m,r}]\Phi^{\alpha}(x), \quad i=0,1,\ldots,m.$$

Therefore, the mentioned operational matrix is achieved as follows [31]:

$$F^{(\nu,\alpha)} = \begin{bmatrix} \sum_{r=0}^{m} \omega_{m0,0,r}^{(\nu,\alpha)} & \sum_{r=0}^{m} \omega_{m0,1,r}^{(\nu,\alpha)} & \cdots & \sum_{r=0}^{m} \omega_{m0,m,r}^{(\nu,\alpha)} \\ \sum_{r=1}^{m} \omega_{m1,0,r}^{(\nu,\alpha)} & \sum_{r=1}^{m} \omega_{m1,1,r}^{(\nu,\alpha)} & \cdots & \sum_{r=1}^{m} \omega_{m1,m,r}^{(\nu,\alpha)} \\ \vdots & \vdots & \cdots & \vdots \\ \sum_{r=m-1}^{m} \omega_{m(m-1),0,r}^{(\nu,\alpha)} & \sum_{r=m-1}^{m} \omega_{m(m-1),1,r}^{(\nu,\alpha)} & \cdots & \sum_{r=m-1}^{m} \omega_{m(m-1),m,r}^{(\nu,\alpha)} \\ \omega_{mm,0,m}^{(\nu,\alpha)} & \omega_{mm,1,m}^{(\nu,\alpha)} & \cdots & \omega_{mm,m,m}^{(\nu,\alpha)} \end{bmatrix}.$$

For example, for $m = 2$, and $\nu = \alpha = \frac{1}{2}$, the operational matrix $F^{(\nu,\alpha)}$ is as follows:

$$F^{(\frac{1}{2},\frac{1}{2})} = \begin{bmatrix} 0.125375 & 0.0940316 & -0.134528 \\ -0.0626877 & 0.376126 & 0.410521 \\ 0.0125375 & -0.0752253 & 0.626877 \end{bmatrix}.$$

4.1.3 Fractional-Order Genocchi Functions and Their Properties

Genocchi numbers and polynomials are introduced by Angelo Genocchi in 1817–1889 [41]. These polynomials are used in some research works such as numerical solution of fractional differential equations [72], etc. The Genocchi polynomials are defined by means of the exponential generating functions as [72, 73]:

$$\frac{2te^{xt}}{e^t+1} = \sum_{m=0}^{\infty} G_m(x)\frac{t^m}{m!}, \qquad (|t| < \pi). \tag{19}$$

Considering above equation, the next relation is derived [41]:

$$\frac{2t}{e^t+1} = \sum_{m=0}^{\infty} g_m \frac{t^m}{m!}, \qquad (|t| < \pi),$$

where, $g_m = G_m(0)$ are Genocchi numbers derived as follows [41]:

$$g_k = 2B_k - 2^{k+1}B_k,$$

and B_k is the well-known Bernoulli number [74]. For example, the first few Genocchi numbers are given as [41]:

$$g_0 = 0, \quad g_1 = 1, \quad g_2 = -1, \quad g_4 = 1, \quad g_6 = -3,$$

in which $g_{2k+1} = 0$, $k = 1, 2, \ldots$. Also, the Genocchi polynomials of degree m is defined as [41]

$$G_m(x) = \sum_{k=0}^{m} \binom{m}{k} g_{m-k} x^k. \tag{20}$$

The first few Genocchi polynomials are given as follows:

$$G_0(x) = 0, \quad G_1(x) = 1, \quad G_2(x) = 2x - 1,$$

$$G_3(x) = 3x^2 - 3x, \quad G_4(x) = 4x^3 - 6x^2 + 1.$$

While many of properties of Genocchi polynomial are close to the corresponding properties of Bernoulli and Euler polynomials, some properties are rather different [75]. The following properties are established for Genocchi polynomials [41]:

- $\int_0^1 G_m(x)G_n(x)dx = \frac{2(-1)^m m!n!}{(m+n)!} g_{m+n}, \quad m, n \geq 1.$
- $\int_a^b G_m(x)dx = \frac{G_{m+1}(b)-G_{m+1}(a)}{m+1}.$
- $\int_0^x G_m(t)dt = \frac{G_{m+1}(x)-g_{m+1}}{m+1}.$

Next, Dehestani et al. defined fractional-order Genocchi functions (FGFs) using Eq. (19) as follows [41]:

$$\frac{2te^{x^{\alpha}t}}{e^t + 1} = \sum_{m=0}^{\infty} G_m^{\alpha}(x) \frac{t^m}{m!}, \quad (|t| < \pi), \quad 0 < \alpha \leq 1, \tag{21}$$

where, FGFs of order $m\alpha$ are denoted by $G_m^{\alpha}(x)$.

The analytic form of these functions is presented by the following formulae [41]:

$$G_m^{\alpha}(x) = \sum_{k=0}^{m} \binom{m}{k} g_{m-k} x^{\alpha k}. \tag{22}$$

In special case, the first few FGFs are proposed as follows:

$$G_0^{\alpha}(x) = 0, \quad G_1^{\alpha}(x) = 1, \quad G_2^{\alpha}(x) = 2x^{\alpha} - 1,$$

$$G_3^{\alpha}(x) = 3x^{2\alpha} - 3x^{\alpha}, \quad G_4^{\alpha}(x) = 4x^{3\alpha} - 6x^{2\alpha} + 1.$$

Moreover, these functions satisfy the following properties [41]:

- $\int_0^1 G_m^{\alpha}(x)G_n^{\alpha}(x)w_{\alpha}(x)dx = \frac{2(-1)^m m!n!}{\alpha(m+n)!} g_{m+n}, \quad m, n \geq 1.$
- $G_m^{\alpha}(x+1) + G_m^{\alpha}(x) = 2mx^{(m-1)\alpha}, \quad m > 1.$

The two-dimensional (2D) fractional-order Genocchi functions are introduced on $\Omega = [0, 1] \times [0, 1]$ as follows [41]:

$$G_{mn}^{\alpha,\beta}(x,t) = G_m^{\alpha}(x)G_n^{\beta}(t), \quad m = 1, 2, \ldots, M, \quad n = 1, 2, \ldots, N, \quad 0 < \alpha, \beta \leq 1. \tag{23}$$

The following property is established for these functions [41]:

$$\int_0^1\int_0^1 w_{\alpha,\beta}(x,t)G_{mn}^{\alpha,\beta}(x,t)G_{ij}^{\alpha,\beta}(x,t)dxdt = \frac{4(-1)^{m+i}m!n!i!j!}{\alpha\beta(m+n)!(i+j)!}g_{m+n}g_{i+j}, \quad m,n,i,j \geq 1.$$

Integral pseudo-operational matrix for FGFs

The pseudo-operational matrix of FGFs is presented using Eq. (22) by Dehestani et al. [41]. For this aim, the following relation is obtained as [41]:

$$\int_0^x G_m^\alpha(\xi)d\xi = \int_0^x\left(\sum_{k=0}^m\binom{m}{k}G_{m-k}\xi^{\alpha k}\right) = \sum_{k=0}^m\binom{m}{k}\frac{G_{m-k}}{\alpha k+1}x^{\alpha k+1}.$$

Now, x^α is approximated by FGFs, and it is substituted in the above equation [41]:

$$\begin{aligned}\int_0^x G_m^\alpha(\xi)d\xi &\simeq x\sum_{k=0}^m\binom{m}{k}\frac{G_{m-k}}{\alpha k+1}\left(\sum_{j=1}^M a_j^k G_j^\alpha(x)\right)\\ &= x\sum_{j=1}^M\left(\sum_{k=0}^m\binom{m}{k}\frac{G_{m-k}a_j^k}{\alpha k+1}\right)G_j^\alpha(x), \end{aligned} \tag{24}$$

where

$$a_j^k = \frac{(2j)!}{2(-1)^j(j!)^2G_{2j}}\sum_{r=1}^j\binom{j}{r}G_{j-r}\frac{1}{(k+r+1)}. \tag{25}$$

Thus, considering above discussion, the following relation is achieved [41]:

$$\begin{aligned}\int_0^x G_m^\alpha(\xi)d\xi &\simeq x\left[\sum_{k=0}^m\binom{m}{k}\frac{G_{m-k}a_1^k}{\alpha k+1}\ \dots\ \sum_{k=0}^m\binom{m}{k}\frac{G_{m-k}a_N^k}{\alpha k+1}\right]G^\alpha(x) \quad (26)\\ &= xP_\alpha^1 G^\alpha(x).\end{aligned}$$

Finally, we get

$$\int_0^x G^\alpha(\xi)d\xi \simeq xP_\alpha^1 G^\alpha(x).$$

For example, for $M = 4$, this matrix is as [41]:

$$P_\alpha^1 = \begin{bmatrix} 1 & 0 & 0 & 0 \\ \frac{-\alpha}{\alpha+1} & \frac{1}{\alpha+1} & 0 & 0 \\ \frac{-3\alpha}{2(2\alpha+1)(\alpha+1)} & \frac{-3\alpha}{2(2\alpha+1)(\alpha+1)} & \frac{1}{2\alpha+1} & 0 \\ \frac{6\alpha^2}{(3\alpha+1)(2\alpha+1)} & \frac{-3\alpha}{(3\alpha+1)(2\alpha+1)} & \frac{-2\alpha}{(3\alpha+1)(2\alpha+1)} & \frac{1}{3\alpha+1} \end{bmatrix}.$$

Also, similar to the above process, the pseudo-operational matrix of integration with respect to t is obtained as [41]:

$$\int_0^t G^{\beta}(\eta)d\eta \simeq t P_{\beta}^{1} G^{\beta}(t). \tag{27}$$

Pseudo-operational matrix of variable-order fractional derivative for FGFs

The variable-order fractional derivative of the vector $G^{\alpha}(x) = \{G_1^{\alpha}(x), G_2^{\alpha}(x), \ldots, G_M^{\alpha}(x)\}$ is obtained as [41]:

$$D^{\nu(x)}\left(x^r G^{\alpha}(x)\right) \simeq x^{r-\nu(x)} \Upsilon_r^{\alpha,\nu} G^{\alpha}(x), \quad r = 0, 1, \ldots, \qquad q-1 < \nu(x) \le q, \tag{28}$$

where $\Upsilon_r^{\alpha,\nu}$ is named the operational matrix of variable order fractional derivative for the FGFs.

To compute the considered operational matrix, the variable-order Caputo's derivative is taken on FGFs, then the following relation is obtained [41]:

$$\begin{aligned} D^{\nu(x)}\left(x^r G_m^{\alpha}(x)\right) &= D^{\nu(x)}\left(\sum_{k=0}^{m} \binom{m}{k} G_{m-k} x^{\alpha k + r}\right) \\ &= \sum_{k=\lceil \frac{r-q}{\alpha} \rceil}^{m} \binom{m}{k} G_{m-k} \frac{\Gamma(\alpha k + r + 1)}{\Gamma(\alpha k + r - \nu(x) + 1)} x^{\alpha k + r - \nu(x)} \\ &= x^{r-\nu(x)} \sum_{k=\lceil \frac{r-q}{\alpha} \rceil}^{m} \binom{m}{k} G_{m-k} \frac{\Gamma(\alpha k + r + 1)}{\Gamma(\alpha k + r - \nu(x) + 1)} x^{\alpha k}. \end{aligned} \tag{29}$$

Now, $x^{\alpha k}$ is approximated by M terms of FGFs, as [41]:

$$x^{\alpha k} \simeq \sum_{j=1}^{M} b_j^k G_j^{\alpha}(x), \tag{30}$$

here, b_j^k is calculated same as a_j^k. By substituting Eq. (30) in Eq. (29), the following equation is achieved [41]:

$$\begin{aligned} D^{\nu(x)}\left(x^r G_m^{\alpha}(x)\right) &\simeq x^{r-\nu(x)} \sum_{k=\lceil \frac{r-q}{\beta} \rceil}^{m} \binom{m}{k} G_{m-k} \frac{\Gamma(\alpha k + r + 1)}{\Gamma(\alpha k + r - \nu(x) + 1)} \left(\sum_{j=1}^{M} b_j^k G_j^{\alpha}(x)\right) \\ &= x^{r-\nu(x)} \sum_{j=1}^{M} \left(\sum_{k=\lceil \frac{r-q}{\alpha} \rceil}^{m} \binom{m}{k} G_{m-k} \frac{\Gamma(\alpha k + r + 1)}{\Gamma(\alpha k + r - \nu(x) + 1)} b_j^k \right) G_j^{\alpha}(x) \end{aligned}$$

$$= x^{r-\nu(x)} \sum_{j=1}^{M} (\sum_{k=\lceil \frac{r-q}{\alpha} \rceil}^{m} \kappa_{j,m,k}^{\alpha,r}) G_j^{\alpha}(x),$$

where

$$\kappa_{j,m,k}^{\alpha,r} = \binom{m}{k} G_{m-k} \frac{\Gamma(\alpha k + r + 1)}{\Gamma(\alpha k + r - \nu(x) + 1)} b_j^k.$$

Therefore, the following relation is derived [41]:

$$D^{\nu(x)}(x^r G_m^{\alpha}(x)) \simeq x^{r-\nu(x)} \left[\sum_{k=\lceil \frac{r-q}{\alpha} \rceil}^{m} \kappa_{1,m,k}^{\alpha,r} \cdots \sum_{k=\lceil \frac{r-q}{\alpha} \rceil}^{m} \kappa_{M,m,k}^{\alpha,r} \right] G^{\alpha}(x).$$

Similar to the above computational process, the pseudo-operational matrix of the variable order fractional derivative of order $\nu(x, t)$ is achieved [41]:

$$D_t^{\nu(x,t)} \left(x^s G^{\beta}(x) \right) \simeq x^{s-\nu(x,t)} \Upsilon_s^{\beta,\nu} G^{\alpha}(x), \quad s = 0, 1, \ldots, \qquad q - 1 < \nu(x, t) \leq q. \tag{31}$$

4.2 *Numerical Methods*

In this subsection, we present a numerical method based on various kinds of functions to obtain approximate solution of Problems (1) and (2).

4.2.1 Description a Numerical Method for Problem 1

Here, a numerical scheme for solving the fractional integro-differential problem given in Eqs. (1)–(2) is proposed.

An arbitrary $f \in L^2[0, 1]$ can be expanded in terms of the fractional-order alternative Legendre functions as follows [31]:

$$f(x) = \sum_{i=0}^{\infty} c_i p_{m,i}^{\alpha}(x), \tag{32}$$

and the coefficients c_i are calculated as [31]:

$$c_i =< f, p_{m,i}^{\alpha} >= (2i + 1)\alpha \int_0^1 f(x) p_{m,i}^{\alpha}(x) x^{\alpha-1} dx, \tag{33}$$

where, $\langle ., . \rangle$ denotes the inner product in $L^2[0, 1]$. If the infinite series in Eq. (32) is truncated, then the following relation is derived [31]:

$$f(x) \simeq \sum_{i=0}^{m} c_i p_{m,i}^{\alpha}(x) = C^T \Phi^{\alpha}(x), \tag{34}$$

where, T indicates transposition, C is $(m+1) \times 1$ vector as

$$C = [c_0, c_1, c_2, \ldots, c_m]^T, \tag{35}$$

and $\Phi^{\alpha}(x)$ is defined in Eq. (14).

For solving the considered problem, first $D^{\nu}u(x)$ is expanded by the FALFs as follows [31]:

$$D^{\nu}u(x) \simeq D^{\nu}u_m(x) = U^T \Phi^{\alpha}(x). \tag{36}$$

Taking I^{ν} on both sides of (36), the following relation is obtained [31]:

$$u(x) \simeq u_m(x) = U^T F^{(\nu,\alpha)} \Phi^{\alpha}(x) + \sum_{k=0}^{n-1} \frac{x^k}{k!} \tau_k, \tag{37}$$

where, $F^{(\nu,\alpha)}$ is FALF operational matrix of fractional integration. Now, Eqs. (36) and (37) are substituted into Problem 1. Then, this problem is rewritten as follows [31]:

$$\begin{aligned} U^T \Phi^{\alpha}(x) = F\Bigg(& x, U^T F^{(\nu,\alpha)} \Phi^{\alpha}(x) + \sum_{k=0}^{n-1} \frac{x^k}{k!} \tau_k, \\ & \int_0^x \kappa_1(x,s) G_1(s, U^T F^{(\nu,\alpha)} \Phi^{\alpha}(s) + \sum_{k=0}^{n-1} \frac{s^k}{k!} \tau_k)) ds, \\ & \int_0^1 \kappa_2(x,s) G_2(s, U^T F^{(\nu,\alpha)} \Phi^{\alpha}(s) + \sum_{k=0}^{n-1} \frac{s^k}{k!} \tau_k)) ds\Bigg) \\ & + Res_m(x), \end{aligned} \tag{38}$$

where, $Res_m(x), x \in [0, 1]$ is residual error. This error made when substituting the approximate solution into the governing equation.
To evaluate the integral in Eq. (38), the Gauss-Legendre numerical integration [77] is utilized. Then, we have [31]:

$$\begin{aligned} U^T \Phi^{\alpha}(x) = F\Bigg(& x, U^T F^{(\nu,\alpha)} \Phi^{\alpha}(x) + \sum_{k=0}^{n-1} \frac{x^k}{k!} \tau_k, \frac{x}{2} \sum_{j=1}^{\tilde{m}} \omega_j \kappa_1(x, \frac{x}{2} + \frac{x}{2}\zeta_j) \\ & G_1(\frac{x}{2} + \frac{x}{2}\zeta_j, U^T F^{(\nu,\alpha)} \Phi^{\alpha}(\frac{x}{2} + \frac{x}{2}\zeta_j) + \sum_{k=0}^{n-1} \frac{(\frac{x}{2} + \frac{x}{2}\zeta_j)^k}{k!} \tau_k), \\ & \frac{1}{2} \sum_{j=1}^{\tilde{m}} \omega_j \kappa_2(x, \frac{1+\zeta_j}{2}) G_2(\frac{1+\zeta_j}{2}, U^T F^{(\nu,\alpha)} \Phi^{\alpha}(\frac{1+\zeta_j}{2}) + \sum_{k=0}^{n-1} \frac{(\frac{1+\zeta_j}{2})^k}{k!} \tau_k)\Bigg) \end{aligned}$$

$$+ E_{\tilde{m}} + Res_m(x), \tag{39}$$

where, ζ_j, $j = 1, 2, \ldots, \tilde{m}$ are zeros of the Legendre polynomial $P_{\tilde{m}}(x)$ and $\omega_j = \frac{-2}{(\tilde{m}+1)P'_{\tilde{m}}(\zeta_j)P_{\tilde{m}+1}(\zeta_j)}$ and $E_{\tilde{m}}$ is the error between the Gauss-Legendre rule and the exact integral given in [77]. Applying the collocation method in above equation at the zeros of the shifted Legendre polynomials $L_{m+1}(x)$; $(x_i, i = 0, 1, \ldots, m)$, the following equation is achieved [31]:

$$\begin{aligned} U^T\Phi^\alpha(x_i) = F\Bigg(& x_i, U^T F^{(\nu,\alpha)}\Phi^\alpha(x_i) + \sum_{k=0}^{n-1}\frac{x_i^k}{k!}\tau_k, \frac{x_i}{2}\sum_{j=1}^{\tilde{m}}\omega_j\kappa_1(x_i, \frac{x_i}{2} + \frac{x_i}{2}\zeta_j) \\ & G_1(\frac{x_i}{2} + \frac{x_i}{2}\zeta_j, U^T F^{(\nu,\alpha)}\Phi^\alpha(\frac{x_i}{2} + \frac{x_i}{2}\zeta_j) \\ & + \sum_{k=0}^{n-1}\frac{(\frac{x_i}{2} + \frac{x_i}{2}\zeta_j)^k}{k!}\tau_k), \frac{1}{2}\sum_{j=1}^{\tilde{m}}\omega_j\kappa_2(x_i, \frac{1+\zeta_j}{2}) \\ & G_2(\frac{1+\zeta_j}{2}, U^T F^{(\nu,\alpha)}\Phi^\alpha(\frac{1+\zeta_j}{2}) + \sum_{k=0}^{n-1}\frac{(\frac{1+\zeta_j}{2})^k}{k!}\tau_k)\Bigg). \end{aligned} \tag{40}$$

For deriving the unknown U, the above nonlinear equations are solved by implementing Newton's iterative method. Next, $u(x)$ is obtained from Eq. (37).

4.2.2 Description a Numerical Method for Problem 2

A continuous and bounded function $f \in L^2(\Omega)$, $\Omega = [0, 1] \times [0, 1]$ can be expanded in terms of the two-dimensional fractional-order Genocchi functions as [41]:

$$f(x,t) \simeq \sum_{m=1}^{M}\sum_{n=1}^{N} c_{mn}G_{mn}^{\alpha,\beta}(x,t) = G^{\alpha T}(x)\widehat{C}G^\beta(t),$$

and the coefficient matrix $\widehat{C}$ is derived as follows [41]:

$$\widehat{C} = D^{-1}\langle\langle f(x,t), G^\alpha(x)\rangle, G^\beta(t)\rangle\tilde{D}^{-1},$$

where

$$D = \langle G^\alpha(x), G^\alpha(x)\rangle, \qquad \tilde{D} = \langle G^\beta(t), G^\beta(t)\rangle.$$

Now, an estimate of the error norm of the best approximation of the two variables smooth function is determined by the 2D FGFs. Suppose that $f(x,t)$ is a sufficiently smooth function on the interval Ω and $Y = span\{G_{mn}^{\alpha,\beta}(x,t); m = 1, 2, \ldots, M,\ n = 1, 2, \ldots, N\}$.

If f^*_{MN} is the best approximation of f out of Y then for any $P^{\alpha,\beta}_{MN} f \in Y$, the following relation is established [41]:

$$\|f - f^*_{MN}\|_2 \leq \|f - P^{\alpha,\beta}_{MN} f\|_2, \tag{41}$$

in which, $P^{\alpha,\beta}_{MN} f$ is the interpolating polynomial to f at points (x_m, t_n), where x_m, $m = 1, 2, \ldots, M$ and t_n, $n = 1, 2, \ldots, N$ are the roots of the M-degree and N-degree of the shifted Chebyshev polynomial, respectively. Then, we get [41]:

$$f(x,t) - P^{\alpha,\beta}_{MN} f(x,t) = \frac{D_x^{M\beta} f(\xi, t)}{\Gamma(M\alpha + 1)} \prod_{i=0}^{M-1} (x - x_i)^{\alpha} + \frac{D_t^{N\beta} f(x, \eta)}{\Gamma(N\beta + 1)} \prod_{j=0}^{N-1} (t - t_j)^{\beta}$$
$$+ \frac{D_{xt}^{M\alpha+N\beta} F(\xi', \eta')}{\Gamma(M\alpha + 1)\Gamma(N\beta + 1)} \prod_{i=0}^{M-1} (x - x_i)^{\alpha} \prod_{j=0}^{N-1} (t - t_j)^{\beta},$$

where $\xi, \eta, \xi', \eta' \in [0, 1]$. Therefore, the following relation is achieved [41]:

$$|f(x,t) - P^{\alpha,\beta}_{MN} f(x,t)| \leq \frac{\max_{(x,t)\in\Omega} \left|D_x^{M\alpha} f(\xi, t)\right|}{\Gamma(M\alpha + 1)} \prod_{i=0}^{M-1} (x - x_i)^{\alpha}$$
$$+ \frac{\max_{(x,t)\in\Omega} \left|D_t^{N\beta} f(x, \eta)\right|}{\Gamma(N\beta + 1)} \prod_{j=0}^{N-1} (t - t_j)^{\beta}$$
$$+ \frac{\max_{(x,t)\in\Omega} \left|D_{xt}^{M\alpha+N\beta} f(\xi', \eta')\right|}{\Gamma(M\alpha + 1)\Gamma(N\beta + 1)} \prod_{i=0}^{M-1} (x - x_i)^{\alpha} \prod_{j=0}^{N-1} (t - t_j)^{\beta}.$$

Suppose that there are real numbers C_1, C_2 and C_3, as [41]:

$$C_1 = \max_{(x,t)\in\Omega} \left|D_x^{M} f(x,t)\right|, \quad C_2 = \max_{(x,t)\in\Omega} \left|D_t^{(N+1)\beta} f(x,t)\right|, \quad C_3 = \max_{(x,t)\in\Omega} \left|D_{xt}^{M+(N+1)\beta} f(x,t)\right|.$$

Then, according to the estimates for Chebyshev interpolation nodes on the interval Ω [78], we obtain [41]:

$$|f(x,t) - P^{\alpha,\beta}_{MN} f(x,t)| \leq \frac{C_1}{2^{(2M-1)\alpha}\Gamma(M\alpha + 1)} + \frac{C_2}{2^{(2N-1)\beta}\Gamma(N\beta + 1)}$$
$$+ \frac{C_3}{2^{(2M-1)\alpha+(2M-1)\beta}\Gamma(M\alpha + 1)\Gamma(N\beta + 1)}.$$

Therefore, using Eq. (41), the following relation is derived [41]:

$$\|f - P_{MN}^{\alpha,\beta} f\|_{L^2_{w_{\alpha,\beta}}(\Omega)} \leq \frac{C_1}{\alpha\beta 2^{(2M-1)\alpha}\Gamma(M\alpha+1)} + \frac{C_2}{\alpha\beta 2^{(2N-1)\beta}\Gamma(N\beta+1)} + \frac{C_3}{\alpha\beta 2^{(2M-1)\alpha+(2M-1)\beta}\Gamma(M\alpha+1)\Gamma(N\beta+1)}. \tag{42}$$

Considering the proposed discussion, it can be seen that with increasing the 2D fractional-order Genocchi functions terms the approximate function converges to exact solution [41].

Problem 2.1:

Here, a numerical technique is proposed for solving the variable-order fractional nonlinear Fredholm-Volterra integro-differential equations in Eqs. (3)–(4). For this purpose, consider [41]:

$$u^{(n)}(x) \simeq U^T G^\alpha(x). \tag{43}$$

By integrating above equation with respect to x and utilizing the pseudo-operational matrix of integration of FGFs, the following relation is derived [41]:

$$\begin{aligned} u^{(n-1)}(x) &\simeq xU^T P_\alpha^1 G^\alpha(x) + \delta_{n-1}, \\ u^{(n-2)}(x) &\simeq x^2 U^T P_\alpha^2 G^\alpha(x) + x\delta_{n-1} + \delta_{n-2}, \\ &\vdots \\ u(x) &\simeq x^n U^T P_\alpha^n G^\alpha(x) + \sum_{i=0}^{n-1} \frac{x^i}{i!}\delta_i = u_M(x), \end{aligned} \tag{44}$$

where

$$\underbrace{\int_0^x \int_0^x \cdots \int_0^x}_{r\ times} G^\alpha(\xi) d\xi d\xi \cdots d\xi \simeq x^r P_\alpha^r G^\alpha(x), \quad r = 1, 2, \ldots, n,$$

and similarly to P_α^1, P_α^r is calculated. Using the above process, we have [41]:

$$\begin{aligned} \psi_f\left(x, x^n U^T P_\alpha^n G^\alpha(x) + \sum_{i=0}^{n-1} \frac{x^i}{i!}\delta_i\right) &\simeq A^T G^\alpha(x), \\ \psi_v\left(x, x^n U^T P_\alpha^n G^\alpha(x) + \sum_{i=0}^{n-1} \frac{x^i}{i!}\delta_i\right) &\simeq B^T G^\alpha(x). \end{aligned} \tag{45}$$

Moreover, using Eq. (44), the following relation is established [41]:

$$D^{\nu(x)}u(x) \simeq D^{\nu(x)}\left(x^n U^T P_\alpha^n G^\alpha(x) + \sum_{i=0}^{n-1}\frac{x^i}{i!}\delta_i\right) \tag{46}$$
$$\simeq x^{n-\nu(x)}U^T P_\alpha^n \Upsilon_n^{\alpha,\nu} G^\alpha(x) + \sum_{i=\lceil n\rceil}^{n-1}\frac{\Gamma(i+1)}{i!\Gamma(i+1-\nu(x))}\delta_i x^{i-\nu(t)}.$$

Then, the kernel functions is approximated using the terms of FGFs as follows [41]:

$$k_f(x,t) \simeq G^{\alpha T}(x)K_f G^\alpha(t), \qquad k_v(x,t) \simeq G^{\alpha T}(x)K_v G^\alpha(t), \tag{47}$$

where

$$K_f = D^{-1}\langle\langle k_f(x,t), G^\alpha(x)\rangle, G^\alpha(t)\rangle D^{-1},$$
$$K_v = D^{-1}\langle\langle k_v(x,t), G^\alpha(x)\rangle, G^\alpha(t)\rangle D^{-1}.$$

Substituting the above equations in Eq. (3), this equation is rewritten as follows [41]:

$$x^{n-\nu(x)}U^T P_\alpha^n \Upsilon_n^{\alpha,\nu} G^\alpha(x) + \sum_{i=\lceil n\rceil}^{n-1}\frac{\Gamma(i+1)}{i!\Gamma(i+1-\nu(x))}\delta_i x^{i-\nu(t)} \tag{48}$$
$$+\sum_{k=0}^{n} p_k(x)\left(x^{n-k}U^T P_\alpha^{n-k} G^\alpha(x) + \sum_{i=0}^{k-1}\frac{x^i}{i!}\delta_i\right)$$
$$= h(x) + \lambda_f G^{\alpha T}(x)K_f H_f A + \lambda_v G^{\alpha T}(x)K_v H_v(x)B,$$

where

$$H_f = \int_0^1 G^\alpha(x)G^{\alpha T}(x)dx, \qquad H_f = [h_{mn}^f], \qquad m,n = 1,2,\ldots,M,$$

$$h_{mn}^f = \sum_{k=0}^{m}\sum_{j=0}^{n}\binom{m}{k}\binom{n}{j}\frac{g_{m-k}g_{n-j}}{\alpha(k+j)+1},$$

$$H_v(x) = \int_0^x G^\alpha(\xi)G^{\alpha T}(\xi)d\xi, \qquad H_v(x) = [h_{mn}^v(x)], \qquad m,n = 1,2,\ldots,M,$$

$$h_{mn}^v(x) = \sum_{k=0}^{m}\sum_{j=0}^{n}\binom{m}{k}\binom{n}{j}\frac{g_{m-k}g_{n-j}x^{\alpha(k+j)+1}}{\alpha(k+j)+1}.$$

Finally, Eqs. (45), (48) are collocated at Newton-Cotes nods [79]. The achieved system of algebraic equations for the unknown vector U can be solved using Newton's iterative method.

Problem 2.2:

Now, we consider the variable-order fractional nonlinear Volterra partial integro-differential equations in Eq. (5).

For deriving a numerical solution for this problem, let [41]:

$$\frac{\partial^3 u}{\partial x^2 \partial t}(x,t) \simeq G^{\alpha T}(x) U G^{\beta}(t), \tag{49}$$

in which

$$U = [u_{mn}], \qquad m = 1, 2, \ldots, M, \qquad n = 1, 2, , \ldots, N.$$

By integrating from Eq. (49) with respect to t, the following relation is obtained [41]:

$$\frac{\partial^2 u}{\partial x^2}(x,t) \simeq t G^{\alpha T}(x) U P_{\beta} G^{\beta}(t) + f''(x). \tag{50}$$

Also, by integrating from Eq. (50) with respect to x of order 2, we get [41]:

$$\frac{\partial u}{\partial x}(x,t) \simeq x t G^{\alpha T}(x) P_{\alpha}^{T} U P_{\beta} G^{\beta}(t) + f'(x) - f'(0) + \frac{\partial u}{\partial x}(0,t), \tag{51}$$

and

$$u(x,t) \simeq x^2 t G^{\alpha T}(x) (P_{\alpha}^{T})^2 U P_{\beta} G^{\beta}(t) + f(x) - f(0) - x f'(0) + x \frac{\partial u}{\partial x}(0,t) + \varphi_0(t). \tag{52}$$

For approximating $\frac{\partial u}{\partial x}(0,t)$ in Eqs. (51) and (52), by integrating from Eq. (51) with respect to x from 0 to 1, we have [41]:

$$\frac{\partial u}{\partial x}(0,t) \simeq \varphi_1(t) - \varphi_0(t) - t S^T P_{\alpha}^{T} U P_{\beta} G^{\beta}(t) - f(1) + f(0) + f'(0) = \frac{\partial u_{MN}}{\partial x}(0,t), \tag{53}$$

where

$$S = \int_0^1 x G^{\alpha T}(x) dx, \quad S = [s_{m1}], \quad s_{m1} = \sum_{k=0}^{m} \binom{m}{k} \frac{G_{m-k}}{\alpha k + 2}, \quad m = 1, 2, \ldots, M.$$

By substituting Eq. (53) into Eqs. (52) and (51), the following equations are established [41]:

$$\begin{aligned} \frac{\partial u}{\partial x}(x,t) \simeq\; & x t G^{\alpha T}(x) P_{\alpha}^{T} U P_{\beta} G^{\beta}(t) + f'(x) - f'(0) + \varphi_1(t) - \varphi_0(t) \\ & - t S^T P_{\alpha}^{T} U P_{\beta} G^{\beta}(t) - f(1) + f(0) + f'(0), \end{aligned} \tag{54}$$

and

$$
\begin{aligned}
u(x,t) &\simeq x^2 t G^{\alpha T}(x)(P_\alpha^T)^2 U P_\beta G^\beta(t) + f(x) - f(0) - xf'(0) \\
&+ x(\varphi_1(t) - \varphi_0(t) - tS^T P_\alpha^T U P_\beta G^\beta(t) - f(1) + f(0) + f'(0)) + \varphi_0(t) \\
&= u_{MN}(x,t).
\end{aligned} \tag{55}
$$

Using the properties of variable-order fractional Caputo operator and Eq. (55), the variable-order fractional derivatives is computed as follows [41]:

$$
\begin{aligned}
D_t^{\nu(x,t)} u(x,t) &\simeq x^2 t^{1-\nu(x,t)} G^{\alpha T}(x)(P_\alpha^T)^2 U P_\beta \Upsilon_1^{\beta,\nu} G^\beta(t) \\
&+ x(D_t^{\nu(x,t)}[\varphi_1(t) - \varphi_0(t)] - t^{1-\nu(x,t)} S^T P_\alpha^T U P_\beta \Upsilon_1^{\beta,\nu} G^\beta(t)) \\
&+ D_t^{\nu(x,t)} \varphi_0(t).
\end{aligned} \tag{56}
$$

In the following, considering above discussion, the differential part of the Eq. (5) is achieved as [41]:

$$
\begin{aligned}
D_t^{\nu(x,t)} u(x,t) &+ \sum_{k=0}^{2} Q_k(x,t) \frac{\partial^k u}{\partial x^k}(x,t) \\
&\simeq x^2 t^{1-\nu(x,t)} G^{\alpha T}(x)(P_\alpha^T)^2 U P_\beta \Upsilon_1^{\beta,\nu} G^\beta(t) \\
&+ x(D_t^{\nu(x,t)}[\varphi_1(t) - \varphi_0(t)] - t^{1-\nu(x,t)} S^T P_\alpha^T U P_\beta \Upsilon_1^{\beta,\nu} G^\beta(t)) + D_t^{\nu(x,t)} \varphi_0(t) \\
&+ Q_0(x,t)(x^2 t G^{\alpha T}(x)(P_\alpha^T)^2 U P_\beta G^\beta(t) + f(x) - f(0) - xf'(0)) \\
&+ x Q_0(x,t)(\varphi_1(t) - \varphi_0(t) - tS^T P_\alpha^T U P_\beta G^\beta(t) - f(1) + f(0) + f'(0)) \\
&+ Q_0(x,t)\varphi_0(t) \\
&+ Q_1(x,t)(xtG^{\alpha T}(x) P_\alpha^T U P_\beta G^\beta(t) + f'(x) - f'(0)) \\
&+ Q_1(x,t)(\varphi_1(t) - \varphi_0(t) - tS^T P_\alpha^T U P_\beta G^\beta(t) - f(1) + f(0) + f'(0)) \\
&+ Q_2(x,t)\left(tG^{\alpha T}(x) U P_\beta G^\beta(t) + f''(x)\right) = \Delta(x,t).
\end{aligned} \tag{57}
$$

Now, let [41]:

$$
\psi(x,t,u(x,t)) \simeq G^{\alpha T}(x) A G^\beta(t), \tag{58}
$$

subject to A is considered the same as matrix U. Then, the approximation of $u(x,t)$ is substituted in Eq. (55). Moreover, the separable kernel $k(x,t,\eta)$ is proposed as [41]:

$$
k(x,t,\eta) = \sum_{i=1}^{l} X_i(x) W_i(t) Z_i(\eta). \tag{59}
$$

The integral part of the Eq. (5) is calculated as [41]:

$$
\begin{aligned}
\int_0^t k(x,t,\eta)\psi(x,t,u(x,t))d\eta &\simeq \sum_{i=1}^{l} X_i(x) W_i(t) \left(\int_0^t Z_i(\eta) G^\beta(\eta) d\eta\right) A^T G^\alpha(x) \\
&= \Psi(x,t).
\end{aligned} \tag{60}
$$

Hence, by considering Eq. (5), Eq. (57) and above equation, the following equation is derived [41]:

$$\Delta(x,t) = h(x,t) + \lambda\Psi(x,t). \tag{61}$$

Consequently, by replacing the Newton-Cotes points [79] as nodes of collocation into Eqs. (57) and (61), the system of equations with an unknown vectors U and A is derived. Thus, by substituting the vector U in Eq. (55), the approximate solution of the proposed technique is achieved.

4.3 Convergence Analysis and Error Estimation

In this subsection, the convergence of the proposed method for solving considered problems is investigated.

4.3.1 Convergence Analysis of the Proposed Method Based on FALFs

Theorem 1 *Suppose that u and u_m be the exact and approximate solution of Eqs. (1)–(2), respectively. Moreover, let [31]:*

(i) $\|u\|_2 < \infty$,
(ii) $K_1 = Max|\kappa_1(x,s)|$, $K_2 = Max|\kappa_2(x,s)|$; $(x,s) \in [0,1] \times [0,1]$,
(iii) F, G_1 *and* G_2 *satisfy Lipschitz conditions with the Lipschitz constants* η, η_1 *and* η_2, *respectively,*
(iiii) $\Gamma(\nu) - \eta - K_1\eta\eta_1 - K_2\eta\eta_2 \neq 0$,
then $\|u - u_m\|_2 \longrightarrow 0$.

Proof Let the error function is denoted by e_m [31]:

$$e_m(x) = u(x) - u_m(x),$$

Considering Problem 1, the following equation is established [31]:

$$\begin{aligned} D^{\nu} e_m(x) = {} & F\left(x, u(x), \int_0^x \kappa_1(x,s)G_1(s,u(s))ds, \int_0^1 \kappa_2(x,s)G_2(s,u(s))ds\right) \\ & - F\left(x, u_m(x), \int_0^x \kappa_1(x,s)G_1(s,u_m(s))ds, \int_0^1 \kappa_2(x,s)G_2(s,u_m(s))ds\right) \\ & - E_{\tilde{m}} - Res_m(x). \end{aligned} \tag{62}$$

Considering the definition of the Riemann-Liouville fractional integral and the Caputo derivative in Eqs. (6)–(7), the above equation can be rewritten as [31]:

$$e_m(x) = I^{\nu}\Bigg(F\Big(x, u(x), \int_0^x \kappa_1(x,s)G_1(s,u(s))ds, \int_0^1 \kappa_2(x,s)G_2(s,u(s))ds\Big) \\ - F\Big(x, u_m(x), \int_0^x \kappa_1(x,s)G_1(s,u_m(s))ds, \int_0^1 \kappa_2(x,s)G_2(s,u_m(s))\Big)ds\Bigg) \\ - I^{\nu}(E_{\tilde{m}}) - I^{\nu}(Res_m(x)). \tag{63}$$

Then, we have [31]:

$$e_m(x) = \Lambda_1(x) - \Lambda_2(x) - \Lambda_3(x), \tag{64}$$

where

$$\Lambda_1(x) = I^{\nu}\Bigg(F\Big(x, u(x), \int_0^x \kappa_1(x,s)G_1(s,u(s))ds, \\ \int_0^1 \kappa_2(x,s)G_2(s,u(s))ds\Big) \\ - F\Big(x, u_m(x), \int_0^x \kappa_1(x,s)G_1(s,u_m(s))ds, \\ \int_0^1 \kappa_2(x,s)G_2(s,u_m(s))ds\Big)\Bigg), \tag{65}$$

$$\Lambda_2(x) = I^{\nu}(E_{\tilde{m}}), \tag{66}$$

$$\Lambda_3(x) = I^{\nu}(Res_m(x)). \tag{67}$$

Now, for estimation of Λ_1, the following relation is derived [31]:

$$|\Lambda_1(x)| = \Bigg|\frac{1}{\Gamma(\nu)}\int_0^x (x-t)^{\nu-1}\Bigg[\Bigg(F\Big(t, u(t), \int_0^t \kappa_1(t,s)G_1(s,u(s))ds, \int_0^1 \kappa_2(t,s)G_2(s,u(s))ds\Big) \\ - F\Big(t, u_m(t), \int_0^t \kappa_1(t,s)G_1(s,u_m(s))ds, \int_0^1 \kappa_2(t,s)G_2(s,u_m(s))ds\Big)\Bigg)\Bigg]dt\Bigg| \\ \leq \frac{1}{\Gamma(\nu)}\int_0^x |x-t|^{\nu-1}\Bigg|\Bigg(F\Big(t, u(t), \int_0^t \kappa_1(t,s)G_1(s,u(s))ds, \int_0^1 \kappa_2(t,s)G_2(s,u(s))ds\Big) \\ - F\Big(t, u_m(t), \int_0^t \kappa_1(t,s)G_1(s,u_m(s))ds, \int_0^1 \kappa_2(t,s)G_2(s,u_m(s))ds\Big)\Bigg)\Bigg|dt. \tag{68}$$

□

Since $|x-t| \leq 1$ and F, G_1 and G_2 satisfy Lipschitz conditions, we have [31]:

$$|\Lambda_1(x)| \leq \frac{1}{\Gamma(\nu)}\int_0^x (\eta + K_1\eta\eta_1 + K_2\eta\eta_2)|u(t) - u_m(t)|dt. \tag{69}$$

Since $0 \leq t \leq x \leq 1$, then [31]:

$$|\Lambda_1(x)| \leq \frac{1}{\Gamma(\nu)} \int_0^1 (\eta + K_1\eta\eta_1 + K_2\eta\eta_2)|e_m(t)|dt. \tag{70}$$

So that

$$\|\Lambda_1\|_2 \leq \frac{(\eta + K_1\eta\eta_1 + K_2\eta\eta_2)}{\Gamma(\nu)} \|e_m\|_2. \tag{71}$$

Also, for Λ_2, Λ_3, the following relations are established [31]:

$$\|\Lambda_2\|_2 \leq \frac{1}{\Gamma(\nu)} \|E_{\tilde{m}}\|_2, \tag{72}$$

$$\|\Lambda_3\|_2 \leq \frac{1}{\Gamma(\nu)} \|Res_m\|_2. \tag{73}$$

From (64), the following relation is derived [31]:

$$\|e_m\|_2 \leq \|\Lambda_1\|_2 + \|\Lambda_2\|_2 + \|\Lambda_3\|_2.$$

So that

$$\|e_m\|_2 \leq \frac{(\eta + K_1\eta\eta_1 + K_2\eta\eta_2)}{\Gamma(\nu)} \|e_m\|_2 + \frac{1}{\Gamma(\nu)} \|E_{\tilde{m}}\|_2 + \frac{1}{\Gamma(\nu)} \|Res_m\|_2. \tag{74}$$

Finally, we get [31]:

$$\|e_m\|_2 \leq \frac{\|E_{\tilde{m}}\|_2 + \|Res_m\|_2}{\Gamma(\nu) - \eta - K_1\eta\eta_1 - K_2\eta\eta_2}. \tag{75}$$

By choosing $\tilde{m}$ sufficiently large, then, $E_{\tilde{m}}$ tends to 0 [77]. Therefore, when Res_m tends to 0 then $\|e_m\|_2 = \|u - u_m\|_2 \longrightarrow 0$ [31]. The numerical results show that Res_m tends to 0 [31].

4.3.2 Convergence Analysis and Error Estimate of the Proposed Method Based on FGFs

Lemma 1 *Let* $g \in L^2[0, 1]$ *is approximated by* g_M *as [41]*

$$g(x) \simeq g_M(x) = \sum_{m=1}^{M} a_m G_m^{\alpha}(x),$$

and consider

$$L_M(g) = \int_0^1 [g(x) - g_M(x)]^2 dx,$$

so that $\lim_{M\to\infty} L_M(g) = 0$.

Proof This lemma is proved similar to process proposed in [66, 76].
The error vector for the pseudo-operational matrix of integration of order n is defined as follows [41]:

$$E_x^{\alpha,n} = \int_0^x \int_0^x \dots \int_0^x G^\alpha(\xi)d\xi d\xi \dots d\xi - x^n P_\alpha^n G^\alpha(x),$$

$$E_x^{\alpha,n} = \left[e_{m1}^{\alpha,n}\right], \quad m = 1, 2, \dots, M.$$

Due to the presented pseudo-operational matrices and Lemma 1, each component of $E_x^{\alpha,n}$ is derived as follows [41]:

$$\begin{aligned}\|e_{m1}^{\alpha,n}\|_2 &= \| \int_0^x \int_0^x \dots \int_0^x G_m^\alpha(\xi)d\xi d\xi \dots d\xi - x^n P_\alpha^n G^\alpha(x)\|_2 \\ &= |\sum_{k=0}^{m} \binom{m}{k} \frac{x^n G_{m-k}}{(\alpha k+1)(\alpha k+2)\dots(\alpha k+n)}| \|x^{\alpha k} - \sum_{j=1}^{M} a_j^k G_j^\alpha(x)\|_2 \\ &\leq \sum_{k=0}^{m} \binom{m}{k} \frac{G_{m-k}}{(\alpha k+1)(\alpha k+2)\dots(\alpha k+n)} \left(\frac{G(x^{\alpha k}, G_1^\alpha(x), \dots, G_M^\alpha(x))}{G(G_1^\alpha(x), G_2^\alpha(x), \dots, G_M^\alpha(x))}\right)^{\frac{1}{2}}.\end{aligned} \tag{76}$$

Consequently, if the terms of the FGFs bases increase, the error vectors $E_x^{\alpha,n}$ tend to zero [41].

Theorem 2 *Let $u(x)$ is the exact solution of variable-order fractional integro-differential equations in Eq. (3). Then, the upper bound of error for the approximate solution is determined by [41]*

$$\|u - u_M\|_2 \leq \|U^T\|_2 \|E_x^{\alpha,n}\|_2. \tag{77}$$

Proof Using Eqs. (44) and (76), the following equation is achieved [41]:

$$\begin{aligned}\|u - u_M\|_2 &= \| \left(U^T[\int_0^x \int_0^x \dots \int_0^x G^\alpha(\xi)d\xi d\xi \dots d\xi] + \sum_{i=0}^{n-1} \frac{x^i}{i!}\delta_i\right) \\ &- \left(x^n U^T P_\alpha^n G^\alpha(x) + \sum_{i=0}^{n-1} \frac{x^i}{i!}\delta_i\right) \|_2 \\ &= \|U^T \left(\int_0^x \int_0^x \dots \int_0^x G^\alpha(\xi)d\xi d\xi \dots d\xi - x^n P_\alpha^n G^\alpha(x)\right) \|_2 \\ &= \|U^T\|_2 \| \int_0^x \int_0^x \dots \int_0^x G^\alpha(\xi)d\xi d\xi \dots d\xi - x^n P_\alpha^n G^\alpha(x)\|_2 \\ &\leq \|U^T\|_2 \|E_x^{\alpha,n}\|_2.\end{aligned}$$

According to the behavior of the error vector $E_x^{\alpha,n}$, as M approaches ∞, the error of the proposed scheme tends to zero.

Theorem 3 *Suppose that $u(x,t)$ is the exact solution of variable-order fractional partial integro-differential equations in Eq. (5). Then, the upper bound of error for the approximate solution is calculated as follows [41]:*

$$\begin{aligned}\|u-u_{MN}\|_2 \le \| &\int_0^x\int_0^x G^{\alpha T}(\xi)d\xi d\xi\|_2\|U\|_2\|E_t^{\beta,1}\|_2 \\ &+\|E_x^{\alpha,2}\|_2\|U\|_2\|P_\beta\|_2\|tG^\beta(t)\|_2 \\ &+\|\int_0^1\int_0^x G^{\alpha T}(\xi)d\xi dx\|_2\|U\|_2\|E_t^{\beta,1}\|_2 \\ &+\|E_x^{\alpha,1}\|_2\|U\|_2\|P_\beta\|_2\|tG^\beta(t)\|_2,\end{aligned} \tag{78}$$

where, $E_t^{\beta,n}$ is achieved same as $E_x^{\alpha,n}$.

Proof Using Eqs. (53) and (55), the following relations is derived [41]:

$$\begin{aligned}u(x,t)=&\left[\int_0^x\int_0^x G^{\alpha T}(\xi)d\xi d\xi\right]U\left[\int_0^t G^\beta(\eta)d\eta\right]\\ &+f(x)-f(0)-xf'(0)+x\left(\frac{\partial u}{\partial x}(x,0)\right)+\varphi_0(t),\end{aligned}$$

and

$$\begin{aligned}\frac{\partial u}{\partial x}(x,0)=\varphi_1(t)-\varphi_0(t)-&\left[\int_0^1\int_0^x G^{\alpha T}(\xi)d\xi dx\right]U\left[\int_0^t G^\beta(\eta)d\eta\right]\\ &-f(1)+f(0)+f'(0).\end{aligned}$$

So that [41]:

$$\begin{aligned}&\|u-u_{MN}\|_2\\ &=\|\left[\int_0^x\int_0^x G^{\alpha T}(\xi)d\xi d\xi\right]U\left[\int_0^t G^\beta(\eta)d\eta\right]-x^2tG^{\alpha T}(x)(P_\alpha^T)^2UP_\beta G^\beta(t)\|_2\\ &+\|\left[\int_0^1\int_0^x G^{\alpha T}(\xi)d\xi dx\right]U\left[\int_0^t G^\beta(\eta)d\eta\right]-t\left[\int_0^1 xG^{\alpha T}(x)dx\right]P_\alpha^T UP_\beta G^\beta(t)\|_2\\ &\le\|\int_0^x\int_0^x G^{\alpha T}(\xi)d\xi d\xi\|_2\|U\|_2\|\int_0^t G^\beta(\eta)d\eta-tP_\beta G^\beta(t)\|_2\\ &+\|\int_0^x\int_0^x G^{\alpha T}(\xi)d\xi d\xi-x^2G^{\alpha T}(x)(P_\alpha^T)^2\|_2\|P_\beta\|_2\|tG^\beta(t)\|_2\\ &+\|\int_0^1\int_0^x G^{\alpha T}(\xi)d\xi dx\|_2\|U\|_2\|\int_0^t G^\beta(\eta)d\eta-tP_\beta G^\beta(t)\|_2\\ &+\|\int_0^1(\int_0^x G^{\alpha T}(\xi)d\xi-xG^{\alpha T}(x)P_\alpha^T)dx\|_2\|U\|_2\|P_\beta\|_2\|tG^\beta(t)\|_2\\ &=\|\int_0^x\int_0^x G^{\alpha T}(\xi)d\xi d\xi\|_2\|U\|_2\|E_t^{\beta,1}\|_2+\left\|E_x^{\alpha,2}\right\|_2\|U\|_2\left\|P_\beta\right\|_2\left\|tG^\beta(t)\right\|_2\end{aligned}$$

$$+\left\|\int_0^1\int_0^x G^{\alpha T}(\xi)d\xi dx\right\|_2 \|U\|_2 \left\|E_t^{\beta,1}\right\|_2 + \left\|E_x^{\alpha,1}\right\|_2 \|U\|_2 \left\|P_\beta\right\|_2 \left\|tG^\beta(t)\right\|_2.$$

Then, the proof is completed. □

The achieved results from the Lemma 1 and Eq. (77) display that by increasing the number of FGFs, the upper bound of error tends to zero [41].

4.4 Numerical Results

To show the effectiveness of proposed methods, several examples are presented for each numerical methods.

4.4.1 Numerical Results Obtained from Numerical Method Based on FALFs

Here, some examples are considered to demonstrate the accuracy and efficiency of the presented method based on FALFs to obtain the numerical solution of Problem 1.

Example 1 Consider the following nonlinear fractional Volterra integro-differential equation as [31]:

$$D^{\frac{5}{3}}u(x) = f(x) - xu(x) + \frac{1}{\Gamma(6.5)}\int_0^x (x-s)^{5.5}u^3(s)ds, \quad 0 \le x \le 1,$$

with the initial conditions $u(0) = u'(0) = 0$ and

$$f(x) = \Gamma(\frac{8}{3}) + x^{\frac{8}{3}} - \frac{0.000252451x^{11.5}}{\Gamma(6.5)}.$$

The exact solution of this problem is $u(x) = x^{\frac{5}{3}}$.

The absolute errors between the exact and approximate solutions, for $m = 5$ and various values of α are shown in Table 2. Considering the reported results in Table 2, we can see that the best case of α for this problem is $\alpha = \frac{5}{3}$. Moreover, CPU time for various values of α are listed in this table [31].

Example 2 Consider the following nonlinear fractional Fredholm integro-differential equation as [31]:

$$D^{\frac{5}{3}}u(x) - \int_0^1 (x+s)^2u^3(s)ds = f(x), \quad 0 \le x \le 1,$$

Table 2 The absolute errors and CPU time with $m = 5$ and different values of α in Ref. [31] for Example 1

x	$\alpha = \frac{1}{4}$	$\alpha = \frac{1}{3}$	$\alpha = \frac{1}{2}$	$\alpha = 1$	$\alpha = \frac{5}{3}$
0	2.56×10^{-3}	1.38×10^{-15}	8.25×10^{-5}	1.05×10^{-3}	3.23×10^{-17}
0.1	1.91×10^{-4}	1.78×10^{-14}	1.26×10^{-5}	1.89×10^{-4}	1.04×10^{-17}
0.2	1.19×10^{-3}	2.75×10^{-14}	1.44×10^{-5}	1.55×10^{-4}	1.39×10^{-17}
0.3	1.78×10^{-3}	3.55×10^{-14}	8.99×10^{-6}	8.65×10^{-5}	5.55×10^{-17}
0.4	1.59×10^{-3}	4.23×10^{-14}	8.93×10^{-6}	8.83×10^{-5}	0
0.5	1.02×10^{-3}	4.80×10^{-14}	1.94×10^{-5}	1.10×10^{-4}	1.11×10^{-16}
0.6	4.51×10^{-4}	5.23×10^{-14}	1.47×10^{-5}	1.54×10^{-5}	1.67×10^{-16}
0.7	2.45×10^{-4}	5.54×10^{-14}	1.25×10^{-6}	1.11×10^{-4}	2.22×10^{-16}
0.8	6.97×10^{-4}	5.71×10^{-14}	1.52×10^{-5}	2.63×10^{-5}	5.55×10^{-16}
0.9	2.06×10^{-3}	5.74×10^{-14}	6.82×10^{-6}	1.26×10^{-4}	3.55×10^{-15}
1	4.55×10^{-3}	5.60×10^{-14}	5.00×10^{-5}	2.91×10^{-4}	4.44×10^{-16}
CPU time	0.016	0.016	0.015	0.0001	0.015

with the initial conditions $u(0) = u'(0) = 0$ and

$$f(x) = \frac{6}{\Gamma(\frac{1}{3})} x^{\frac{1}{3}} - \frac{x^2}{7} - \frac{x}{4} - \frac{1}{9}.$$

The exact solution of this equation is $u(x) = x^2$.

The absolute errors achieved by the proposed method based on FALFs are compared with $m = 6$ for various values of α in Table 3. Also, CPU time for various values of α are shown in Table 3 [31].

Example 3 Consider the following nonlinear fractional Volterra integro-differential equation as [31]:

$$D^{\frac{1}{2}} u(x) = f(x)u(x) + g(x) + \sqrt{x} \int_0^x u^2(s)ds, \quad 0 \leq x \leq 1,$$

with the initial condition $u(0) = 0$, and

$$f(x) = 2\sqrt{x} + 2x^{\frac{3}{2}} - (\sqrt{x} + x^{\frac{3}{2}})Ln(1 + x),$$

$$g(x) = \frac{2Arcsinh\sqrt{x}}{\sqrt{\pi}\sqrt{1 + x}} - 2x^{\frac{3}{2}}.$$

The exact solution of this problem is $u(x) = Ln(1 + x)$ [31].

Table 4 demonstrates the absolute errors achieved between the approximate solutions and exact solution, for $m = 12$ and various values of α. In addition, the absolute

Table 3 The absolute errors and CPU time with $m = 6$ and different values of α in Ref. [31] for Example 2

x	$\alpha = \frac{1}{3}$	$\alpha = \frac{1}{2}$	$\alpha = 1$	$\alpha = \frac{5}{3}$	$\alpha = 2$
0	5.45×10^{-16}	1.03×10^{-6}	1.03×10^{-4}	2.93×10^{-3}	1.08×10^{-2}
0.1	1.12×10^{-14}	5.13×10^{-5}	2.96×10^{-4}	5.54×10^{-5}	6.04×10^{-3}
0.2	2.11×10^{-14}	8.40×10^{-5}	4.73×10^{-4}	1.25×10^{-3}	3.04×10^{-3}
0.3	3.23×10^{-14}	1.16×10^{-4}	6.61×10^{-4}	1.25×10^{-3}	6.08×10^{-3}
0.4	4.53×10^{-14}	1.51×10^{-4}	8.60×10^{-4}	7.17×10^{-4}	3.78×10^{-4}
0.5	6.03×10^{-14}	1.88×10^{-4}	1.07×10^{-3}	1.68×10^{-4}	7.36×10^{-3}
0.6	7.75×10^{-14}	2.29×10^{-4}	1.28×10^{-3}	4.72×10^{-4}	2.72×10^{-3}
0.7	9.74×10^{-14}	2.74×10^{-4}	1.53×10^{-3}	2.16×10^{-3}	1.02×10^{-2}
0.8	1.20×10^{-13}	3.24×10^{-4}	1.82×10^{-3}	3.37×10^{-3}	8.72×10^{-3}
0.9	1.45×10^{-13}	3.80×10^{-4}	2.15×10^{-3}	8.49×10^{-4}	1.26×10^{-2}
1	1.74×10^{-13}	4.44×10^{-4}	2.46×10^{-3}	5.70×10^{-3}	1.50×10^{-2}
CPU time	0.110	0.063	0.063	0.078	0.078

Table 4 The absolute errors and CPU time with $m = 12$ and different values of α in Ref. [31] for Example 3

x	$\alpha = \frac{1}{4}$	$\alpha = \frac{1}{2}$	$\alpha = 1$	$\alpha = \frac{3}{2}$
0	3.31×10^{-7}	3.97×10^{-11}	1.29×10^{-3}	1.74×10^{-2}
0.1	1.47×10^{-6}	2.51×10^{-5}	2.08×10^{-4}	6.37×10^{-3}
0.2	1.44×10^{-6}	2.59×10^{-9}	4.05×10^{-5}	2.49×10^{-3}
0.3	1.67×10^{-6}	2.92×10^{-9}	1.65×10^{-4}	3.62×10^{-3}
0.4	1.91×10^{-6}	3.73×10^{-9}	9.88×10^{-5}	8.66×10^{-3}
0.5	2.78×10^{-6}	5.18×10^{-9}	2.35×10^{-4}	1.91×10^{-2}
0.6	4.48×10^{-6}	7.95×10^{-9}	2.97×10^{-4}	2.98×10^{-2}
0.7	7.27×10^{-6}	1.32×10^{-8}	5.10×10^{-4}	4.03×10^{-2}
0.8	1.29×10^{-5}	2.39×10^{-8}	9.40×10^{-4}	4.77×10^{-2}
0.9	2.59×10^{-5}	4.71×10^{-8}	1.85×10^{-3}	5.63×10^{-2}
1	5.50×10^{-5}	1.01×10^{-7}	4.08×10^{-3}	3.95×10^{-2}
CPU time	0.907	0.468	0.453	0.469

errors for different values of m obtained using the present method with $\alpha = \frac{1}{2}$ [31] are compared with the results of method based on the Jacobi polynomials of order M_1 [80] in Table 5.

Example 4 Consider the following mixed Fredholem-Volterra fractional integro-differential equation as [31]:

$$D^{2.3}u(x) = f(x) + \frac{1}{4}\int_0^x (x-s)u(s)ds + \frac{1}{2}\int_0^1 xsu(s)ds, \quad 0 \le x \le 1,$$

Table 5 The absolute errors with $\alpha = \frac{1}{2}$ in Ref. [31] for Example 3

Method	Absolute error
Method in [80]	
$M_1 = 4$	1×10^{-2}
$M_1 = 6$	1×10^{-2}
Method based on FALFs [31]	
$m = 4$	2.68×10^{-2}
$m = 6$	8.74×10^{-4}
$m = 8$	9.74×10^{-5}
$m = 10$	1.21×10^{-5}
$m = 12$	1.01×10^{-7}

Table 6 Comparison of the absolute errors with the other methods for $m = 10$ and $\alpha = \frac{1}{2}$, in Ref. [31] for Example 4

x	ADM [31]	LWM [71]	FALFs [31]
1/8	1.02×10^{-5}	6.64×10^{-6}	1.55×10^{-7}
2/8	4.20×10^{-5}	4.53×10^{-5}	3.82×10^{-7}
3/8	9.24×10^{-5}	3.14×10^{-5}	6.46×10^{-7}
4/8	4.17×10^{-4}	7.37×10^{-5}	9.40×10^{-7}
5/8	8.16×10^{-4}	2.44×10^{-4}	1.26×10^{-6}
6/8	2.31×10^{-3}	3.81×10^{-4}	1.60×10^{-6}
7/8	8.07×10^{-3}	6.02×10^{-4}	1.96×10^{-6}
CPU time	–	–	0.125

with the initial conditions $u(0) = u'(0) = u''(0) = 0$ and

$$f(x) = \frac{\Gamma(4.5)}{\Gamma(2.2)} x^{1.2} - \frac{1}{99} x^{5.5} - \frac{1}{11} x.$$

The exact solution of this problem is $u(x) = x^{\frac{7}{2}}$.

Table 6 demonstrates the absolute errors achieved for various values of x using the Adomian decomposition method (ADM) with $n = 5$, the Legendre wavelet method (LWM) with $k = 5$, $M = 2$ [71] and the proposed technique based on FALFs with $m = 10, \alpha = \frac{1}{2}$ [31].

Example 5 Consider the following mixed Fredholem-Volterra fractional integro-differential equation as [31]:

$$D^{2.2}u(x) = f(x) + \frac{1}{3}\int_0^x (x+s)u^2(s)ds + \frac{1}{4}\int_0^1 (x-s)u^3(s)ds, \quad 0 \le x \le 1,$$

Table 7 The l_∞ norm of absolute errors with $\alpha = \frac{1}{3}$ for Example 5

	CAS [31]		LWM [81]		FALFs [31]	
	$m = 16$	$m = 32$	$m = 16$	$m = 32$	$m = 9$	$m = 12$
l_∞	7.84×10^{-2}	5.24×10^{-3}	4.43×10^{-3}	2.14×10^{-3}	1.25×10^{-6}	2.42×10^{-7}
CPU time	–	–	–	–	0.079	0.219

with the initial conditions $u(0) = u'(0) = u''(0) = 0$ and

$$f(x) = \frac{\Gamma(4)}{\Gamma(1.8)}x^{0.8} - \frac{5}{56}x^8 - \frac{1}{40}x + \frac{1}{44}.$$

The exact solution of this equation is $u(x) = x^3$.
To compare the numerical results obtained by the CAS wavelet method [31], LWM [81] and the present method [31], we use l_∞ norm which is given by

$$\|e_m\|_\infty = max_{1\le i\le m}\{|u(x_i) - u_m(x_i)|\}, \qquad x_i = \frac{i}{m}, i = 0, 1, \ldots, m-1,$$

where $u(x)$ is the exact solution and $u_m(x)$ is the numerical solution.
Table 7 shows the l_∞ norm of absolute errors derived by applying CAS wavelet method, LWM and the present method with $m = 9, 12$ and $\alpha = \frac{1}{3}$.

Example 6 Consider the following mixed Fredholem-Volterra fractional integro-differential equation as [31]:

$$D^{0.15}u(x) = f(x) + \frac{1}{4}\int_0^x \frac{u(s)}{(x-s)^{\frac{1}{2}}}ds + \frac{1}{7}\int_0^1 e^{x+s}u(s)ds, \quad 0 \le x \le 1,$$

with the initial condition $u(0) = 0$ and

$$f(x) = \frac{\Gamma(3)}{\Gamma(2.85)}x^{1.85} - \frac{\Gamma(2)}{\Gamma(1.85)}x^{0.85} - \frac{\sqrt{\pi}x^{\frac{5}{2}}\Gamma(3)}{4\Gamma(\frac{7}{2})} + \frac{\sqrt{\pi}x^{\frac{3}{2}}\Gamma(2)}{4\Gamma(\frac{5}{2})} - \frac{e^{x+1} - 3e^x}{7}.$$

The exact solution of this equation is $u(x) = x(x-1)$ [31].
The obtained solutions using the proposed method [31] with the method of Legendre wavelets operational matrix method (LWM) and Legendre wavelets collocation method (LWCM) [82] are compared, in Table 8.

Example 7 Consider linear system of fractional integro-differential equation as [31]:

$$\begin{cases} D^\nu u_1(x) = u_1(x) + u_2(x), & 0 \le x \le 1, 0 < \nu \le 1, \\ D^\nu u_2(x) = -u_1(x) + u_2(x), \end{cases}$$

Table 8 The approximate solution at various points with $\alpha = 1$ for Example 6

x	LWCM [82]		LWM [82]		FALFs [31]		Exact solution
	$\hat{m} = 8$	$\hat{m} = 16$	$\hat{m} = 8$	$\hat{m} = 16$	$m = 8$	$m = 12$	
0	0.0025	0.0011	0.0016	0.0006	−0.0003	−0.0003	0.0000
1/8	−0.1007	−0.1023	−0.1014	−0.1058	−0.1079	−0.1084	−0.1094
2/8	−0.1769	−0.1813	−0.1774	−0.1829	−0.1851	−0.1859	−0.1875
3/8	−0.2224	−0.2257	−0.2231	−0.2293	−0.2308	−0.2319	−0.2344
4/8	−0.2431	−0.2442	−0.2382	−0.2447	−0.2455	−0.2470	−0.2500
5/8	−0.2230	−0.2281	−0.2227	−0.2292	−0.2290	−0.2306	−0.2344
6/8	−0.1804	−0.1829	−0.1765	−0.1827	−0.1815	−0.1835	−0.1875
7/8	−0.1011	−0.1055	−0.0990	−0.1052	−0.1033	−0.1051	−0.1084
CPU time	–	–	–	–	0.063	0.235	–

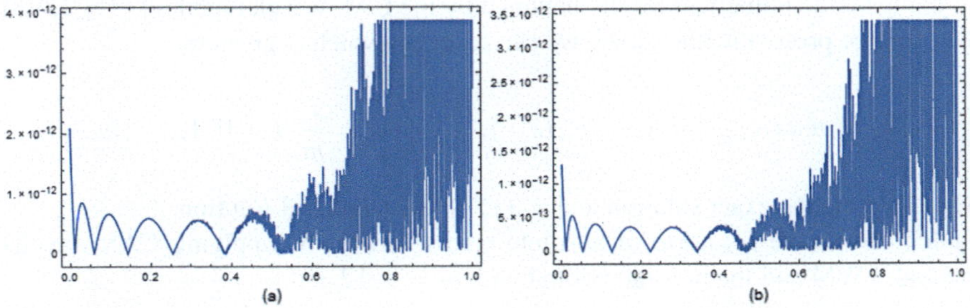

Fig. 7 The absolute errors with $m = 10$ and $\alpha = \nu = 1$ for **a** $u_1(x)$, **b** $u_2(x)$ of Example 7 in Ref. [31]

with the initial conditions

$$u_1(0) = 0, \ u_2(0) = 1,$$

and the exact solution for $\nu = 1$ is

$$u_1(x) = e^x \sin(x), \qquad u_2(x) = e^x \cos(x).$$

The absolute errors achieved between the approximate solutions and the exact solutions at $\nu = \alpha = 1$ and $m = 10$ for $u_1(x)$ and $u_2(x)$ are shown in Fig. 7. We can see that the approximate solutions are in high agreement with the exact solution. Moreover, the obtained results for $u_1(x)$ and $u_2(x)$ with $m = 5$, and $\nu = 0.7, 0.8, 0.9, 1$ are plotted in Fig. 8.

Example 8 Consider the following linear system of fractional integro-differential equation as [31]:

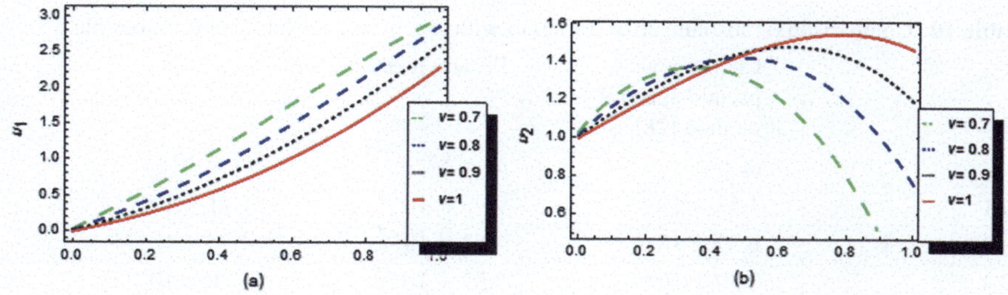

Fig. 8 Plots of **a** $u_1(x)$ and **b** $u_2(x)$ with $m = 5$ and different values of ν in Ref. [31] for Example 7

Table 9 Comparison of absolute errors for $u_1(x)$ with $\alpha = \nu = 1$ in Ref. [31] for Example 8

x	Chebyshev pseudo-spectral method [28]	Present method [31]	
	$m = 5$	$m = 5$	$m = 10$
0.1	1.06×10^{-5}	9.95×10^{-7}	8.58×10^{-13}
0.2	6.45×10^{-5}	3.83×10^{-7}	8.40×10^{-13}
0.3	9.29×10^{-5}	7.57×10^{-7}	5.36×10^{-13}
0.4	7.23×10^{-5}	1.43×10^{-7}	3.20×10^{-13}
0.5	1.50×10^{-5}	7.78×10^{-7}	1.16×10^{-12}
0.6	4.63×10^{-5}	2.58×10^{-7}	1.51×10^{-13}
0.7	7.66×10^{-5}	7.03×10^{-7}	2.60×10^{-12}
0.8	5.68×10^{-5}	4.75×10^{-7}	2.29×10^{-12}
0.9	5.14×10^{-6}	9.65×10^{-7}	1.77×10^{-11}
CPU time	–	0.063	0.469

$$\begin{cases} D^{\nu}u_1(x) = f_1(x) - u_2(x) - \int_0^x [u_1(s) + u_2(s)]ds, & 0 < \nu \leq 1, \\ D^{\nu}u_2(x) = f_2(x) + u_1(x) - \int_0^x [u_1(s) - u_2(s)]ds, \end{cases}$$

with the initial conditions $u_1(0) = 1,\ u_2(0) = -1$ and

$$f_1(x) = 1 + x + x^2, \qquad f_2(x) = -1 - x.$$

The exact solutions for $\nu = 1$ are $u_1(x) = x + e^x,\ u_2(x) = x - e^x$. The absolute errors obtained for various values of x by using the proposed scheme [31] and the Chebyshev pseudo-spectral method [28] are reported in Tables 9 and 10. For more investigation, the numerical results for $u_1(x)$ and $u_2(x)$ with $\nu = 0.7, 0.8, 0.9, 1$ and $m = 5$ are shown in Fig. 9 [31]. This figure shows that, as ν approaches 1, the numerical solution by FALFs, converges to the exact solution, i.e. the solution of fractional order differential equation approaches the solution of integer order differential equation [31].

Table 10 Comparison of absolute error for $u_2(x)$ with $\alpha = \nu = 1$ in Ref. [31] for Example 8

x	Chebyshev pseudo-spectral method [28]	Present method [31]	
	$m = 5$	$m = 5$	$m = 10$
0.1	1.58×10^{-5}	9.95×10^{-7}	1.41×10^{-13}
0.2	6.32×10^{-6}	3.83×10^{-7}	4.90×10^{-13}
0.3	1.29×10^{-5}	7.57×10^{-7}	1.30×10^{-12}
0.4	4.96×10^{-6}	1.43×10^{-7}	2.43×10^{-12}
0.5	3.81×10^{-5}	7.78×10^{-7}	4.22×10^{-12}
0.6	6.71×10^{-5}	2.58×10^{-7}	6.69×10^{-12}
0.7	7.25×10^{-5}	7.03×10^{-7}	9.97×10^{-12}
0.8	4.58×10^{-5}	4.75×10^{-7}	1.48×10^{-11}
0.9	3.12×10^{-6}	9.65×10^{-7}	2.00×10^{-11}
CPU time	–	0.063	0.469

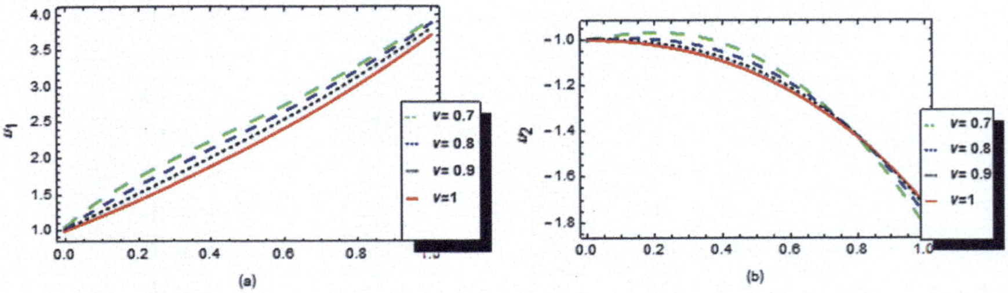

Fig. 9 Plots of **a** $u_1(x)$ and **b** $u_2(x)$ with $m = 5$ and various values of ν in Ref. [31] for Example 8

Example 9 Consider the fractional population growth model of a species. The model is characterized by the nonlinear fractional Volterra integro-differential equation as [83]:

$$D^{\nu}u(x) = au(x) - bu^2(x) - cu(x)\int_0^x u(s)ds, \quad 0 \le \nu \le 1,$$

with the initial condition $u(0) = 0.1$. In this model, u is the population of identical individuals at time x, $a > 0$ is the birth rate coefficient, $b > 0$ is the crowding coefficient, and $c > 0$ is the toxicity coefficient.

The exact solution of considered equation is not available. Then, to display accuracy of the proposed scheme, the residual function is defined as follows [31]:

$$Rex(x) = D^{\nu}u_m(x) - au_m(x) + bu_m^2(x) + cu_m(x)\frac{x}{2}\sum_{j=1}^{\tilde{m}}\omega_j u_m(\frac{x}{2} + \frac{x}{2}\zeta_j).$$

Table 11 Comparison of residual functions for $a = b = c = 5$ for Example 9

x	Bessel functions collocation method [83]		Present method [31]	
	$\nu = 0.5$	$\nu = 0.75$	$\nu = 0.5$	$\nu = 0.75$
0.1	4.95×10^{-3}	9.53×10^{-3}	2.04×10^{-5}	9.22×10^{-6}
0.2	1.50×10^{-3}	2.92×10^{-4}	2.66×10^{-4}	4.88×10^{-5}
0.3	1.31×10^{-3}	3.34×10^{-3}	1.61×10^{-4}	1.71×10^{-5}
0.4	1.12×10^{-4}	9.04×10^{-4}	6.76×10^{-5}	3.91×10^{-6}
0.5	2.22×10^{-4}	2.77×10^{-5}	8.99×10^{-10}	9.09×10^{-12}
0.6	1.75×10^{-4}	9.92×10^{-4}	5.89×10^{-5}	3.53×10^{-7}
0.7	9.37×10^{-4}	8.13×10^{-4}	1.09×10^{-4}	1.05×10^{-7}
0.8	6.79×10^{-4}	1.42×10^{-4}	1.27×10^{-4}	3.98×10^{-7}
0.9	1.76×10^{-3}	2.76×10^{-3}	7.93×10^{-6}	2.98×10^{-8}

The values of residual function calculated using the present technique for $m = 12, \alpha = \nu$ with the method based on Bessel functions collocation method [83] are listed in Table 11.

4.4.2 Numerical Results Obtained from Numerical Method Based on FGFs

Here, we present several examples by applying the proposed method based on fractional-order Genocchi functions. The errors which are used in some examples are defined as follows [41]:

$$L_2 = \left(\sum_{i=1}^{M} \sum_{j=1}^{N} |u(x_i, t_j) - u_{MN}(x_i, t_j)|^2 \right)^{\frac{1}{2}}.$$

Example 10 Consider the following variable-order fractional Volterra integro-differential equation as [40, 41]:

$$D^{\nu(x)}u(x) + 6\int_0^x u(\xi)d\xi + 2tu'(x) + u(x) = h(x), \qquad 1 < \nu(x) \leq 2,$$

with the initial condition $u(0) = 0$ and

$$h(x) = \frac{10}{\Gamma(3 - \nu(x))} x^{2-\nu(x)} + 10x^3 + 70x^2 + 45x.$$

The exact solution is $u(x) = 5x^2 + 15x$ [41]. According to the proposed technique based on FGFs, for $M = 3, \alpha = 1$ and $\nu(x) = 2$, the unknown vector U is achieved

Table 12 Comparison of the absolute errors obtained by the present method [41] for various values of $\nu(x)$ and $\alpha = 1$, $M = 3$ with method in [40] of Example 10

x	Present method [41]		Method in [40]
	$\nu(x) = \frac{3}{5}(\sin(x) + \cos(x))$	$\nu(x) = \frac{1+\sin(x)}{4}$	$\nu(x) = \frac{3}{5}(\sin(x) + \cos(x))$
0.125	1.6059×10^{-17}	5.1517×10^{-18}	0
0.375	4.3361×10^{-16}	1.3909×10^{-16}	8.8820×10^{-16}
0.625	2.0074×10^{-15}	6.4396×10^{-16}	8.8817×10^{-16}
0.875	5.5085×10^{-15}	1.7670×10^{-15}	0

as follows [41]:

$$U = \begin{bmatrix} 20 & 5 & -3.461871477881021 \times 10^{-16} \end{bmatrix}.$$

Then, the numerical solution is calculated as

$$u(x) \simeq 15x - 3.4618 \times 10^{-16}x^3 + 5x^2.$$

Also, by implementing the algorithm based on FGFs, the following equation is derived as [41]:

$$E_x^{\alpha,1} \leq \begin{bmatrix} 2.0775 \times 10^{-17} \\ 1.0388 \times 10^{-17} \\ 1.0388 \times 10^{-17} \end{bmatrix}.$$

Therefore, considering Theorem 2, the bound of error is obtained as [41]:

$$\begin{aligned} \|u - u_3\|_2 &\leq \|U\|_2 \|E_x^{\alpha,1}\|_2 \\ &\leq 20.6155 \times (2.5444 \times 10^{-17}) = 5.2454 \times 10^{-16}. \end{aligned}$$

In Table 12, the numerical results achieved using the proposed technique [41] are compared with the results of shifted Legendre Gauss-Lobatto collocation method [40].

Example 11 Consider the following variable-order fractional Fredholm integro-differential equation as [41]:

$$D^{\nu(x)}u(x) = x\exp(x) + \exp(x) - x + \int_0^1 xu(x)dx, \qquad 0 < \nu(x) \leq 1,$$

with the initial condition $u(0) = 0$. For $\nu(x) = 1$, the exact solution of this problem is $u(x) = x\exp(x)$ [41]. Figure 10 displays the behavior of approximate solutions for the different choices of $\nu(x)$.

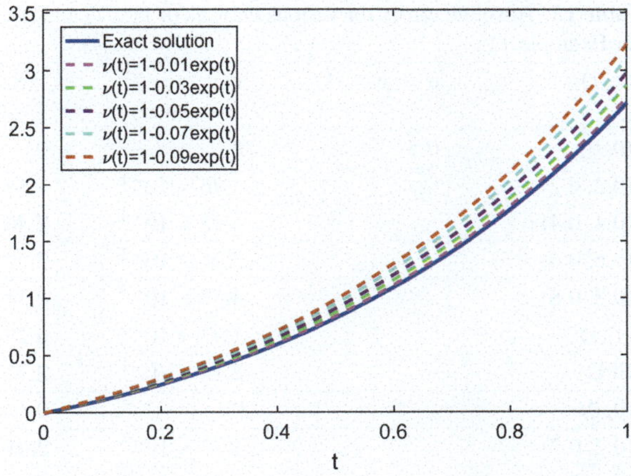

Fig. 10 Approximate solutions for different values of $\nu(x)$ with $\alpha = 1, M = 5$ in Ref. [41] for Example 11

Example 12 Consider the following variable-order fractional nonlinear partial Volterra integro-differential equation as [41, 84]:

$$D_t^{\nu(x,t)}u(x,t) = h(x,t) + (1 + x^2)\frac{\partial^2 u}{\partial x^2}(x,t) + \int_0^t (x^2 + tu^2(x,\eta))d\eta,$$

with the initial condition

$$u(x,0) = \sinh(x) + 1, \qquad x \in [0,1],$$

and the boundary conditions

$$u(0,t) = 1, \qquad u(1,t) = \sinh(1) + \exp(-t), \qquad t \in [0,1].$$

where, $0 < \nu(x,t) \leq 1$ and $h(x,t)$ is selected as the exact solution is $u(x,t) = \sinh(x) + \exp(-xt)$ for $\nu(x,t) = 1$ [41]. The absolute errors for various choices of $\nu(x,t)$ and α, β with $M = N = 3$ are reported in Table 13.

5 Conclusion

In this chapter, we first reported integro-differential equations and spectral method from 1970 to 3rd March 2020 with quantitative and qualitative approaches. Then, due to the great use of considered equations in various branches of sciences, we have provided examples of their applications in physics, biology and other sciences. Here, the following classes of integro-differential equations with arbitrary order are considered:

Table 13 Absolute errors for various choices of $\nu(x,t)$ and $\alpha=\beta$ with $M=N=3$ in Ref. [41] for Example 12

(x,t)	$\alpha=\beta$	$\nu(x,t)=$ $1-\sin(xt)$	$\nu(x,t)=$ $1-0.1\sin(xt)$	$\nu(x,t)=$ $1-0.3\sin(xt)$
(0, 0)	0.5	0	0	0
(0.2, 0.2)		7.76×10^{-4}	1.33×10^{-3}	8.04×10^{-4}
(0.4, 0.4)		6.92×10^{-4}	2.46×10^{-3}	1.93×10^{-3}
(0.6, 0.6)		2.24×10^{-3}	2.99×10^{-3}	3.61×10^{-3}
(0.8, 0.8)		4.53×10^{-3}	2.34×10^{-3}	3.95×10^{-3}
(1, 1)		2.97×10^{-41}	6.20×10^{-41}	6.20×10^{-41}
CPU		2.14×10^{-1}	2.17×10^{-1}	2.13×10^{-1}
(0, 0)	1	0	0	0
(0.2, 0.2)		1.77×10^{-3}	2.01×10^{-4}	5.58×10^{-4}
(0.4, 0.4)		2.94×10^{-3}	2.23×10^{-4}	7.74×10^{-4}
(0.6, 0.6)		2.27×10^{-4}	1.61×10^{-4}	4.06×10^{-4}
(0.8, 0.8)		2.92×10^{-3}	6.26×10^{-4}	1.73×10^{-3}
(1, 1)		2.97×10^{-41}	2.97×10^{-41}	2.97×10^{-41}
CPU		2.15×10^{-1}	2.12×10^{-41}	2.13×10^{-1}

- Fractional integro-differential equations
- Variable order fractional integro-differential equations and variable order fractional partial integro-differential equations.

A numerical method is presented for solving each class of considered equations. These methods are based on different sets of basic functions included fractional-order alternative Legendre functions and fractional-order Genocchi functions.

For each set of basic functions, their properties are expressed and numerical techniques to obtain the approximate solution of these equations are proposed. Convergence analysis of each methods is presented. Finally, to demonstrate the effectiveness of the proposed methods, several numerical examples are proposed and the numerical results are compared with the results obtained by applying some existing methods. For future works, the following cases are suggested:

- Stability analysis of the suggested methods for numerical approximation of arbitrary-order integro-differential equations
- Utilizing other basic functions such as wavelets and hybrid functions for solving arbitrary-order integro-differential equations
- Presenting new numerical methods to solve arbitrary-order integro-differential equations with weakly singular kernel
- Applying spectral methods for solving arbitrary-order integro-differential equations in infinity interval
- Solving various mathematical models arising in different branches of sciences, numerically.

Acknowledgements Part of the contents of Sect. 3 was published in "Elsevier, 365, P. Rahimkhani, Y. Ordokhani, Approximate solution of nonlinear fractional integro-differential equations using fractional alternative Legendre functions, Journal of Computational and Applied Mathematics, 112365, Copyright Elsevier (2020)" and in "Springer Nature, H. Dehestani, Y. Ordokhani, M. Razzaghi, Pseudo-operational matrix method for the solution of variable-order fractional partial integro-differential equations, Engineering with Computers, 1–16, Copyright Springer Nature (2020)."

Moreover, the authors would like to thank the referees for their precise remarks, which improved the presentation of this chapter.

References

1. Singh, H., Baleanu, D., Srivastava, H.M., Dutta, H., Jha, N.K.: Solution of multi-dimensional Fredholm equations using Legendre scaling functions. Appl. Numer. Math. **150**, 313–324 (2020)
2. Babolian, E., Shamloo, A.S.: Numerical solution of Volterra integral and integro-differential equations of convolution type by using operational matrices of piecewise constant orthogonal functions. J. Comput. Appl. Math. **214**(2), 495–508 (2008)
3. Razzaghi, M., Ordokhani, Y.: Solution of differential equations via rationalized Haar functions. Math. Comput. Simul. **56**(3), 235–246 (2001)
4. Sabermahani, S., Ordokhani, Y., Yousefi, S.A.: Fractional-order general Lagrange scaling functions and their applications. BIT Numer. Math. **60**, 101–128 (2020)
5. Sabermahani, S., Ordokhani, Y., Lima, P.M.: A novel Lagrange operational matrix and Tau-collocation method for solving variable-order fractional differential equations. Iran. J. Sci. Technol. Trans. A Sci. **44**(1), 127–135 (2020)
6. Sabermahani, S., Ordokhani, Y., Yousefi, S.A.: Fractional-order Lagrange polynomials: an application for solving delay fractional optimal control problems. Trans. Inst. Measure. Control **41**(11), 2997–3009 (2019)
7. Rabiei, K., Ordokhani, Y., Babolian, E.: Fractional-order Boubaker functions and their applications in solving delay fractional optimal control problems. J. Vibrat. Control **24**(15), 3370–3383 (2018)
8. Sabermahani, S., Ordokhani, Y., Yousefi, S.A.: Two-dimensional M'untz-Legendre hybrid functions: theory and applications for solving fractional-order partial differential equations. Comput. Appl. Math. **39**(111) (2020). https://doi.org/10.1007/s40314-020-1137-5
9. Guo, B.: Spectral Methods and their Applications. World Scientific (1998)
10. Gottlieb, D., Orszag, S.A.: Numerical Analysis of Spectral Methods: Theory and Applications, vol. 26. Siam (1977)
11. Boyd, J.P.: Chebyshev and Fourier Spectral Methods. Courier Corporation (2001)
12. Wazwaz, A.M.: Linear and Nonlinear Integral Equations, vol. 639. Springer, Berlin (2011)
13. Lakshmikantham, V.: Theory of Integro-Differential Equations, vol. 1. CRC Press (1995)
14. Guo, D.: Existence of solutions for nth-order integro-differential equations in Banach spaces. Comput. Math. Appl. **41**(5–6), 597–606 (2001)
15. Agarwal, R.P., Ntouyas, S.K., Ahmad, B., Alhothuali, M.S.: Existence of solutions for integro-differential equations of fractional order with nonlocal three-point fractional boundary conditions. Adv. Diff. Eq. **2013**(1), 128 (2013)
16. Wu, J., Liu,Y.: Existence and uniqueness of solutions for the fractional integro-differential equations in Banach spaces. Electron. J. Diff. Eq. (EJDE) [electronic only] (2009)
17. Matar, M.M.: Existence and uniqueness of solutions to fractional semilinear mixed Volterra-Fredholm integrodifferential equations with nonlocal conditions. Electron. J. Diff. Eq. (EJDE) [electronic only] (2009)
18. Jin, C., Luo, J.: Stability of an integro-differential equation. Comput. Math. Appl. **57**(7), 1080–1088 (2009)

19. Drozdov, A.: Stability of integro-differential equations with periodic operator coefficients. Quart. J. Mech. Appl. Math. **49**(2), 235–260 (1996)
20. Agarwal, R.P., Domoshnitsky, A., Goltser, Y.: Stability of partial functional integro-differential equations. J. Dyn. Control Syst. **12**(1), 1–31 (2006)
21. Zhao, J., Meng, F.: Stability analysis of solutions for a kind of integro-differential equations with a delay. Math. Prob. Eng. (2018)
22. Mittal, R.C., Nigam, R.: Solution of fractional integro-differential equations by Adomian decomposition method. Int. J. Appl. Math. Mech. **4**(2), 87–94 (2008)
23. Abazari, R., Klcman, A.: Application of differential transform method on nonlinear integro-differential equations with proportional delay. Neural Comput. Appl. **24**(2), 391–397 (2014)
24. Enright, W.H., Hu, M.: Continuous Runge-Kutta methods for neutral Volterra integro-differential equations with delay. Appl. Numer. Math. **24**(2–3), 175–190 (1997)
25. AlHabees, A., Maayah, B., Bushnaq, S.: Solving fractional proportional delay Integro differential equations of first order by reproducing kernel Hilbert space method. Global J. Pure Appl. Math. **12**(4), 3499–3516 (2016)
26. Xu, Y., Suat, E.V.: A finite difference technique for solving variable-order fractional integro-differential equations 699–712 (2014)
27. Moghaddam, B.P., Machado, J.A.T.: A computational approach for the solution of a class of variable-order fractional integro-differential equations with weakly singular kernels. Fract. Calculus Appl. Anal. **20**(4), 1023–1042 (2017)
28. Khader, M.M., Sweilam, N.H.: On the approximate solutions for system of fractional integro-differential equations using Chebyshev pseudo-spectral method. Appl. Math. Model. **37**(24), 9819–9828 (2013)
29. Saeedi, H., Moghadam, M.M.: Numerical solution of nonlinear Volterra integro-differential equations of arbitrary order by CAS wavelets. Comm. Nonlinear Sci. Numer. Simul. **16**(3), 1216–1226 (2011)
30. Saadatmandi, A., Dehghan, M.: A Legendre collocation method for fractional integro-differential equations. J. Vibrat. Control **17**(13), 2050–2058 (2011)
31. Rahimkhani, P., Ordokhani, Y.: Approximate solution of nonlinear fractional integro-differential equations using fractional alternative Legendre functions. J. Comput. Appl. Math. **365**, 112365 (2020)
32. Sheng, C., Wang, Z., Guo, B.: An hp-spectral collocation method for nonlinear Volterra functional integro-differential equations with delays. Appl. Numer. Math. **105**, 1–24 (2016)
33. Aghazadeh, N., Khajehnasiri, A.A.: Solving nonlinear two-dimensional Volterra integro- differential equations by block-pulse functions. Math. Sci. **7**(1), 3 (2013)
34. Sabermahani, S., Ordokhani, Y.: A new operational matrix of Müntz-Legendre polynomials and Petrov-Galerkin method for solving fractional Volterra-Fredholm integro-differential equations. Comput. Methods Diff. Eq. (2020). https://doi.org/10.22034/cmde.2020.32623.1515
35. Brunner, H., Hu, Q., Lin, Q.: Geometric meshes in collocation methods for Volterra integral equations with proportional delays. IMA J. Numer. Anal. **21**(4), 783–798 (2001)
36. Wei, Y., Chen, Y.: Legendre spectral collocation methods for pantograph Volterra delay-integro-differential equations. J. Sci. Comput. **53**(3), 672–688 (2012)
37. Yüzbasi, S.: Laguerre approach for solving pantograph-type Volterra integro-differential equations. Appl. Math. Comput. **232**, 1183–1199 (2014)
38. Zhao, J., Cao, Y., Xu, Y.: Sinc numerical solution for pantograph Volterra delay-integro-differential equation. Int. J. Comput. Math. **94**(5), 853–865 (2017)
39. Sedaghat, S., Ordokhani, Y., Dehghan, M.: On spectral method for Volterra functional integro-differential equations of neutral type. Numer. Funct. Anal. Optimizat. **35**(2), 223–239 (2014)
40. Doha, E.H., Abdelkawy, M.A., Amin, A.Z.M., Baleanu, D.: Spectral technique for solving variable-order fractional volterra integro-differential equations. Numer. Methods Partial Diff. Eq. **34**(5), 1659–1677 (2018)
41. Dehestani, H., Ordokhani, Y., Razzaghi, M.: Pseudo-operational matrix method for the solution of variable-order fractional partial integro-differential equations. Eng. Comput. 1–16 (2020)

42. Singh, C.S., Singh, H., Singh, V.K., Singh, O.P.: Fractional order operational matrix methods for fractional singular integro-differential equation. Appl. Math. Model. **40**(23–24), 10705–10718 (2016)
43. Ordokhani, Y., Moosavi, S., Shahrezaee, M.: Tau numerical solution of the Volterra-Fredholm Hammerstein integro-differential equations with the Bernstein multi-scaling functions. J. Inf. Comput. Sci. **8**(4), 243–255 (2013)
44. Aria, M., Cuccurullo, C.: Bibliometrix: an R-tool for comprehensive science mapping analysis. J. Inf. **11**(4), 959–975 (2017)
45. Jenner, A., Yun, C.O., Yoon, A., Kim, P.S., Coster, A.C.: Modelling heterogeneity in viral-tumour dynamics: the effects of gene-attenuation on viral characteristics. J. Theoret. Biol. **454**, 41–52 (2018)
46. Durdiev, D.K., Rahmonov, A.A.: Inverse problem for a system of integro-differential equations for SH waves in a visco-elastic porous medium: global solvability. Theoret. Math. Phys. **195**(3), 923–937 (2018)
47. Akgül, A., Khan, Y.: A novel simulation methodology of fractional order nuclear science model. Math. Methods Appl. Sci. **40**(17), 6208–6219 (2017)
48. Jama, M.A., Zhao, W., Ahmad, W., Buffo, A., Alopaeus, V.: Analytical time-stepping solution of the discretized population balance equation. Comput. Chem. Eng. **135**, 106741 (2020)
49. Cushing, J.M.: Integrodifferential Equations and Delay Models in Population Dynamics, vol. 20. Springer Science & Business Media (2013)
50. Becker, M., Levy, M.D.: Modeling the transfer of drug resistance in solid tumors. Bull. Math. Biol. **79**(10), 2394–2412 (2017)
51. Babolian, E., Biazar, J.: Solving the problem of biological species living together by Adomian decomposition method. Appl. Math. Comput. **129**(2–3), 339–343 (2002)
52. Maddalena, L.: Analysis of an integro-differential system modeling tumor growth. Appl. Math. Comput. **245**, 152–157 (2014)
53. Greene, J.M., Levy, D., Fung, K.L., Souza, P.S., Gottesman, M.M., Lavi, O.: Modeling intrinsic heterogeneity and growth of cancer cells. J. Theoret. Biol. **367**, 262–277 (2015)
54. Dbowski, M., Lachowicz, M., Szymaska, Z.: Microscopic description of DNA thermal denaturation. Appl. Math. Comput. **361**, 47–60 (2019)
55. Hossain, M.E.: Numerical investigation of memory-based diffusivity equation: the integro-differential equation. Arab. J. Sci. Eng. **41**(7), 2715–2729 (2016)
56. Koley, S., Sahoo, T.: An integro-differential equation approach to study the scattering of water waves by a floating flexible porous plate. Geophys. Astrophys. Fluid Dyn. **112**(5), 345–356 (2018)
57. Lienert, M.: Direct interaction along light cones at the quantum level. J. Phys. A Math. Theoret. **51**(43), 435302 (2018)
58. Pecina, P.: An analytical theory of a scattering of radio waves on meteoric ionization II. Solution of the integro-differential equation in case of backscatter. Month. Not. R Astronom. Soc. **463**(2), 1185–1198 (2016)
59. Pinheiro, I.F., Sphaier, L.A., Alves, L.S. de B.: Integral transform solution of integro-differential equations in conduction-radiation problems. Numer. Heat Transfer Part A Appl. **73**(2), 94–114 (2018)
60. Zhou, L.Q., Meleshko, S.V.: Symmetry groups of integro-differential equations for linear thermoviscoelastic materials with memory. J. Appl. Mech. Tech. Phys. **58**(4), 587–609 (2017)
61. Vlasov, V.V., Rautian, N.A.: A study of operator models arising in problems of hereditary mechanics. J. Math. Sci. **244**(2), 170–182 (2020)
62. Elghribi, M.: Stochastic calculus in a risk model with stochastic return on investments. Stochastics 1–20 (2020)
63. Zsaki, A., Bitar, D.: Modeling the behavior of a reinforced granular layer overlying soft soil using an integro-differential equation. Int. J. Geotech. Eng. **5**(1), 45–60 (2011)
64. Marinov, M., Rangelov, T.: Integro-differential equations for anti-plane cracks in inhomogeneous piezoelectric plane. Comptes rendus de l'Academie bulgare des Sciences **64**, 1669–1678 (2011)

65. Avaji, M., Dehcheshmeh, S.S., Hafshejani, J.S., Ghahfarokhi, D.F.: On the solution of linear and nonlinear fractional integro-differential equations in nano-transistors. J. Comput. Theoret. Nanosci. **10**(2), 510–513 (2013)
66. Rahimkhani, P., Ordokhani, Y., Babolian, E.: Fractional-order Bernoulli wavelets and their applications. Appl. Math. Model. **40**, 8087–8107 (2016)
67. Xu, Y., He, Z.: Existence and uniqueness results for Cauchy problem of variable-order fractional differential equations. J. Appl. Math. Comput. **43**(1–2), 295–306 (2013)
68. Chelyshkov, V.S.: Alternative orthogonal polynomials and quadratures. Electron. Trans. Numer. Anal. **25**(7), 17–26 (2006)
69. Bazm, S.: Solution of nonlinear Volterra-Hammerstein integral equations using alternative Legendre collocation method. Sahand Comm. Math. Anal. **4**(1), 57–77 (2016)
70. Bazm, S., Hosseini, A.: Numerical solution of nonlinear integral equations using alternative Legendre polynomials. J. Appl. Math. Comput. **56**(1–2), 25–51 (2018)
71. Meng, Z., Yi, M., Huang, J., Song, L.: Numerical solutions of nonlinear fractional differential equations by alternative Legendre polynomials. Appl. Math. Comput. **336**, 454–464 (2018)
72. Roshan, S., Jafari, H., Baleanu, D.: Solving FDEs with Caputo-Fabrizio derivative by operational matrix based on Genocchi polynomials. Math. Methods Appl. Sci. **41**, 1–8 (2018)
73. Araci, S.: Novel identities involving Genocchi numbers and polynomials arising from applications of umbral calculus. Appl. Math. Comput. **233**, 599–607 (2014)
74. Hosseininia, M., Heydari, M.H., Avazzadeh, Z., Ghaini, F.M.: Two-dimensional Legendre wavelets for solving variable-order fractional nonlinear advection-diffusion equation with variable coefficients. Int. J. Nonlinear Sci. Numer. Simul. **19**(7–8), 793–802 (2018)
75. Horadam, A.F.: Genocchi polynomials. In: Applications of Fibonacci Numbers, pp. 145–166. Springer, Dordrecht (1991)
76. Kreyszig, E.: Introductory Functional Analysis with Applications. Wiley, New York (1978)
77. Stoer, J., Bulirsch, R.: Introduction to Numerical Analysis, 2nd edn. Springer, Berlin (2002)
78. Rahimkhani, P., Ordokhani, Y.: Numerical solution a class of 2D fractional optimal control problems by using 2D Müntz-Legendre wavelets. Optimal Control Appl. Methods **36**(9), 1916–1934 (2018)
79. Phillips, G.M., Taylor, P.J.: Theory and Applications of Numerical Analysis. Elsevier (1996)
80. Dehghan, M., Manafian, J., Saadatmandi, A.: Solving nonlinear fractional partial differential equations using the homotopy analysis method. Numer. Methods Part. Diff. Eq. **26**(2), 448–479 (2010)
81. Meng, Z., Wang, L., Li, H., Zhang, W.: Legendre wavelets method for solving fractional integro-differential equations. Int. J. Comput. Math. **92**(6), 1275–1291 (2015)
82. Yi, M., Wang, L., Jun, H.: Legendre wavelets method for the numerical solution of fractional integro-differential equations with weakly singular kernel. Appl. Math. Model. **40**(4), 3422–3437 (2016)
83. Parand, K., Nikarya, M.: Application of Bessel functions for solving differential and integro-differential equations of the fractional order. Appl. Math. Model. **38**(15–16), 4137–4147 (2014)
84. Avazzadeh, Z., Beygi Rizi, Z., Maalek Ghaini, F.M., Loghmani, G.B.: A numerical solution of nonlinear parabolic-type Volterra partial integro-differential equations using radial basis functions. Eng. Anal. Bound. Elements **36**, 881–893 (2012)

An Efficient Numerical Algorithm to Solve Volterra Integral Equation of Second Kind

Ram K. Pandey and Harendra Singh

Abstract In this chapter, our aim to construct exponential fitting (ef) of Volterra Runge-Kutta (VRK) method to find the numerical solution of Volterra integral equation (VIE) of the second kind. Volterra integral equations have several areas of applications in science and engineering. Sometimes, the traditional Runge-Kutta method produces a large amount of error to solve VIE having exponential/trigonometric solutions. This gives the motivation to construct the fitting version of VRK method so that the new exponentially fitted VRK method would be best tuned to solve the Volterra integral equation of second kind having exponential/trigonometric solutions. The development is based on the assumption that the linear operators corresponding to the internal and external stages annihilate the set of basis exponential functions with unknown frequencies where the optimum values of frequencies are computed by minimizing the local truncation error.

Keywords Volterra integral equation · Volterra-Runge-Kutta method · Exponential-fitting · Order conditions · Truncation error

1 Introduction

The Integral equation is one of the important mathematical tools which are used to model several phenomena in physics, heat flow, fluid flow, vibration, chemical reaction, nuclear reaction and several areas of economics and finance. In this paper, we shall be focused on the Volterra integral equation of the second kind and is given by

R. K. Pandey (✉)
Department of Mathematics and Statistics, Dr. H.S. Gour Vishwavidyalaya, Sagar 470003, M.P., India
e-mail: pandeywavelet@gmail.com

H. Singh
Department of Mathematics, P. G. College, Ghazipur, U.P., India
e-mail: harendra059@gmail.com

H. Singh et al. (eds.), *Topics in Integral and Integro-Differential Equations*,
Studies in Systems, Decision and Control 340,
https://doi.org/10.1007/978-3-030-65509-9_8

$$y(x) = f(x) + \int_0^x k(x, s, y(s))ds, \quad x \in [0, X] \tag{1.1}$$

Here, we assume that the kernel $k(x, s, y(s))$ is continuous (at least) on $S \times \mathbb{R}^n$ where $S = \{(x, s) : x_0 \leq x \leq s \leq X\}$, and that the solution $y(x)$ exists uniquely and also continuous on $[0, X]$.

Volterra integral equations (VIE) have wide area of applications in various fields of sciences and engineering such as, heat conduction problem [1], the concrete problem of mechanics or physics [2], the unsteady poiseuille flow in a pipe [3], diffusion problems [4], electroelastic dynamics [5], contact problems [6], plasma physics [7], the image deblurring problem and its regularization [8], axisymmetric contact problems for bodies with complex rheology [9], diffraction theory [10] and the electrochemical behaviour of an inlaid microband electrode for the case of equal diffusion coefficients [11]. These types of integral equations also occur as reformulations from some mixed boundary value problems arising in several branches of applied sciences; for example, solid and fluid mechanics, electrostatics, heat transfer, diffraction and scattering of waves, etc.

Several authors contributed by various research papers to find the numerical solution of Volterra and Fredholm integral equations. The piecewise polynomial collocation methods [12, 13], spline collocation and iterated collocation methods [14], Galerkin methods [15], Adomian decomposition methods [16], product integration methods [17], hybrid collocation methods [18], Walsh functions [19], Haar wavelets [20], variable transformations [21], extrapolation methods [22], the block by block method [23] and the differential transform method [24] have been applied to solve one-dimensional Volterra integral equations. Moreover, the iterated discrete Galerkin method [25], the iterated collocation method [26], the Galerkin method with spline functions as the basis [27], the Nyström method [28], two-dimensional rationalized Haar (RH) functions [29], the two-dimensional differential transform method [30], the degenerate method [31], the piecewise polynomial projection methods [32], the Nyström method [33] and the Gauss product quadrature rules [34] have been utilized to obtain an approximate solution of two-dimensional Volterra integral equations. Assari and Dehghan [35] constructed a numerical scheme to solve the nonlinear Volterra integral equations of the second kind using the radial basis functions. Recently, Singh et al. [36] proposed a method to find the approximate solution of multi-dimensional Fredholm integral equation of the second kind using the n-dimensional Legendre scaling functions. The use of operational matrix approach, to solve integral equations and integro-differential equations of integer and fractional order can be found in [37] and the references cited therein.

Volterra-Ruge-Kutta (VRK) methods for the numerical solution of (1.1) was first introduced by Bellen et al. [38], and further investigated in [39, 40]. They include the special cases of VRK as previously developed in literature Pouzet-type methods [41], Bel'tyukov-type methods [42]. Cote et al. [43] reanalysed and introduced natural VRK method and also introduced some A and V stable VRK methods. Recently,

Abdi et al. [44] proposed the new code based on VRK methods to solve VIEs. They have constructed the variable step size formulation of Natural VRK method.

In case of ordinary differential equations (ODEs), the traditional RK family formulae fails to solve initial value problems (IVPs) having special characteristics like periodicity, energy conservation, oscillation, phase conservation, etc. To solve the IVP having periodic and oscillatory solutions, a first theoretical foundation was given by Gautschi [45] and Lyche [46] where they proposed a technique called functional fitting. The study of exponentially-fitted RK method is a modern development and is widely useful to solve equations with specific solution behaviour. A lot of researches have been done to construct functionally-fitted numerical methods. The initial value problems (IVPs) having exponential solutions can be solved exactly with the help of an exponential fitting approach. Berghe et al. [47–49], Paternoster [50], Simos [51] have first combined the exponential-fitting with the existing RK methods and thus developed various types of exponential-fitted RK methods. The approach of Simos was different with the approach of Berghe in the sense that Berghe uses the additional parameter in internal stages and this certainly improves the rate of convergence of exponential-fitted Runge-Kutta method (ef-RKM) which was shown in the numerical experiments given in Berghe et al. [47, 48]. The principle of the fitting is based on the annihilation of the linear integral operators associated with internal stages and external stages provided the existing RK method has been converted into a functional form using the approach of Albrecht [52]. Ixaru et al. [53] discovered the algorithm to compute the weights of exponentially-fitted multi-step algorithms for ODEs. The development of exponential fitting of various numerical integrators is covered in the excellent monograph (see [54]). Ozawa [55] made the functional fitting of RKM with variable coefficients.

The main objective of the present paper is to derive the exponential-fitting of Volterra-Runge-Kutta (VRK) method to find the numerical solution of linear and nonlinear Volterra integral equations of the second kind. The fitting approach is carried out based on the annihilation of internal and external integral operators associated with VRK method. Here, we assume that the internal and external integral operators associated with VRKM annihilate the set functions $\{e^{\pm\omega x}, e^{\pm\nu x} y(s)\}$ with unknown frequencies $\omega, \ \nu \in \mathbb{R}$ (or $\omega, \ \nu \in i\mathbb{R}$). The optimum values of frequencies $\omega, \ \nu$ are obtained by minimizing the term of local truncation error.

2 Derivation of the Method

In this section, we first introduce previously developed Volterra-Runge-Kutta method in the first subsection and in next subsection we have constructed its exponential fitting. The construction of the fitting process is similar to the Berghe's approach [47–49] (see also Ixaru [54]). The motivation behind the fitting is that ef-version of VRKM can be used to solve VIE of the second kind having exponential/trigonometric solutions more efficiently. In the case of limit, when frequencies tending to zero, exponential-fitted VRK method becomes standard VRK method.

2.1 Volterra Runge-Kutta Method (VRK)

In this section, we give the formulation of VRK method proposed previously by several researchers [38–40]. In order to introduce VRKM for nonlinear VIE (1.1), we take $x_n = x_0 + nh$, $n = 0, 1, \dots N-1$ with $h = \frac{b-a}{N}$.

We denote y_n as the numerical approximation of the exact solution $y(x_n)$. We denote

$$F_n(x) = f(x) + \int_{x_n}^{x} k(x, s, y(s))ds, x \geq x_n \ (n = 0, 1, \dots N-1) \tag{2.1}$$

This is called the lag-term. And suppose $\tilde{F}_n(x)$ be an approximation of $F_n(x)$. An m-stage implicit Runge-Kutta method (VRK method) for VIE (1.1) is given by

$$Y_i^{[n]} = \tilde{F}_n(x_n + c_i h) + h\sum_{j=1}^{m} a_{ij} k\left(x_n + d_{ij}h, x_n + e_{ij}h, Y_j^{[n]}\right), \ i = 1, 2, \dots, m \tag{2.2}$$

$$y_{n+1} = \tilde{F}_n(x_n + h) + h\sum_{i=1}^{m} b_i k\left(x_n + e_i h, x_n + c_i h, Y_i^{[n]}\right). \tag{2.3}$$

where, (2.2) and (2.3) respectively called as internal and external stages. And $\tilde{F}_n(x_n + c_i h)$ is the approximation of lag-term (tail) and is defined in Eq. (2.1). The tail $\tilde{F}_n(x_n + c_i h)$ is the approximation to $F_n(x_n + c_i h)$ and it should be computed using quadrature formula of high order in such a way that it preserves the order of convergence of VRK method (2.2–2.3).

The first approach to compute tail approximation was proposed by Bellen et al. [38] and reanalysed in Conte et al. [43]. The approach of computing $\tilde{F}_n(x)$ is based on natural continuous extensions of the numerical solution by the interpolation formula of degree $d \leq p$.

$$u(x_n + \theta h) = \sum_{j=1}^{\mu} w_j(\theta) Y_j^{[n]}, \ n = 0, 1, \dots N-1, \ \theta \in [0, 1]. \tag{2.4}$$

where $w_j(\theta)$ are the polynomials of degree d with $\left\lfloor \frac{p}{2} \right\rfloor \leq d \leq \min\{\mu - 1, p\}$. This leads to the following tail approximation:

$$\tilde{F}_n(x) = f(x) + h\sum_{l=1}^{n}\sum_{j=1}^{\mu} \gamma_j k\left(x, x_{l-1} + \xi_j h, Y_j^{[l-1]}\right), \tag{2.5}$$

with γ_j as weights and ξ_j are abscissas (nodes), $j = 1, 2, \ldots, \mu$ which corresponds to a quadrature rule of order greater than or equal to p. The method corresponding to the tail approximation (2.4) was reported in the literature as an extended VRK method and in Conte et al. [43], it is referred to as natural VRK (NVRK) method.

If in Eqs. (2.2–2.3), $a_{ij} = 0,\ i \le j$ then corresponding VRKM is called explicit and if $a_{ij} = 0,\ i < j$ then VRKM is called diagonally implicit.

If $d_{ij} = c_i,\ e_i = 1,\ \theta_i = c_i$ the VRK is of Pouzet-type (see Pouzet [41]). If $d_{ij} = e_j,\ \theta_i = c_i$, then corresponding VRK will be designated as Bel'tyukov type [42].

2.2 Exponentially-Fitted Volterra-Runge-Kutta (ef-VRK) Method

To propose exponential-fitting of VRK method, we assume the kernel $k(x, s, y(s))$ is independent of s. Like in Conte et al. [43], Bellen et al. [38], Brunner et al. [40], we can take VIE (1.1) in the following simpler form:

$$y(x) = f(x) + \int_0^x k(x, y(s))ds, \quad x \in [0, X]. \tag{2.6}$$

We rewrite the VRK in the following form:
The internal stage as

$$Y_i^{[n]} = \tilde{F}_n(x_n + c_i h) + h \sum_{j=1}^{m} a_{ij} k\left(x_n + d_{ij} h, Y_j^{[n]}\right), \ i = 1, 2, \ldots, m \tag{2.7}$$

whereas, the external stage (objective) of the solution is

$$y_{n+1} = \tilde{F}_n(x_n + h) + h \sum_{i=1}^{m} b_i k\left(x_n + e_i h, Y_i^{[n]}\right). \tag{2.8}$$

The tail approximation $\tilde{F}_n(x)$, corresponding to the tail $F_n(x)$ is given by

$$\tilde{F}_n(x) = f(x) + h \sum_{l=1}^{n} \sum_{j=1}^{\mu} \gamma_j \ k\left(x, Y_j^{[l-1]}\right). \tag{2.9}$$

To compute the tail approximation in (2.9), we use the same ef-VRKM as in external (objective) stage of solution (2.8).

Following the approach proposed by Albrecht [52], for RK methods for ordinary differential equations (ODEs), we write the internal, external stages in functional form. To write this, we approximate Y_i by $y(x_n+c_ih)$ and $\tilde{F}_n(x_n+c_ih)$ by $F_n(x_n+c_ih)$ in the external and internal stages (2.7–2.8).

To construct the exponential-fitting, we follow the approach proposed by of Berghe's et al. [47–49] and Ixaru and Berghe [54] for ordinary differential equations (ODEs).

We define the linear integral operator corresponding to the external stage (2.8) as

$$L(y(x), k; h) = \int_{x}^{x+h} k(x+h, y(x))ds - h\sum_{i=1}^{m} b_i k(x+h, y[x+c_i h]). \tag{2.10}$$

And linear integral operator corresponding to the internal stage (2.7) is defined as

$$L_i(Y_i(x), k; h) = \int_{x}^{x+c_i h} k(x+c_i h, y(s))ds - h\sum_{j=1}^{m} a_{ij} k\big(x+c_j h, y[x+c_j h]\big), \tag{2.11}$$

where, $i = 1, 2, \ldots, m$ and in case of explicit ef-VRKM we take $m = i - 1$ in (2.7).

We assume that the integral operators defined in Eqs. (2.10–2.11) annihilate the set of functions

$$\big\{1, x, y(x) = e^{\pm\omega x}, k(x, y(s)) = e^{\pm v x} y(s)\big\}. \tag{2.12}$$

Then the annihilation of set of functions (2.12) by the operators L and L_i gives $L\big(e^{\pm\omega x}, e^{\pm v x} y(s); h\big) = 0$ and $L_i\big(e^{\pm\omega(x+c_i h)}, e^{\pm v(x+c_i h)} y((x+c_i h)); h\big) = 0$.

This gives the following order conditions for the exponential-fitting of explicit VRK method as:

$$\frac{e^{hv}\big(-1+e^{hw}\big)}{w} - h\sum_{i=1}^{m} e^{hv+hwc_i} b_i = 0, \tag{2.13}$$

$$\frac{e^{hv}\big(-1+e^{hw}\big)}{w} - h\sum_{i=1}^{m} e^{hv+hwc_i} b_i = 0, \tag{2.14}$$

$$\frac{e^{hwc_i}\big(-1+e^{hvc_i}\big)}{v} - h\sum_{j=1}^{i-1} a_{ij} e^{hvc_j+hwc_j} = 0. \tag{2.15}$$

Here, we restrict ourselves for $m = 2$ and we proposed the exponential fitting of 2-stage explicit VRKM whose coefficients are given by:

$$b_1 = -\frac{e^{hwc_1}\big(-1+e^{hw}\big)\big(-1+e^{hv+h(-v-w)+2hwc_2}\big)}{\big(e^{2hwc_1}-e^{2hwc_2}\big)hw}, \tag{2.16}$$

$$b_2 = -\frac{e^{hwc_2}\left(-1+e^{hw}\right)\left(-1+e^{hv+h(-v-w)+2hwc_1}\right)}{\left(-e^{2hwc_1}+e^{2hwc_2}\right)hw}, \tag{2.17}$$

$$a_{21} = \frac{e^{-hvc_1-hwc_1+hwc_2}\left(-1+e^{hvc_2}\right)}{hv}. \tag{2.18}$$

where, ω, ν are the unkonwn frequencies to be determined during the fitting process. From Eqs. (2.16–2.18) it is clear that under the limits $\omega \to 0$, $\nu \to 0$ with pre-assigned nodes $c_1 = 0$, $c_2 = 1$, one can get $b_1 = \frac{1}{2}$, $b_2 = \frac{1}{2}$ and $a_{21} = 1$ which is the standard Runge-Kutta method.

Here, we use the same ef-VRK method to compute the tail approximation of $\tilde{F}_n(x_n + c_i h)$.

3 Local Truncation Error

In the similar way, as in [44, 54], we define the local error for explicit 2-stage ef-VRK method at x_{n+1} is given by

$$LE_{n+1} = \left| \int_{x_n}^{x_n+h} k(x_n + h, y_n(s))ds - h\sum_{i=1}^{m} b_i k\left(x + h, Y_i^{[n]}\right)\right|, \tag{3.1}$$

where values of parameters b_1, b_2 and a_{21} are given in Eqs. (2.16–2.18) and

$$Y_i^{[n]} = \tilde{F}_n(x_n + c_i h) + h\sum_{j=1}^{i-1} a_{ij} k\left(x_n + d_{ij}h, Y_j^{[n]}\right),\ i = 1, 2 \tag{3.2}$$

and

$$y_n(x) = f(x) + \tilde{F}_n(x) + \int_{x_n}^{x} k(x, y_n(s)),\quad x \in [x_n, x_{n+1}],\ n \geq 0. \tag{3.3}$$

Following Taylor's approximation (like in the case of standard RK method), one can derive that $LE_{n+1} = O\left(h^3\right)$ i.e. the proposed ef-VRKM is of order two.

In the similar way, as by Ixaru and Berghe [54] (in the case of IVP in ODE), by minimizing the expression of (3.1), we can find the optimum value of frequencies ω, ν of solution and kernel respectively.

4 Numerical Experiments

Here, we apply ef-VRK method proposed in Sect. 2.2, to two VIEs given in Examples 1 and 2. We have also compared the results in Tables 1 and 2. In Tables 1 and 2, we report the norm of end point errors and compared it with the results obtained by previously developed method given in [43, 44].

Example 1 First, we consider the following non linear VIE of the second kind:

$$y(x) = e^{-x} + \int_0^x e^{s-x}\left(y(s) + e^{-y(s)}\right)ds, \quad x \in [0, 20] \tag{4.1}$$

with the exact solution $y(x) = \log(x + e)$.

We have implemented the ef-VRKM (Sect. 2.2) and standard RK (VRK) (Sect. 2.1) to solve numerically the VIE (4.1) for the step size $h = \frac{X-x_0}{N}$ with $N = 2^i$, $i = 6, 7, \ldots, 10$.

The absolute values of end point errors are compared in Table 1.

Like in Conte et al. [43], we denote

$e_n = |y(X) - y_N|$ as absolute endpoint error,

Table 1 Comparison of absolute end point error and order of the method for Example 1 with frequencies $\omega = 1, \ \nu = 1$

N	Our method (ef-VRKM)		Standard RK method (RKM)		NVRKM (Conte et al. [43])	
	cdy_N	$\tilde{p}$	cdy_N	$\tilde{p}$	cdy_N	$\tilde{p}$
64	3.15	–	1.58	–	2.05	–
128	3.75	2.0	2.18	2.0	3.16	3.69
256	4.35	2.0	2.78	2.0	4.32	3.84
512	4.96	2.02	3.38	2.0	5.50	3.92
1024	5.56	2.0	3.99	2.02	6.69	3.96

Table 2 Comparison of absolute end point error ($X = 5$) for Example 2, with frequencies $\omega = -2, \ \nu = -\frac{1}{2}$

Tol.	Step size	Our method (ef-VRKM)			NVRKM (Abdi et al. [44])		
		$\lvert y(X) - y_N\rvert$	cdy_N	p	$\lvert y(X) - y_N\rvert$	cdy_N	p
10^{-4}	0.0590	1.79×10^{-5}	4.74	–	5.83×10^{-4}	–	–
10^{-6}	0.0125	8.12×10^{-7}	6.09	1.99	2.67×10^{-5}	3.69	4.91
10^{-8}	0.0027	3.78×10^{-8}	7.42	2.00	1.54×10^{-6}	3.84	2.99
10^{-10}	0.0006	1.87×10^{-9}	8.72	1.99	6.62×10^{-8}	3.92	3.81
10^{-12}	0.0001	–	–	–	1.23×10^{-9}	3.96	3.96

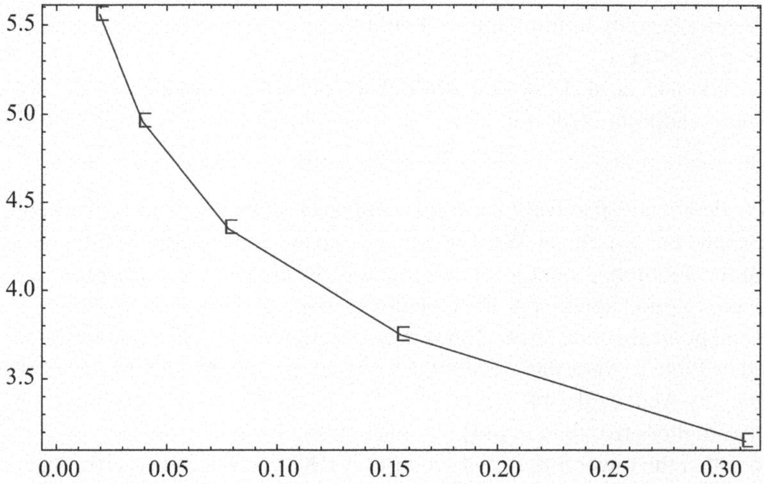

Fig. 1 Comparison of correct digits of the solution ($X = 20$) versus step size h obtained by 2nd order ef-VRKM (Sect. 2.2) for Example 1

$cdy_N = -\log_{10}(|y(X) - y_N|)$ as correct digits in endpoint solution,

and $\tilde{p} = \frac{(cdy_N - cdy_{N/2})}{\log_{10}(2)}$ as estimated order of the method.

The comparison of cdy_N and order $\tilde{p}$ are in good agreement with the errors given in Conte et al. [43]. It is observed from Table 1, that almost equal or less error is achieved by second order ef-VRK method where as NVRK (proposed in Conte et al. [43]) of order 4 is used to achieve the same accuracy. In other words, 2nd order ef-VRKM produces almost equal error with the errors obtained by 4th order NVRKM (Conte et at. [43]), which shows the great advantageous feature of our proposed development. Figure 1, is plotted to show the relationship between step sizes and correct digits in solution (logarithm of endpoint absolute error).

Example 2 Next, we consider the following linear VIE of second kind [44]

$$y(x) = \frac{1}{2}x^2 e^{-x} + \int_0^x \frac{(x-s)^2}{2} e^{s-x} y(s)ds, \quad x \in [0, 5]. \tag{4.2}$$

The exact solution of VIE (4.2) is given by [44]

$$y(x) = \frac{1}{3}\left(1 - e^{-\frac{3}{2}x}\left[\left(\cos\left(\frac{\sqrt{3}}{2}x\right) + \sqrt{3}\sin\left(\frac{\sqrt{3}}{2}x\right)\right)\right]\right). \tag{4.3}$$

We apply the ef-VRK method proposed in Sect. (2.2) to the VIE (4.2) and computed results are reported in Table 2, for varying step sizes h. Here, the step size h is calculated by minimizing the tolerance as the local error at $x_1 = x_0 + h$ and is denoted by $\text{est}(x_1) = |y(x_1) - y_1|$ [44].

Like in Conte et al. [43], and Abdi et al. [44], we denote $e_n = |y(X) - y_N|$ as absolute endpoint error and $cdy_N = -\log_{10}(|y(X) - y_N|)$ as correct digits in endpoint solution and $p = \frac{(cdy_{N_2} - cdy_{N_1})}{\log_{10}\left(\frac{ns_2}{ns_1}\right)}$ as estimated order of the method where ns_2, ns_1 denotes respectively the total number of steps involved in the solution at next level and previous level. Whereas cdy_{N_2}, cdy_{N_1} respectively denotes the correct digits in the solution at next level and previous level. The comparison of cdy_N and order p are in good agreement with the errors given in Abdi et al. [44].

The end point absolute error, with varying tolerance and respective step sizes h are reported in Table 2. Also, these endpoint errors are compared with the errors obtained previously by Abdi et al. [44]. From Table 2, it is inferred us that the present method produces smaller error than in [44]. Another strong benefit of our proposed ef-VRK method is that the lower order (2nd order) ef-VRKM produces less error than higher order (almost 4th order) NVRKM (proposed by Abdi et al. [44]). This fact certainly indicates the cost minimization of the method. In other words, it is inferred that our proposed ef-VRKM is cost efficient. As to implement or propose higher order methods, one would require large number of parameters, large number of times steps. Figure 2, is plotted to show the effect of step size on end point absolute error. For both Examples 1 and 2, the CPU time (in seconds) is reported in Table 3 for varying step-sizes. It is observed in Table 3, the time consumption of the proposed ef-VRNM is some fraction of minutes. This fact assures the cost-efficiency and superiority of the proposed method.

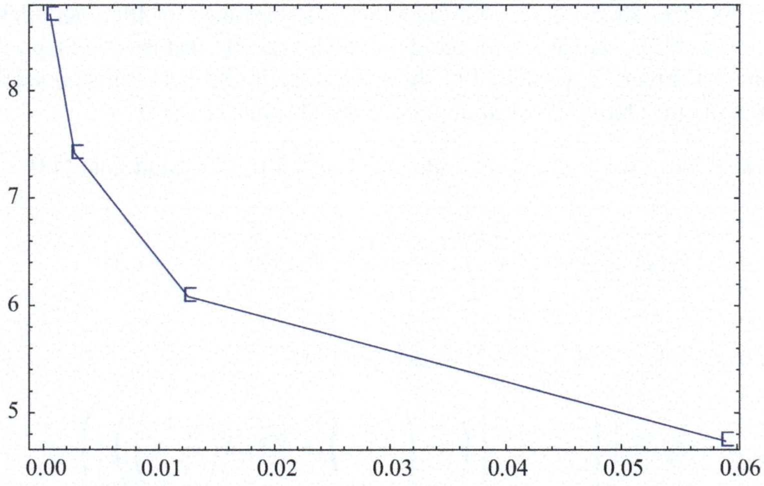

Fig. 2 Comparison of correct digits of the solution ($X = 5$) versus step size h obtained by 2nd order ef-VRKM (Sect. 2.2) for Example 2

Table 3 CPU Time for Execution of ef-VRKM for Examples 1 and 2

Step size h	Example 1	Example 2
	CPU time (In Seconds)	CPU time (In Seconds)
0.3125	1.685	1.498
0.15625	2.325	1.623
0.078125	5.179	2.076
0.0390625	16.397	3.839
0.01953125	59.077	10.688

5 Conclusion

In this paper, we successfully constructed the exponential-fitting of Volterra-Runge-Kutta (VRK) method to find the numerical solution of Volterra integral equation (VIE) of the second kind. The proposed method has tested on two numerical experiments of nonlinear and linear VIE of the second kind. The results are reported in Tables 1, 2 and Figs. 1, 2. A comparison of correct digits (logarithm of absolute endpoint errors) in the solution for Examples 1, 2 are made in Tables 1 and 2 respectively. The results, in Tables, show that the present method is very appropriate and best tuned to solve VIEs of the second kind. A great benefit of present ef-VRKM is that only 2nd order ef-VRKM is sufficient to produce very less endpoint error whereas to achieve the same accuracy, fourth or higher-order NVRKM (previously developed) is required. In Table 3, CUP time for Examples 1 and 2 is shown for varying step-sizes and it is found to be some fraction of minutes. This fact shows the cost-effectiveness of the proposed method. Figures 1 and 2 are plotted for step size versus the correct digits in the numerical solution for Examples 1 and 2 respectively. Another feature of the exponentially-fitted version is that it converted into standard RK method under the limits $\omega \to 0, \nu \to 0$. These methods have a wide scope of applicability to solve Volterra integral equations which occurred frequently in science and engineering such as diffusion problems, electrostatics, plasma physics, mechanics etc.

To improve the accuracy of the ef-VRK method, we can develop these methods for variable step sizes. In other words, the step size is chosen as per the pre-assigned tolerance of the numerical scheme. Like the recent research paper, Abdi et al. [44], the construction of exponentially-fitting of natural Volterra-Runge-Kutta method with variable step size will be the next aim of the research in this direction. Another attractive future development will include the construction of exponentially-fitted RK type method to solve Volterra integro-differential equations having exponential/trigonometric solution.

References

1. Bartoshevich, M.A.: A heat-conduction problem. J. Eng. Phys. **28**(2), 240–244 (1975)
2. Yousefi, S.A.: Numerical solution of Abel's integral equation by using Legendre wavelets. Appl. Math. Comput. **175**(1), 574–580 (2006)
3. Galdi, G.P., Pileckas, K., Silvestre, A.L.: On the unsteady poiseuille flow in a pipe. Z. Angew. Math. Phys. **58**(6), 994–1007 (2007)
4. McKee, S., Cuminato, J.A.: Nonlocal diffusion, a Mittag-Leffler function and a two-dimensional Volterra integral equation. J. Math. Anal. Appl. **423**(1), 243–252 (2015)
5. Ding, H.J., Wang, H.M., Chen, W.Q.: Analytical solution for the electroelastic dynamics of a nonhomogeneous spherically isotropic piezoelectric hollow sphere. Arch. Appl. Mech. **73**(1–2), 49–62 (2003)
6. Kit, G.S., Maksymuk, A.V.: The method of Volterra integral equations in contact problems for thin-walled structural elements. J. Math. Sci. **90**(1), 1863–1867 (1998)
7. Farengo, R., Lee, Y.C., Guzdar, P.N.: An electromagnetic integral equation: Application to microtearing modes. Phys. Fluids **26**(12), 3515–3523 (1983)
8. Hansen, P.C., Jensen, T.K.: Large-scale methods in image deblurring, vol. 4699. LNCS Lect. Notes. Comput, Sc (2007)
9. Manzhirov, A.V.: On a method of solving two-dimensional integral equations of axisymmetric contact problems for bodies with complex rheology. J. Appl. Math. Mech. **49**(6), 777–782 (1985)
10. Radlow, J.: A two-dimensional singular integral equation of diffraction theory. Bull. Amer. Math. Soc. **70**(4), 596–599 (1964)
11. Mirkin, M.V., Bard, A.J.: Multidimensional integral equations. part 1. A new approach to solving microelectrode diffusion problems. J. Electroad. Chem. **323**(1–2), 1–27 (1992)
12. Diogo, T.: Collocation and iterated collocation methods for a class of weakly singular Volterra integral equations. J. Comput. Appl. Math. **229**(2), 363–372 (2009)
13. Brunner, H.: Collocation Methods for Volterra Integral and Related Functional Differential Equations. Cambridge University Press, New York (2004)
14. Brunner, H., Blom, J.G.: The numerical solution of nonlieaner Volterra integral equations of the second kind by collocation and iterated collocation methods. SIAM J. Sci. Stat. Comput. **8**, 806–830 (1987)
15. Zhang, S., Lin, Y., Rao, M.: Numerical solutions for second-kind Volterra integral equations by Galerkin methods. Appl. Math. **45**(1), 19–39 (2000)
16. Wazwaz, A.M.: Linear and Nonlinear Integral equations: Methods and Applications. Higher Education Press and Springer Verlag, Heidelberg (2011)
17. Linz, P.: Product integration methods for Volterra integral equations of the first kind. BIT **11**(4), 413–421 (1971)
18. Rebelo, M., Diogo, T.: A hybrid collocation method for a nonlinear Volterra integral equation with weakly singular kernel. J. Comput. Appl. Math. **234**(9), 2859–2869 (2010)
19. Sloss, B.G., Blyth, W.F.: Corrington's walsh function method applied to a nonlinear integral equation. J. Integral Equ. Appl. **6**(2), 239–256 (1994)
20. Aziz, I., Siraj-Ul-Islam: New algorithms for the numerical solution of nonlinear Fredholm and Volterra integral equations using Haar wavelets, J. Comput. Appl. Math. **239**(1), 333–345 (2013)
21. Galperin, E.A., Kansa, E.J., Makroglou, A., Nelson, S.A.: Variable transformations in the numerical solution of second kind Volterra integral equations with continuous and weakly singular kernels; extensions to Fredholm integral equations. J. Comput. Appl. Math. **115**(1–2), 193–211 (2000)
22. Lima, P., Diogo, T.: Numerical solution of a nonuniquely Solvable Volterra integral equation using extrapolation methods. J. Comput. Appl. Math. **140**(1–2), 537–557 (2002)
23. Katani, R., Shahmorad, S.: Block by block method for the systems of nonlinear Volterra integral equations. Appl. Math. Model. **34**(2), 400–406 (2010)

24. Odibat, Z.M.: Differential transform method for solving Volterra integral equation with separable kernels. Math. Comput. Model. **48**(7–8), 1144–1149 (2008)
25. Chen, W., Lin, W.: Galerkin trigonometric wavelet methods for the natural boundary integral equations. Appl. Math. Comput. **121**(1), 75–92 (2001)
26. Guoqiang, H., Jiong, W.: Richardson extrapolation of iterated discrete Galerkin solution for two-dimensional Fredholm integral equations. J. Comput. Appl. Math. **139**, 49–63 (2002)
27. Carutasu, V.: Numerical solution of two-dimensional nonlinear Fredholm integral equations of the second kind by spline functions. General. Math. **9**, 31–48 (2001)
28. Guoqiang, H., Jiong, W.: Extrapolation of Nyström solution for two dimensional nonlinear Fredholm integral equations. J. Comput. Appl. Math. **134**(1–2), 259–268 (2001)
29. Babolian, E., Bazm, S., Lima, P.: Numerical solution of nonlinear two-dimensional integral equations using rationalized Haar functions. Commun. Nonlinear. Sci. Numer. Simulat. **16**(3), 1164–1175 (2011)
30. Tari, A., Rahimi, M.Y., Shahmorad, S., Talati, F.: Solving a class of two-dimensional linear and nonlinear Volterra integral equations by the differential transform method. J. Comput. Appl. Math. **228**(1), 70–76 (2009)
31. Abdou, M.A., Badr, A.A., Soliman, M.B.: On a method for solving a two-dimensional nonlinear integral equation of the second kin. J. Comput. Appl. Math. **235**(12), 3589–3598 (2011)
32. Pedas, A., Vainikko, G.: Product integration for weakly singular integral equations in m dimensional space. In: Bertram. B., Constanda, C., Struthers, A. (Eds.) Integral Methods in Science and Engineering, pp. 280–285. Chapman and Hall/CRC (2000)
33. Atkinson, K.E.: The numerical evaluation of fixed points for completely continuous operators. SIAM J. Numer. Anal. **10**, 799–807 (1973)
34. Bazm, S., Babolian, E.: Numerical solution of nonlinear two-dimensional Fredholm integral equations of the second kind using Gauss product quadrature rules. Commun. Nonlinear. Sci. Numer. Simulat. **17**(3), 1215–1223 (2012)
35. Assari, P., Dehghan, M.: The approximate solution of nonlinear Volterra integral equations of the second kind using radial basis functions. Appl. Numer. Math. **131**, 140–157 (2018)
36. Singh, H., Baleanu, D., Srivastava, H.M., Dutta, H., Jha, N.K.: Solution of multi-dimensional Fredholm equations using Legendre scaling functions. Appl. Numer. Math. **150**, 313–324 (2020)
37. Singh, C.S., Singh, H., Singh, V.K., Singh, O.P.: Fractional order operational matrix methods for fractional singular integro-differential equation. Appl. Math. Model. **40**, 10705–10718 (2016)
38. Bellen, A., Jackiewicz, Z., Vermiglio, R., Zennaro, M.: Natural continuous extensions of Runge-Kutta methods for Volterra integral equations of the second kind and their applications. Math. Comput. **52**, 49–63 (1989)
39. Bellen, A., Jackiewicz, Z., Vermiglio, R., Zennaro, M.: Stability analysis of Runge-Kutta methods for Volterra integral equations of the second kind. IMA J. Numer. Anal. **10**, 103–118 (1990)
40. Brunner, H., Hairer, E., Nørsett, S.P.: Runge-Kutta theory for Volterra integral equations of the second kind. Math. Comput. **39**, 147–163 (1982)
41. Pouzet, P.: Etude en vue de leur traitement numérique des équations intégrales de type Volterra. Rev. Français Traitement Information (Chiffres) **6**, 79–112 (1963)
42. Bel'tyukov, B.A.: An analogue of the Runge-Kutta method for the solution of nonlinear integral equations of Volterra type. Differ. Equ. **1**, 417–433 (1965)
43. Conte, D., D'Ambrosio, R., Izzo, G., Jackiewicz, Z.: Natural Volterra Runge-Kutta methods. Numer. Algorithms **65**, 421–445 (2014)
44. Abidi, A., Hojjati, G., Jackiewicz, Z., Madi, H.: A new codde for Volterra integral equations based on natural Runge-Kutta methods. Appl. Numer. Math. **143**, 35–50 (2019)
45. Gautschi, W.: Numerical integration of ordinary differential equations based on trigonometric polynomials. Numer. Math. **3**, 381–397 (1961)
46. Lyche, T.: Chebyshevian multistep methods for ordinary differential equations. Numer. Math. **19**, 65–75 (1972)

47. Berghe, G.V., De Meyer, H., Van Daele, M., Van Hecke, T.: Exponentially-fitted explicit Runge-Kutta methods. Comput. Phys. Comm. **123**, 7–15 (1999)
48. Berghe, G.V., De Meyer, H., Van Daele, M., Van Hecke, T.: Exponentially-fitted Runge-Kutta methods. J. Comput. Appl. Math. **125**, 107–115 (2000)
49. Berghe, G.V., Ixaru, L.G., Van Daele, M.: Optimal implicit exponentially-fitted Runge-Kutta methods. Comp. Phys. Comm. **140**, 346–357 (2001)
50. Paternoster, B.: Runge-Kutta(-Nyström) methods for ODEs with periodic solutions based on trigonometric polynomials. Appl. Numer. Math. **28**, 401–412 (1998)
51. Simos, T.E.: An exponentially-fitted Runge-Kutta method for the numerical integration of initial value problems with periodic or oscillating solutions. Comp. Phys. Comm. **115**, 1–8 (1998)
52. Albrecht, P.: A new theoretical approach to RK methods. SIAM J. Numer. Anal. **24**, 391–406 (1987)
53. Ixaru, L.G., Berghe, G.V., De Meyer, H.: Frequency evaluation in exponential fitting multistep algorithms for ODEs. J. Compt. Appl. Math. **140**, 423–434 (2002)
54. Ixaru, L.G., Vanden Berghe, G.: Exponential Fitting. Kluwer Academic Publishers, Dordrecht (2004)
55. Ozawa, K.: A functional fitting Runge-Kutta-Nyström method with variable coefficients. Jpn. J. Ind. Appl. Math. **18**, 105–130 (2001)

An Integral Equation Method for Wave Scattering by a Pair of Horizontal Porous Plates

Sanjib Naskar, Souvik Kundu, and R. Gayen

Abstract A mathematical method comprising of solutions to a pair of coupled hypersingular integral equations for studying the interaction of water waves with a pair of horizontal porous plates submerged at different depths in case of finite as well as infinite depth water domain is developed in the context of linear water wave theory. The porosities of the two plates are arbitrary but uniform across the plates. In practice, as the still seawater level may change significantly due to tide, the use of dual horizontal plates instead of a single horizontal plate has been proposed by researchers. So far two plates of equal length placed at different depths have been considered in such a manner that one plate fully overlaps the other one i.e. one of the plates lies exactly above or below the other plate. Mostly eigenfunction expansion and matching technique has been employed to study such problems. The purpose of the present work is to examine the effect of a pair of horizontal permeable plates of different orientations situated at different depths on the propagation of surface waves. Here we study the hydrodynamic performances of a dual porous plates system for the three cases viz. when the plates are kept in a step-like manner, when there is a partial overlap between the plates and when the plates are fully overlapped. This enables us to compare the effectiveness of different models mentioned above for the three different cases. The assumption of Darcy's law as the boundary condition for fluid across the plates and the appropriate use of Green's integral technique in the water domain reduce the corresponding boundary value problem into two coupled hypersingular integral equations involving the unknown potential difference functions across the plates. The coupled integral equations are solved numerically and with the help of these solutions, the numerical estimates for the reflection coefficient,

S. Naskar · S. Kundu · R. Gayen (✉)
Department of Mathematics, Indian Institute of Technology Kharagpur,
Kharagpur 721302, India
e-mail: rupanwita.gayen@gmail.com

S. Naskar
e-mail: sanjibnaskar62@gmail.com

S. Kundu
e-mail: kundusouvik1988@gmail.com

H. Singh et al. (eds.), *Topics in Integral and Integro-Differential Equations*,
Studies in Systems, Decision and Control 340,
https://doi.org/10.1007/978-3-030-65509-9_9

the transmission coefficient, vertical wave forces acting on the plates and the energy-loss coefficient are computed. The correctness of the numerical results is checked through an energy identity relation for porous plates and by comparing the present computed results with the previous results available in the literature. New results related to different orientations of the plates, different submergence depths, varying lengths and different values of permeability of the plates are depicted.

Keywords Water wave scattering · Horizontal porous plates · Coupled hypersingular integral equations

1 Introduction

A submerged horizontal plate can be used as an offshore breakwater for protecting coastal structures and shorelines. The advantage of using a horizontal plate as a breakwater is that it is less dependent on the depth of the local sea bed and its geometrical condition. Also, it is cost-effective. Moreover, if the horizontal breakwater is submerged, it does not obstruct the sea-view and allows the free exchange of seawater between the sheltered and the open regions which prevents water pollution. A submerged horizontal plate also serves as an efficient energy device [1, 2]. Thus the use of a submerged horizontal plate as a breakwater is environmental-friendly and suitable for many purposes.

Water wave scattering by a submerged horizontal rigid plate was initially studied by Heins [3] and Burke [4] employing Weiner-Hopf technique. The hydrodynamic performances of a submerged horizontal plate have been investigated by many researchers (cf. [5–9]). From all these studies, it has been realized that the incident waves are reflected effectively by a submerged horizontal rigid plate, but the vertical force on the plate is relatively high. Thus the use of a porous plate was proposed instead of a rigid plate in order to reduce the vertical wave load on the plate. Yu and Chwang [10] employed the boundary element method to investigate the problem of water waves over a submerged horizontal porous plate and they exhibited that the plate with proper permeability can significantly reduce the wave force acting on the plate keeping transmission at a low level. The related studies on the horizontal porous plate can be found in the works of Chwang and Wu [11], Cho and Kim [12]. They employed a standard matched eigenfunction expansion method together with complex dispersion relation for considering the effect of submerged porous plate, whereas Liu and Li [13], Cho and Kim [14] studied the wave motion over a submerged horizontal porous plate based on the eigenfunction expansion method without finding the complex roots of the dispersion relation. The study of Liu et al. [15], where they investigated water wave scattering by a submerged horizontal porous plate of finite thickness, represents the fact that with larger width of the plate, more wave energy can be dissipated but with a larger thickness of the plate, the value of the transmission coefficient may not remain at a low level.

All the studies mentioned in the previous paragraph refer to a single porous plate. Apart from studying the effect of a single plate, many researchers investigated the combined effects of a horizontal plate with other different structures as well as the wave interactions with multiple barriers. By means of the eigenfunction matching method, Wu et al. [16] studied wave reflection by a submerged horizontal porous plate attached to a wall. Water wave motion by a perforated wall breakwater with an internal horizontal plate was investigated by Yip and Chwang [17]. Neelamani and Gayathri [18] experimentally examined the wave motion through dual horizontal solid plates fixed on the free surface and supported by piles. Liu et al. [19] studied water wave scattering by a perforated wall breakwater with a submerged horizontal porous plate based on the method of eigenfunction expansion. Later, Liu et al. [20] also studied the hydrodynamic performance of a submerged breakwater with a lower horizontal solid plate and an upper horizontal porous plate. They revealed the fact that with suitable permeability of the upper plate, the vertical wave force acting on both the plates can be reduced significantly. Cho et al. [21] studied the problem of oblique wave scattering by dual submerged horizontal porous plates. They developed both the multi-domain boundary element method solution and an analytic solution. They also carried out an experiment and compared their results with the experimental data. They found that there exist optimal values for the porous effect parameter, plate width, submergence depth of the upper plate and inclination angle of the incoming waves to enhance the performance of the dual plates. They also showed that when the gap between the plates is greater than 10% of the water domain's depth, no significant effect can be found by adding a lower porous plate. Liu and Li [22] carried out further experiments to provide some more useful data in terms of numerical results. Sasmal and De [23] investigated the problem of diffraction of oblique waves by two vertical porous barriers with non-identical submerged gaps by using Havelock's expansion and single-term Galerkin approximations. The study of Sasmal et al. [24], where they examined the effect of porosity on oblique wave diffraction by two unequal vertical porous barriers by using one-term Galerkin approximations, shows that in presence of porosity, the distance and length between the plates play an important role in the scattering behavior of the surface waves.

Integral equation is an important tool to solve boundary value problems arising in the field of mathematical physics. In the recent past, the development of different analytical and numerical techniques of solving integral and integro-differential equations can be found in the works of researchers working in different fields related to the applications of integral equations (cf. [25–29]). The integral equations that arise in the studies of water wave scattering are mostly singular. They may contain weakly singular kernels [30–32], Cauchy kernel [33, 34] or hypersingular kernel [35]. The involvement of coupled singular integral equations can be found in [36–40]. The hypersingular integral approach is an efficient way to tackle plate problems with arbitrary geometry. The use of this technique can be found in the works of [41–45] and others.

In this present study, the hydrodynamic performance of a pair of horizontal porous plates submerged at different depths is investigated in the context of linear water wave potential theory by reducing the governing boundary value problem to a set of hypersingular integral equations. So far two plates of equal lengths placed at different depths have been considered in such a manner that one plate fully overlaps the other one, that is, one of the plates lies exactly above or below the other plate. Mostly eigenfunction expansion and matching technique has been employed to study such problems. The purpose of the present work is to apply an alternative method to examine the effect of a pair of horizontal permeable plates of different orientations situated at different depths on the propagation of surface waves. Here we study the hydrodynamic performances of a dual porous plates system for the three cases viz. when the plates are kept in a step-like manner, when there is a partial overlap between the plates and when the plates are (almost) fully overlapped. This enables us to compare the effectiveness of different models mentioned above for the three different cases. To find the analytical solution for the present problem, we apply a methodology similar to [39]. We first obtain an expression for the velocity potential at an arbitrary point of the fluid by using standard technique of application of Green's integral theorem in the fluid region. The use of boundary conditions on the plates then produces two coupled hypersingular integral equations involving potential difference functions across the plates. These unknown functions are determined numerically using Chebyshev expansions and collocation method and using these solutions, we compute various hydrodynamic quantities. We validate our present numerical results with published results and new results related to different orientations of the plates are presented and analyzed through several figures.

2 Mathematical Formulation

Three different problems, according to the orientations of the plates as shown in Fig. 1, are considered in the two-dimensional Cartesian coordinate system. The x-axis is taken along the free surface and the y-axis is directed vertically downwards. Two thin horizontal porous plates Γ_1 and Γ_2 of lengths a_1 and a_2 respectively are submerged in water at two different depths d_1 and d_2 respectively. Here the permeability of the plates are assumed to be arbitrary but uniform across the plates. The water is homogeneous, inviscid, incompressible and the motion is assumed to be irrotational, so that the velocity potential $\Phi(x, y, t)$ exists. Assuming a time-harmonic motion of

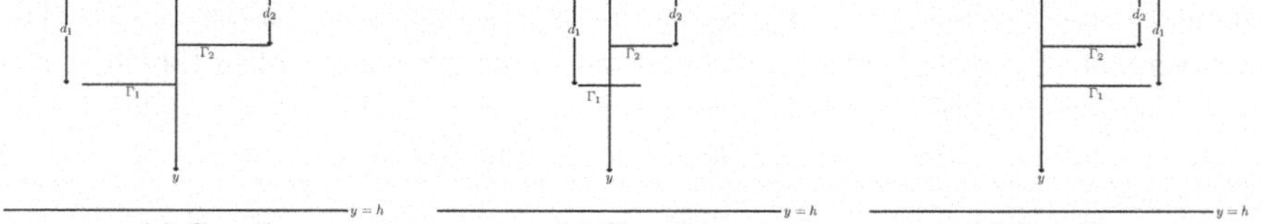

(a) Case 1 (b) Case 2 (c) Case 3

Fig. 1 Schematic diagrams of three different orientations of the plates

frequency σ, the velocity potential can be written as $\Phi(x, y, t) = \Re\{\frac{g^2}{\sigma^3}\phi(x, y)e^{-i\sigma t}\}$. Then $\phi(x, y)$ satisfies the Laplace equation

$$\frac{\partial^2\phi}{\partial x^2} + \frac{\partial^2\phi}{\partial y^2} = 0, \text{ in the water domain,} \tag{1}$$

along with the linearized free surface boundary condition

$$K\phi + \phi_y = 0 \text{ on } y = 0, \tag{2}$$

where $K = \sigma^2/g$ and the bottom boundary condition

$$\begin{aligned} \phi_y = 0, \quad &\text{on} \quad y = h \quad \text{for finite depth water (FD)}, \\ \nabla\phi \to 0, \quad &\text{as} \quad y \to \infty \quad \text{for deep water (ID)}. \end{aligned} \tag{3}$$

$\phi(x, y)$ also satisfies the radiation conditions

$$\phi(x, y) \to \begin{cases} T\phi_{inc}(x, y) & \text{as } x \to \infty, \\ \phi_{inc}(x, y) + R\phi_{inc}(-x, y) & \text{as} x \to -\infty, \end{cases} \tag{4}$$

where R and T are the unknown reflection and transmission coefficients and the incident wave potential $\phi_{inc}(x, y)$ is given by

$$\phi_{inc}(x, y) = F(y; \mu)e^{i\mu x}, \tag{5}$$

where

$$F(y; \mu) = \begin{cases} \dfrac{\cosh \mu(h - y)}{\cosh \mu h} & \text{for FD}, \\ e^{-\mu y} & \text{for ID}, \end{cases} \tag{6}$$

and $\mu = k_0$ or K according as the water is of uniform finite depth or deep and k_0 is the unique positive real root of the equation $K = k \tanh kh$.

Finally, the boundary condition on the plates are presented as (cf. [46])

$$\frac{\partial\phi}{\partial y}(\Gamma_i{}^+) = \frac{\partial\phi}{\partial y}(\Gamma_i{}^-) = -i\mu G_i[\phi](p_i), \; p_i \in \Gamma_i \quad (\text{for} \quad i = 1, 2), \tag{7}$$

where $[\phi](p_i)$ denotes the potential difference across the plate Γ_i and G_i represents the porous-effect parameter defined as

$$G_i = \frac{\gamma_i(f_i + iS_i)}{\mu b(f_i^2 + S_i^2)},$$

with γ_i denoting the porosity of the plate Γ_i; f_i and S_i indicate the resistance and the inertial force coefficients respectively and b stands for the thickness of the porous medium.

At the ends of the plates, the velocity potential behaves as

$$\nabla\phi \to O(r^{-1/2}) \text{ as } r \to 0, \tag{8}$$

where r denotes the distance of any point in the fluid from either edge of the plates.

3 Method of Solution

In this section, we reduce the above boundary problem to a system of coupled integral equations and find their solutions numerically. For that, we express the fundamental potential function $\mathcal{G}(x, y; \xi, \eta)$ due to a line source submerged at a point $q = (\xi, \eta)$ as

$$\mathcal{G}(x, y; \xi, \eta) = \frac{1}{2\pi}\ln\frac{r}{r'} - \frac{1}{\pi}\int_0^\infty P_1(y, \eta; k)\cos k(x-\xi)\mathrm{d}k$$
$$-\frac{1}{\pi}\oint_0^\infty \frac{F(y; k)F(\eta; k)}{\Delta(k)}\cos k(x-\xi)\mathrm{d}k, \tag{9}$$

where

$$P_1(y, \eta; k) = \begin{cases} \dfrac{\mathrm{e}^{-kh}\sinh ky \sinh k\eta}{k\cosh kh} & \text{for FD,} \\ 0 & \text{for ID,} \end{cases} \tag{10}$$

and

$$\Delta(k) = \begin{cases} \dfrac{k\sinh kh - K\cosh kh}{\cosh kh} & \text{for FD,} \\ k - K & \text{for ID,} \end{cases} \tag{11}$$

and the path of the integration is taken along the positive k axis indented below the pole at $k = \mu$ and $r, r' = \{(x-\xi)^2 + (y \mp \eta)^2\}^{1/2}$.

We apply Green's integral theorem to the source potential $\mathcal{G}(x, y; \xi, \eta)$ and the scattered potential $\phi_{sc}(x, y) = \phi(x, y) - \phi_{inc}(x, y)$ in the fluid region bounded externally by the lines $y = 0, -X \le x \le X$; $x = -X, 0 \le y \le Y$; $y = Y, -X \le x \le X$; $x = X, 0 \le y \le Y$; and internally by a small circle of radius ϵ centered at (ξ, η) and the contours enclosing the plate Γ_i and ultimately making $X \to \infty$, $Y \to h$ or ∞ according as the fluid is of uniform finite depth or deep and $\epsilon \to 0$, we get

$$\phi(\xi,\eta) = \phi_{inc}(\xi,\eta) - \int_{\Gamma_1} [\phi](q_1)\frac{\partial \mathcal{G}}{\partial y}(q_1;q)\mathrm{d}s_{q_1} - \int_{\Gamma_2} [\phi](q_2)\frac{\partial \mathcal{G}}{\partial y}(q_2;q)\mathrm{d}s_{q_2}, \quad q_i \in \Gamma_i, \tag{12}$$

where $[\phi](q_i)$ denotes the potential difference across the plate Γ_i.

Now taking normal derivative on both sides of the Eq. (12) at an arbitrary point $p_1 = (\xi_1, \eta_1)$ on Γ_1 and using the boundary condition on the plate Γ_1 as given in Eq.(7), we obtain an integro-differential equation

$$\frac{\partial \phi_{inc}}{\partial \eta}(p_1) = \frac{\partial}{\partial \eta}\int_{\Gamma_1} [\phi](q_1)\frac{\partial \mathcal{G}}{\partial y}(q_1;p_1)\mathrm{d}s_{q_1} + \frac{\partial}{\partial \eta}\int_{\Gamma_2} [\phi](q_2)\frac{\partial \mathcal{G}}{\partial y}(q_2;p_1)\mathrm{d}s_{q_2}$$
$$- \mathrm{i}\mu G_1[\phi](p_1), \quad p_1 \in \Gamma_1. \tag{13}$$

Here we note that in the above equation the kernel $\frac{\partial \mathcal{G}}{\partial y}(q_1; p_1)$ is singular, whereas the kernel $\frac{\partial \mathcal{G}}{\partial y}(q_2; p_1)$ is non-singular. Thus in order to change the order of integration and differentiation in the first term of right-hand side of the Eq. (13), the integral is treated as a Hadamard finite part integral. Hence we obtain the following hypersingular integral equation

$$\frac{\partial \phi_{inc}}{\partial \eta}(p_1) = \oint_{\Gamma_1} [\phi](q_1)\frac{\partial^2 \mathcal{G}}{\partial y \partial \eta}(q_1;p_1)\mathrm{d}s_{q_1} + \int_{\Gamma_2} [\phi](q_2)\frac{\partial^2 \mathcal{G}}{\partial y \partial \eta}(q_2;p_1)\mathrm{d}s_{q_2}$$
$$- \mathrm{i}\mu G_1[\phi](p_1), \quad p_1 \in \Gamma_1. \tag{14}$$

It may be noticed that in the above integral equation two unknown functions $[\phi](q_1)$ and $[\phi](q_2)$ are present. Thus Eq. (14) alone is not sufficient for finding the two unknown functions. In order to obtain a second integral equation, we take normal derivative on both sides of the Eq. (12) at a point $p_2 = (\xi_2, \eta_2)$ on Γ_2 and using the boundary condition on the plate Γ_2 as given in Eq. (7), we obtain the second hypersingular integral equation as given below.

$$\frac{\partial \phi_{inc}}{\partial \eta}(p_2) = \int_{\Gamma_1} [\phi](q_1)\frac{\partial^2 \mathcal{G}}{\partial y \partial \eta}(q_1;p_2)\mathrm{d}s_{q_1} + \oint_{\Gamma_2} [\phi](q_2)\frac{\partial^2 \mathcal{G}}{\partial y \partial \eta}(q_2;p_2)\mathrm{d}s_{q_2}$$
$$- \mathrm{i}\mu G_2[\phi](p_2), \quad p_2 \in \Gamma_2. \tag{15}$$

Now for the different orientations of the plates, we parameterize the points on the plate Γ_i (for $i = 1, 2$) in the following ways:

Case 1. When the plates are kept in a step-like manner:
In this case $\Gamma_1 = [-a_1, 0]$ and $\Gamma_2 = [0, a_2]$. Here the parametric representations of points $p_i \equiv (\xi_i, \eta_i)$ and $q_i \equiv (x_i, y_i)$ on $\Gamma_i (i = 1, 2)$ are given by

$$x_i = \frac{a_i}{2}(v + (-1)^i), \quad y_i = d_i, \quad -1 \leq v \leq 1;$$
$$\xi_i = \frac{a_i}{2}(s + (-1)^i), \quad \eta_i = d_i, \quad -1 \leq s \leq 1. \tag{16}$$

Case 2. When there is a partial overlap between the plates:
In this case $\Gamma_1 = [-a_1/2, a_1/2]$ and $\Gamma_2 = [0, a_2]$ and the parametric representations of the points $p_i \equiv (\xi_i, \eta_i)$ and $q_i \equiv (x_i, y_i)$ on $\Gamma_i (i = 1, 2)$ are given by

$$x_i = \frac{a_i}{2}(v + 1/2(1 + (-1)^i)), \quad y_i = d_i, \quad -1 \le v \le 1;$$
$$\xi_i = \frac{a_i}{2}(s + 1/2(1 + (-1)^i)), \quad \eta_i = d_i, \quad -1 \le s \le 1. \tag{17}$$

Case 3. When one of the plates almost overlaps the other one:
In this case $\Gamma_1 = [0, a_1]$ and $\Gamma_2 = [0, a_2]$. The parametric representations of the points $p_i \equiv (\xi_i, \eta_i)$ and $q_i \equiv (x_i, y_i)$ on $\Gamma_i (i = 1, 2)$ are given by

$$x_i = \frac{a_i}{2}(v + 1), \quad y_i = d_i, \quad -1 \le v \le 1; \quad \xi_i = \frac{a_i}{2}(s + 1), \quad \eta_i = d_i, \quad -1 \le s \le 1. \tag{18}$$

According to the different orientations of the plates, we consider the corresponding parameterization on the plates. With these parameterizations, the integral equations (14) and (15) transform to

$$\tau_i(s) = -\frac{1}{2\pi}\sum_{j=1}^{2} \int_{-1}^{1} g_j(v)\frac{\delta_{ij}}{(v - s)^2}\mathrm{d}v + \sum_{j=1}^{2}\int_{-1}^{1} g_j(v)\mathcal{K}_{ij}(s, v)\mathrm{d}v$$
$$- \frac{1}{2}\mathrm{i}\mu a_i G_i g_i(s), \quad i = 1, 2, \tag{19}$$

where

$$\tau_i(s) = -\frac{1}{2}a_i\mu H(\eta_i; \mu)\mathrm{e}^{\mathrm{i}\mu\xi_i}, \quad [\phi](q_j) \equiv g_j(v), \quad \delta_{ij} = \begin{cases} 1 \; i = j, \\ 0 \; i \ne j, \end{cases} \quad i, j = 1, 2,$$

and

$$\mathcal{K}_{ij}(s, v) = \frac{a_i a_j}{8\pi}\left[\delta_{ij}\left\{\frac{Z_{ij}^2 - X_{ij}^2}{Z_{ij}^2 + X_{ij}^2} - 2M(X_{ij}, \eta_i, y_j)\right\}\right.$$
$$\left. + (1 - \delta_{ij})\left\{\frac{Y_{ij}^2 - X_{ij}^2}{Y_{ij}^2 + X_{ij}^2} + \frac{Z_{ij}^2 - X_{ij}^2}{Z_{ij}^2 + X_{ij}^2} - 2M(X_{ij}, \eta_i, y_j)\right\}\right], \tag{20}$$

with

$$X_{ij} = x_j - \xi_i, \; Y_{ij} = y_j - \eta_i, \; Z_{ij} = y_j + \eta_i,$$

and

$$M(X_{ij}, \eta_i, y_j) = \int_0^\infty k^2 P_2(y_j, \eta_i; k) \cos kX_{ij} \mathrm{d}k + \oint_0^\infty \frac{k^2}{\Delta(k)} H(\eta_i; k) H(y_j; k) \cos kX_{ij} \mathrm{d}k, \quad (21)$$

where

$$P_2(y, \eta; k) = \begin{cases} \dfrac{\mathrm{e}^{-kh} \cosh ky \cosh k\eta}{k \cosh kh} & \text{for FD,} \\ 0 & \text{for ID,} \end{cases} \quad (22)$$

and

$$H(y; k) = \begin{cases} \dfrac{\sinh k(h-y)}{\cosh kh} & \text{for FD,} \\ \mathrm{e}^{-ky} & \text{for ID.} \end{cases} \quad (23)$$

We note that $g_j(\pm 1) = 0$. We solve the coupled integral equations numerically by approximating the unknown function $g_j(v)$ for $(j = 1, 2)$. For that, we take

$$g_j(v) = \sqrt{1 - v^2} \sum_{n=0}^{N} b_n^j U_n(v), \quad j = 1, 2, \quad (24)$$

where b_n^j are unknown coefficients to be determined and $U_n(v)$ is the Chebyshev polynomial of the second kind of order n. Substitution of the above approximation for $g_j(v)$ into the integral equations (19) and collocation at the points $u = u_k (k = 0, 1, 2, ..., N)$ result into the system of $2(N + 1)$ equations with $2(N + 1)$ unknowns, given as:

$$\sum_{j=1}^{2} \sum_{n=0}^{N} b_n^j B_{ij}^n(u_k) = \tau_i(u_k), \ k = 0, 1, 2, ..., N, \ \ i = 1, 2 \quad (25)$$

where u_k are the roots of the Chebyshev polynomial of the first kind given as $u_k = \cos \frac{2k+1}{2N+2}\pi$ and the coefficients $B_{ij}^n(u_k)$ are given by

$$B_{ij}^n(u_k) = \frac{(n+1)\delta_{ij}}{2} U_n(u_k) + \int_{-1}^{1} \sqrt{1 - v^2} U_n(v) \mathcal{K}_{ij}(u_k, v) \mathrm{d}v - \frac{1}{2} \mathrm{i}\mu a_i G_i \delta_{ij} \sqrt{1 - v^2} U_n(u_k). \quad (26)$$

It may be mentioned that the above approach of solving a hypersingular integral equation by expanding in terms of Chebyshev polynomials multiplied by an appropriate weight function and then collocating at discrete points is a well established method (cf. [41]). Its convergence and error analyses were given in [47, 48].

4 Physical Quantities

4.1 Reflection Coefficient and Transmission Coefficient

To obtain the expressions for the reflection and the transmission coefficients, we use the far-field behavior of the function $\mathcal{G}(x, y; \xi, \eta)$ as given by (cf. [49])

$$\mathcal{G}(x, y; \xi, \eta) \to -\mathrm{i}C_o F(y; \mu) F(\eta; \mu) \mathrm{e}^{\pm \mathrm{i}\mu(x-\xi)} \quad \text{as } x \to \pm\infty, \tag{27}$$

where

$$C_0 = \begin{cases} \dfrac{2\cosh^2(k_0 h)}{2k_0 h + \sinh(2k_0 h)} & \text{for FD,} \\ 1 & \text{for ID.} \end{cases} \tag{28}$$

Using the asymptotic behavior of the Green's function $\mathcal{G}(x, y; \xi, \eta)$ as given in expression (27) and the radiation condition as given in expression (4), we deduce the analytical expressions of the reflection and transmission coefficients as given below.

$$R = -\frac{\mathrm{i}}{2} C_0 \mu \sum_{j=1}^{2} \sum_{n=0}^{N} a_j b_n^j \int_{-1}^{1} \sqrt{1-v^2}\, U_n(v) H(y_j; \mu) \mathrm{e}^{\mathrm{i}\mu x_j} \mathrm{d}v, \tag{29}$$

and

$$T = 1 - \frac{\mathrm{i}}{2} C_0 \mu \sum_{j=1}^{2} \sum_{n=0}^{N} a_j b_n^j \int_{-1}^{1} \sqrt{1-v^2}\, U_n(v) H(y_j; \mu) \mathrm{e}^{-\mathrm{i}\mu x_j} \mathrm{d}v. \tag{30}$$

4.2 Energy Identity

The present study deals with the problems of wave scattering by dual submerged horizontal porous plates and the porous medium imposes some resistance to the fluid. As a result of this when a train of incoming waves interacts with the porous plates, some wave energy dissipates. To find an energy balance relation, we apply Green's integral theorem to $\phi(x, y)$ and its complex conjugate $\bar{\phi}(x, y)$ in the fluid region bounded by $x = \pm X$, $0 \le y \le Y$; $y = 0$, $-X \le x \le X$; $y = Y$, $-X \le x \le X$ and the contours enclosing both the plates and ultimately making $X \to \infty$, $Y \to h$ or ∞ according as the water is of finite depth or infinite depth and shrinking the contours enclosing the plates along their sides, we obtain

$$J = 1 - |R|^2 - |T|^2, \tag{31}$$

where

$$J = \mu C_0 \sum_{i=1}^{2} \Re\{G_i\} a_i \int_{-1}^{1} | \sqrt{1-v^2} \sum_{n=1}^{N} a_n^i U_n(v) |^2 \,\mathrm{d}v. \tag{32}$$

Here J indicates the energy loss coefficient which signifies the amount of wave energy dissipated at the surfaces of the porous plates.

4.3 Hydrodynamic Forces

The pressure $p(x, y, t)$ at any point (x, y) in the fluid region is given by

$$p(x, y, t) = -\rho \frac{\partial \Phi}{\partial t}(x, y, t) + \rho g y = \Re \left\{ \frac{\mathrm{i}\rho g^2}{\sigma^2} \phi(x, y) \mathrm{e}^{-\mathrm{i}\sigma t} \right\} + \rho g y, \tag{33}$$

where ρ is the density of the fluid. The expression of the hydrodynamic force per unit length of the plate Γ_i is given by

$$\mathcal{F}_i = \frac{\rho g}{K} \int_{\Gamma_i} \Re\{\mathrm{i}[\phi](q_i)\mathrm{e}^{-\mathrm{i}\sigma t}\} \mathrm{d}s_{q_i}, \quad i = 1, 2. \tag{34}$$

Now we first use the parameterization on the plate Γ_i. Then the approximate form of the potential difference $\big([\phi](q_i) \equiv g_i(v)\big)$ given in Eq. (24) is substituted into the expression (34). This yields the dimensionless wave force F_i acting on the plate Γ_i in the following form:

$$F_i = \left| \frac{K\mathcal{F}_i}{\rho g a_i} \right| = \left| \frac{1}{2} \Re \left\{ \mathrm{i} \sum_{n=0}^{N} b_n^i \int_{-1}^{1} \sqrt{1-v^2} U_n(v) \mathrm{d}v \right\} \right|, \quad i = 1, 2. \tag{35}$$

5 Numerical Results and Discussions

In this section, we depict the numerical values of different physical quantities against the dimensionless wavenumber for different values of the governing physical parameters in case of finite as well as infinite depth water. The parameters are made dimensionless with respect to the uniform finite depth h when we consider finite depth water, whereas a characteristic length L (cf. [36]) is introduced for this purpose when deep water is considered. The first two tables (Tables 1 and 2) represent convergence analysis and support our consideration for the truncation number N for all numerical computations. The next two tables (Table 3 and Table 4) represent validation of numerical values of the reflection and transmission coefficients against the energy identity relation in case of finite depth and deep water respectively. Again,

Table 1 Convergence of $|R|$ with truncation number N in case of finite depth water

N	$k_0h = 1$	$k_0h = 1.5$	$k_0h = 2$
6	0.1245	0.2031	0.5314
8	0.1321	0.2111	0.5193
10	0.1311	0.2121	0.5182
12	0.1311	0.2121	0.5182

Table 2 Convergence of $|R|$ with truncation number N in case of deep water

N	$KL = 1$	$KL = 1.5$	$KL = 2$
6	0.0247	0.4231	0.4614
8	0.0324	0.4411	0.4713
10	0.0317	0.4421	0.4782
12	0.0317	0.4421	0.4782

Table 3 A check on energy identity relation in case of finite depth water

k_0h	$\|R\|$	$\|T\|$	J	$1 - \|R\|^2 - \|T\|^2$
1	0.2801	0.6732	0.4679	0.4683
1.5	0.2472	0.4612	0.7259	0.7262
2	0.0908	0.5308	0.7110	0.7100

Table 4 A check on energy identity relation in case of deep water

KL	$\|R\|$	$\|T\|$	J	$1 - \|R\|^2 - \|T\|^2$
1	0.3186	0.4452	0.7008	0.7006
1.5	0.0815	0.4372	0.8024	0.8022
2	0.2626	0.6123	0.5562	0.5561

through the first set of figures of this section (see Fig. 2), we compare our results based on the present analysis with those available in the literature.

In case of finite depth water, a convergence analysis for the values of $|R|$ with increasing values of N is presented in Table 1 for the Case 1 for three different values of the dimensionless wavenumber k_0h. Here the fixed parametric values are $d_1/h = 0.25$, $d_2/h = 0.15$, $a_1/h = 1$, $a_2/h = 1$, $G_1 = 0.5$, $G_2 = 1$. A similar convergence analysis for the case of deep water is presented in Table 2 for the Case 1 with the parametric values $d_1/L = 0.3$, $d_2/L = 0.1$, $G_1 = 0.5$, $G_2 = 1$, $a_1/L = a_2/L = 1$. Table 1 and Table 2 show that the truncation number $N = 10$ is sufficient for getting results correct up to four decimal places.

To establish the correctness of our numerical results, we first check the energy identity relation (31) for few values of dimensionless wavenumbers in case of finite as well as infinite depth water. In case of finite depth water, the results are listed in

Table 3 for the partially overlapped plates. In this case, the fixed values of dimensionless physical parameters are $d_1/h = 0.25$, $d_2/h = 0.15$, $a_1/h = 1$, $a_2/h = 1$, $G_1 = 0.5$, $G_2 = 1$. In case of deep water, the results are listed in Table 4 for the partially overlapped plates with fixed choices of parametric values as $d_1/L = 0.3$, $d_2/L = 0.1$, $G_1 = 0.5$, $G_2 = 1$, $a_1/L = a_2/L = 1$. Here the reflection and transmission coefficients are determined using expressions (29) and (30) respectively, while the energy loss J is determined from the expression (32).

Through Figs. 2a–c, we validate the present method by comparing the numerical results for particular cases with the results of [9, 14, 20]. Figure 2 shows the comparison of present results for the values of $|R|$, $|T|$ with the results of [20] where they considered the case of complete overlapping of barriers with an upper horizontal porous plate and a lower horizontal solid plate. Figure 2 is generated considering $d_1/h = 0.35$, $d_2/h = 0.1$, $a_1/h = 1$, $a_2/h = 1$, $G_1 = 0$, $G_2 = 0.7958$. Figure 2b shows the comparison of $|R|$ between the present results and the results of [9] where he studied the wave interaction with a single submerged horizontal thin rigid plate. In order to compare our results with those in [9], we select the parameters as $d_1/h = 0.1$, $a_1/h = 1$, $a_2/h = 0.001$, $G_1 = 0$. Consideration of negligible upper plate length allows us to compare our result with that for a single submerged horizontal plate. In Fig. 2c, the values of energy loss coefficients are compared between the present results and the results of [14] where they studied the problem of wave interaction with a submerged horizontal porous plate. In Fig. 2c, the parametric values are $d_1/h = 1$, $a_1/h = 1$, $a_2/h = 0.001$, $G_1 = 5/\pi$. Figures 2a–c show that the present results agree quite well with the results obtained by [9, 14, 20] which strongly establish the validity of the present analysis.

To obtain the new results, three cases have been investigated depending on three different orientations of the plates as depicted in Fig. 1.

5.1 *Results Related to Finite Depth Water*

In this subsection, we present the new results related to finite depth water. Figures 3a–c show the effect of dimensionless gap length $(d_1 - d_2)/h$ between the plates on the reflection coefficients for Case 1, Case 2 and Case 3 respectively. Here the lower plate Γ_1 is fixed at a depth $d_1/h = 0.25$ and the submergence depth of the upper plate Γ_2 is gradually increased. The other fixed parametric values in each case are $G_1 = 1$, $G_2 = 1$, $a_1/h = 1$, $a_2/h = 1$. Figure 3a and b show that for Case 1 and Case 2, as the gap between the plates decreases, that is, as the submergence depth of the upper plate increases, the reflection coefficient decreases. One obvious reason behind this phenomenon is that when the submergence depth of the upper porous plate increases, the system of dual porous plates allows more amount of incident waves to transmit on the other side. On the other hand, reverse behaviour can be observed in Fig. 3c that is for Case 3. Though in this case, the peak values of the reflection coefficient are significantly lower than the other two cases, but as the submergence depth of the upper plate increases keeping the lower plate at a fixed submergence depth, the

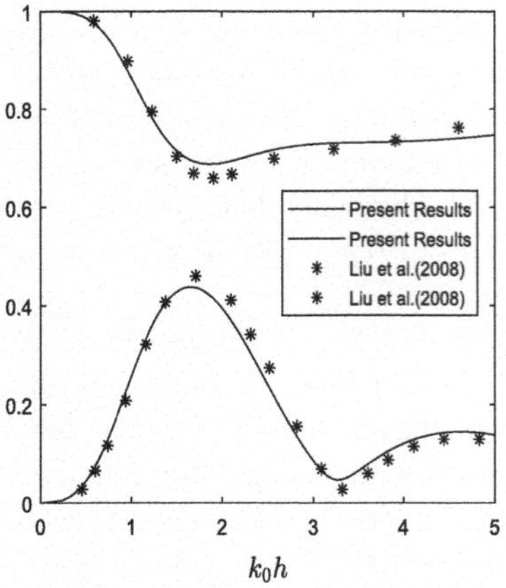

(a) Comparison of our results with Liu et al. [20] for $d_1/h = 0.35, d_2/h = 0.1, a_1/h = 1, a_2/h = 1, G_1 = 0, G_2 = 0.7958$.

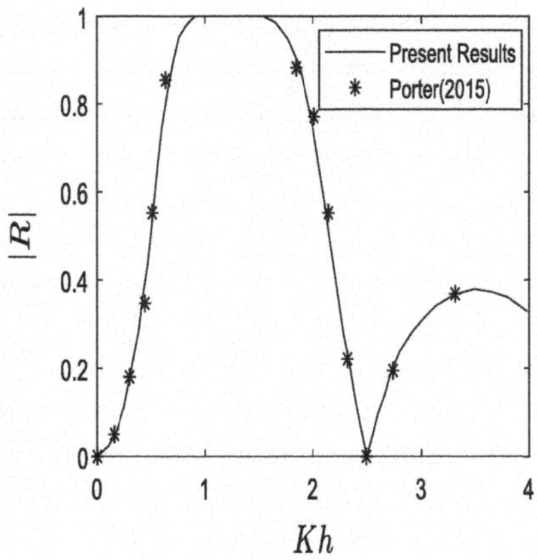

(b) Comparison between results related to the present method and that of Porter [9] with $d_1/h = 0.1, a_1/h = 1, a_2/h = 0.001, G_1 = 0$.

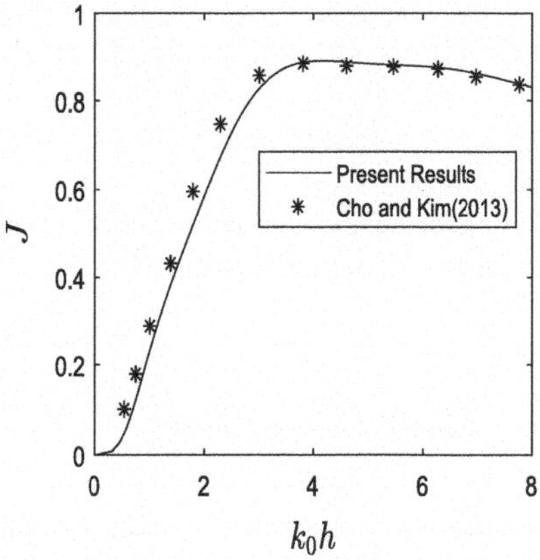

(c) Comparison between the results derived by our method and that of Cho and Kim [14] for $d_1/h = 1, a_1/h = 1, a_2/h = 0.001, G_1 = 5/\pi$.

Fig. 2 Comparison between present results and other published results

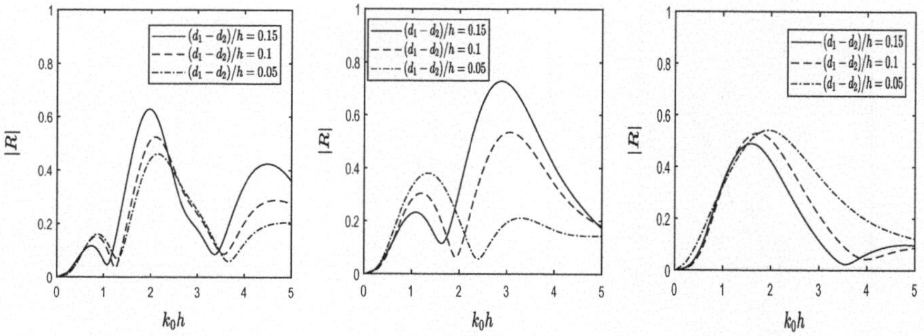

(a) Case 1 (b) Case 2 (c) Case 3

Fig. 3 $|R|$ versus dimensionless wavenumber k_0h for different values of $(d_1 - d_2)/h$ with $d_1/h = 0.25$, $a_2/h = 1$, $a_1/h = 1$, $G_1 = 1$, $G_2 = 1$

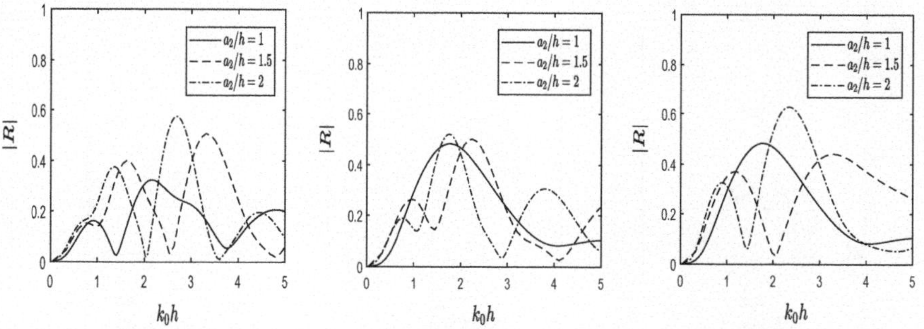

(a) Case 1 (b) Case 2 (c) Case 3

Fig. 4 $|R|$ versus dimensionless wavenumber k_0h for different values of a_2/h with $d_1/h = 0.25$, $d_2/h = 0.15$, $a_1/h = 1$, $G_1 = 1$, $G_2 = 1$.

reflection coefficient increases for higher values of incident wave frequency. This dissimilar behaviour for the Case 3 may occur due to the fact that the decreasing gap length between two fully overlapped symmetric porous plates creates a confined zone in between them. Because of multiple interactions of incoming waves with dual porous plates in this confined region, the effect of dual plates in reflecting incoming waves becomes prominent.

Figures 4a–c and 5a–c depict the impact of upper plate length on the absolute values of the reflection and the transmission coefficients respectively for different orientations of the plates. Fixed parametric values are $d_1/h = 0.25$, $d_2/h = 0.15$, $G_1 = G_2 = 1$, and $a_1/h = 1$. From Figs. 4a–c, it is almost clear that as the length of the upper plate increases, the peak value and the oscillating nature of the reflection coefficient also increases. As the upper plate is situated nearer the free surface, a longer upper plate puts hindrance on the propagation of surface waves resulting in greater reflection. The corresponding transmission coefficients are depicted in Figs. 5a–c. It can be seen from these figures that for a low-frequency region ($k_0h < 2$),

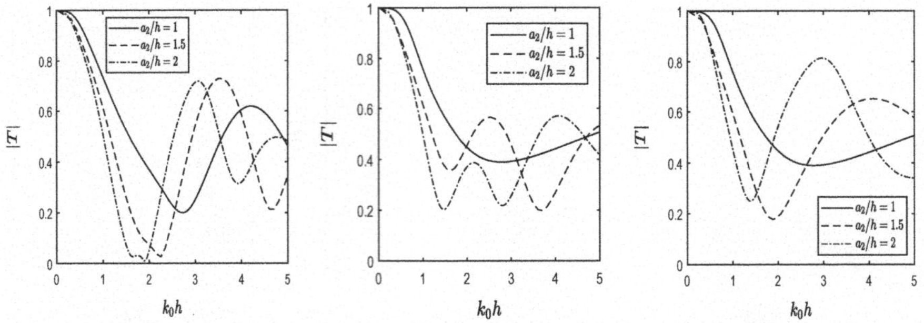

(a) Case 1 (b) Case 2 (c) Case 3

Fig. 5 $|T|$ versus dimensionless wavenumber k_0h for different values of a_2/h with $d_1/h = 0.25, d_2/h = 0.15, a_1/h = 1, G_1 = 1, G_2 = 1$

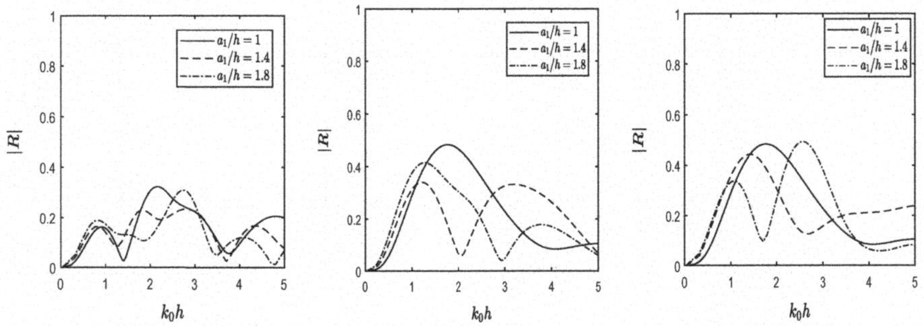

(a) Case 1 (b) Case 2 (c) Case 3

Fig. 6 $|R|$ versus dimensionless wavenumber k_0h for different values of a_1/h with $d_1/h = 0.25, d_2/h = 0.15, a_2/h = 1, G_1 = 1, G_2 = 1$

as the length of the upper plate increases, the value of the transmission coefficient decreases. Beyond this range of wave frequency, the graphs of the transmission coefficient start showing an oscillating nature in almost all three cases.

Figures 6a–c and 7a–c demonstrate the effects of the lower plate length on the reflection and the transmission coefficients respectively. Here the values of the fixed parameters in all three cases are $d_1/h = 0.25$, $d_2/h = 0.15$, $G_1 = 1$, $G_2 = 1$, $a_2/h = 1$. From Figs. 6a–c, it can be observed that for lower values of dimensionless wavenumber ($k_0h \leq 1$), the reflection coefficient increases as the plate length increases. But for higher values of wavenumber, there is no uniform pattern of increasing or decreasing nature of the reflection coefficient. It can be noticed from Figs. 7a–c that when $k_0h < 2$, with the increasing lower plate length and with increasing value of wavenumber the transmission coefficient decreases monotonically. But after that range of k_0h, a fluctuating nature of these curves can be observed.

In Figs. 8a–c, the absolute values of reflection coefficient are plotted against the dimensionless wavenumber k_0h for three different values of $G_1 (= 1, 2, 3)$ for Case 1, Case 2 and Case 3 respectively. Here the other parametric values for all the three

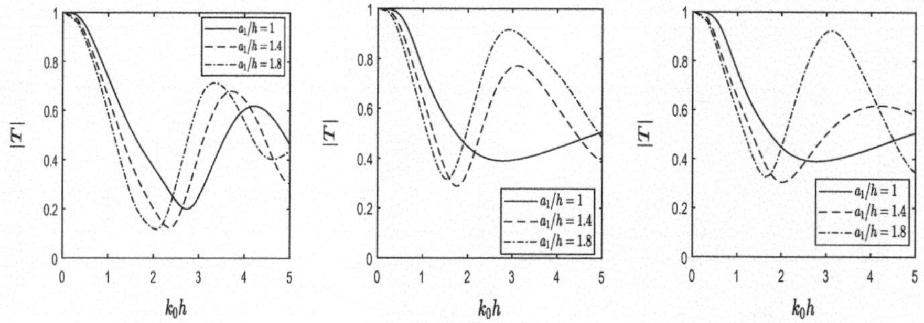

(a) Case 1 (b) Case 2 (c) Case 3

Fig. 7 $|T|$ versus dimensionless wavenumber k_0h for different values of a_1/h with $d_1/h = 0.25, d_2/h = 0.15, a_2/h = 1, G_1 = 1, G_2 = 1$

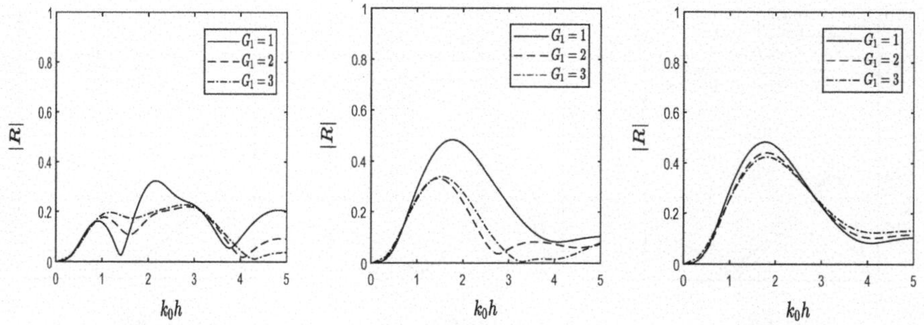

(a) Case 1 (b) Case 2 (c) Case 3

Fig. 8 $|R|$ versus dimensionless wavenumber k_0h for different values of G_1 with $d_1/h = 0.25, d_2/h = 0.15, G_2 = 1, a_1/h = 1, a_2/h = 1$

cases are $d_1/h = 0.25, d_2/h = 0.15, G_2 = 1, a_1/h = 1, a_2/h = 1$. In all the three figures (Figs. 8a–c), it can be noticed that for a lower frequency region ($k_0h \leq 1$), the effect of porous parameter G_1 of the lower plate is insignificant. In the first two cases (Fig. 8a and c), for a higher frequency region, as the permeability of the lower plate increases, reflection coefficient decreases. This is because of the fact that as the permeability of the lower plate increases, more amount of waves get absorbed by the plates and as a result of that reflection coefficient decreases. But, from Fig. 8c, it is clear that in case of fully overlapped dual porous plates, the effect of the permeability of the lower plate on reflecting wave energy remains insignificant throughout the depicted range of wavenumber. From Fig. 8a–c, it is also visible that if the two porous plates are placed in a step-like manner, the reflection by the system of plates in this arrangement is the least compared to the cases when the plates are either partially or fully overlapped.

Figure 9a–c display the effect of permeability of the upper porous plate on the absolute values of the reflection coefficient for three different values of $G_2 (= 1, 2, 3)$ in all three cases (Case 1, Case 2 and Case 3). Here in all the three cases, we choose

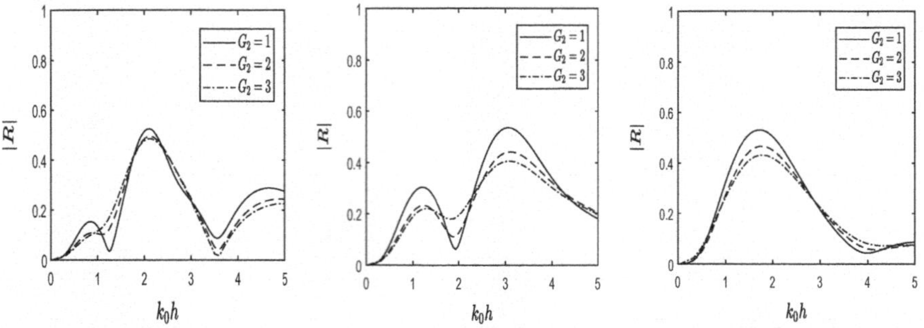

(a) Case 1 (b) Case 2 (c) Case 3

Fig. 9 $|R|$ versus dimensionless wavenumber k_0h for different values of G_2 with $d_1/h = 0.25, d_2/h = 0.15, G_1 = 0.5, a_1/h = 1, a_2/h = 1$

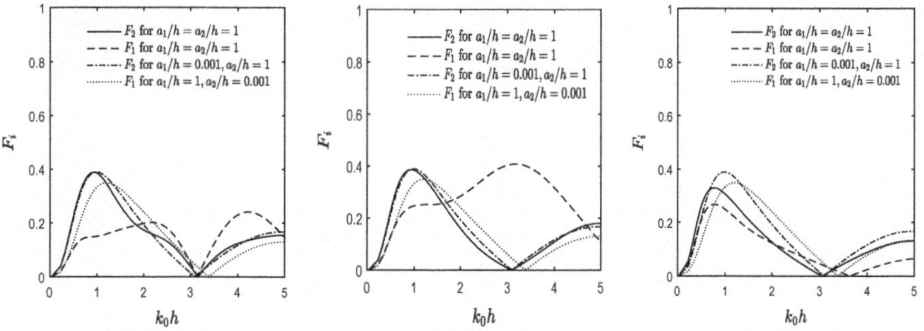

(a) Case 1 (b) Case 2 (c) Case 3

Fig. 10 Comparison of hydrodynamic forces acting on single and dual plate system in finite depth water for $d_1/h = 0.25, d_2/h = 0.15, G_1 = 1, G_2 = 1$

$G_1 = 0.5$ and other parametric values are the same as in Fig. 8. Reference [10] had presented the fact that as the permeability of a single submerged horizontal porous plate increases, the reflection coefficient decreases monotonically. Here in Fig. 9a–c, we observe a similar trend with an exception that the decreasing nature is not uniform.

Comparisons between hydrodynamic forces on dual horizontal porous plates of different orientations and that on a single horizontal porous plate are depicted in Fig. 10a–c. In each of these figure, the solid curve and the dashed curve denote the values of hydrodynamic forces acting on the upper and lower plates respectively for the dual porous plate system with $d_1/h = 0.25, d_2/h = 0.15, G_1 = 1, G_2 = 1, a_1/h = 1, a_2/h = 1$, whereas the dash-dot curves denote the values of hydrodynamic force acting on the upper plate with $d_1/h = 0.25, d_2/h = 0.15, G_1 = 1, G_2 = 1, a_1/h = 0.001, a_2/h = 1$ (i.e., in absence of the lower plate) and the dot-dot lines denote the values of hydrodynamic force acting on the lower plate with $d_1/h = 0.25, d_2/h = 0.15, G_1 = 1, G_2 = 1, a_1/h = 1, a_2/h = 0.001$ (i.e., there is no upper plate). From Fig. 10a and b, it is observed that for Case 1 and Case 2, the

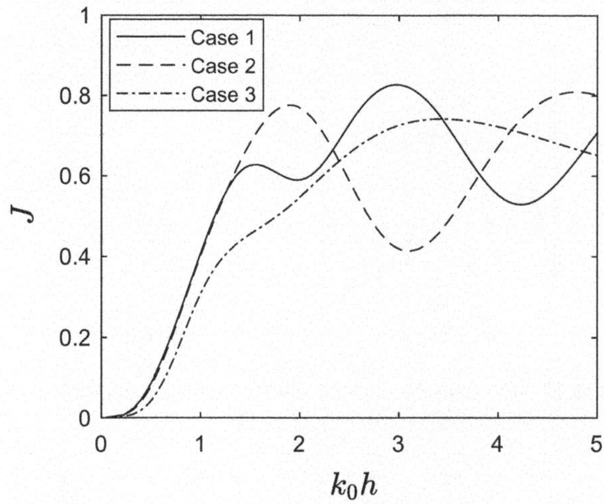

Fig. 11 Energy loss coefficient J versus dimensionless wavenumber k_0h in finite depth water for $G_1 = 0.5, G_2 = 2$, $d_1/h = 0.25, d_2/h = 0.15, a_1/h = 1, a_2/h = 1$

hydrodynamic force acting on the upper plate of the dual porous plates system is not much affected by the addition of a lower plate. On the contrary, the addition of a lower plate significantly reduces the amount of force on the upper plate for the case of two fully overlapped plates (Case 3, Fig. 10c). Figure 10c also shows that for moderate values of dimensionless wavenumber ($k_0h > 1$), the presence of the upper plate reduces the hydrodynamic force acting on the lower plate because the upper plate provides a shelter to the lower plate.

In Fig. 11, the energy loss coefficient is plotted as a function of dimensionless wavenumber k_0h for three different cases. The following fixed parametric values are $d_1/h = 0.25$, $d_2/h = 0.15$, $G_1 = 0.5$, $G_2 = 2$, $a_1/h = 1$, $a_2/h = 1$. It can be seen from Fig. 11 that in Case 3, as the dimensionless wavenumber k_0h increases, the energy loss coefficient monotonically increases until it reaches the maximum value and then it monotonically decreases. However, in Case 1 and Case 2, for an initial range of k_0h, the energy loss coefficient monotonically increases with increasing dimensionless wavenumber, but after that range ($k_0h > 1.5$), it shows an oscillatory nature.

5.2 *Results Related to Deep Water*

In this subsection, we present all the results related to deep water domain. Figure 12a–c show the effect of dimensionless gap length $(d_1 - d_2)/L$ between the lower and upper plates on the reflection coefficient in Case 1, Case 2 and Case 3 respectively. Here the upper plate is fixed at a submergence depth $d_2/L = 0.1$ and the depth of the lower plate is gradually increasing. The fixed values of other parameters are $G_1 = 0.5$, $G_2 = 1$, $a_1/L = 1$, $a_2/L = 1$. From Fig. 12a–c, it is visible that for each

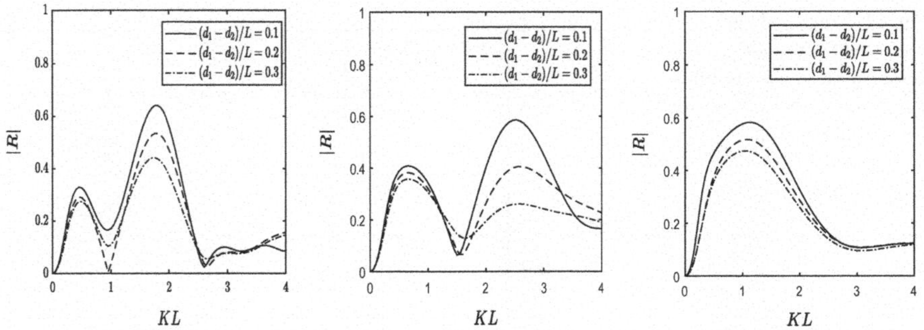

(a) Case 1 (b) Case 2 (c) Case 3

Fig. 12 $|R|$ versus dimensionless wavenumber KL for different values of $(d_1 - d_2)/L$ with $d_2/L = 0.1$, $G_1 = 0.5$, $G_2 = 1$, $a_1/L = 1$, $a_2/L = 1$

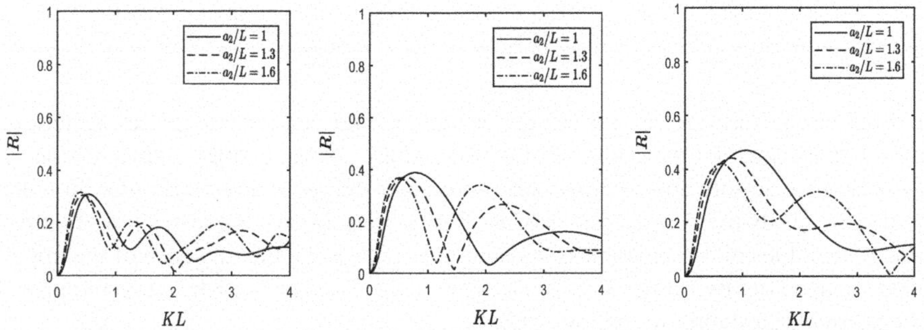

(a) Case 1 (b) Case 2 (c) Case 3

Fig. 13 $|R|$ versus dimensionless wavenumber KL for different values of a_2/L with $d_1/L = 0.3$, $d_2/L = 0.1$, $G_1 = 1.5$, $G_2 = 1$, $a_1/L = 1$

case the absolute values of the reflection coefficient decrease with the increasing dimensionless gap length between the porous plates. This is because of the fact that as the lower plate moves deeper into the water, its effect diminishes and as a whole, the system of dual porous plates put less obstruction to the surface waves. Also, it can be observed from Fig. 12c that the graphs are more regular (oscillation free) in nature compared to the other two cases.

The effects of the length of the upper plate on $|R|$ and $|T|$ with respect to dimensionless wavenumber KL are depicted in Fig. 13a–c and in Fig. 13a–c respectively for the following parametric values: $d_1/L = 0.3$, $d_2/L = 0.1$, $G_1 = 1.5$, $G_2 = 1$, $a_1/L = 1$. From Fig. 14a–c, it can be observed that in all the three cases, within an initial range of the dimensionless wavenumber, the value of $|R|$ increases with increasing upper plate length. Though this increasing nature is not clearly distinguishable. But, after achieving peak values $|R|$ start decreasing in an oscillating manner. Similarly, for increasing upper plate length the values of $|T|$ decrease significantly

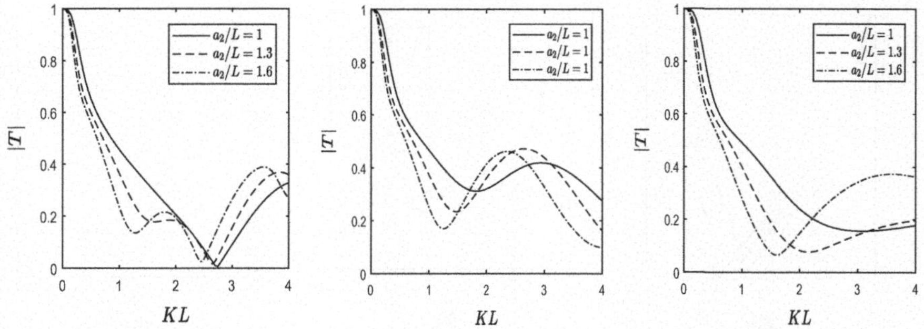

(a) Case 1 (b) Case 2 (c) Case 3

Fig. 14 $|T|$ versus dimensionless wavenumber KL for different values of a_2/L with $d_1/L = 0.3, d_2/L = 0.1, G_1 = 1.5, G_2 = 1, a_1/L = 1$

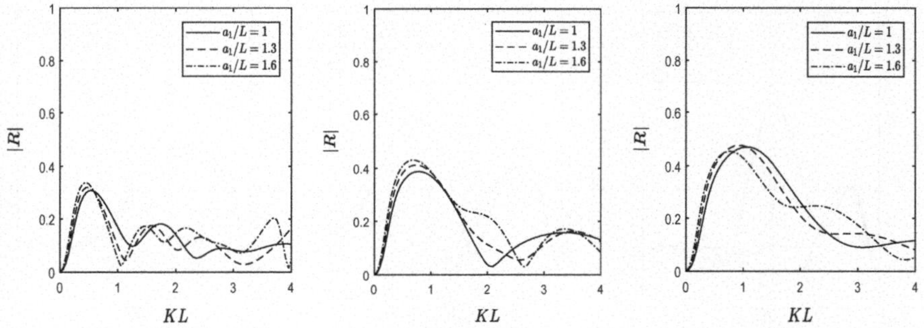

(a) Case 1 (b) Case 2 (c) Case 3

Fig. 15 $|R|$ versus dimensionless wavenumber KL for different values of a_1/L with $d_1/L = 0.3, d_2/L = 0.1, G_1 = 1.5, G_2 = 1, a_2/L = 1$

(see Fig. 14a–c) up to a certain range of wavenumber, beyond which almost all graphs of $|T|$ start showing oscillatory nature.

The variations of $|R|$ and $|T|$ with respect to the dimensionless wavenumber KL for different values of the lower plate length are presented in Fig. 15a–c and Fig. 16a–c respectively. The calculating conditions are $d_1/L = 0.3, d_2/L = 0.1, G_1 = 1.5, G_2 = 1, a_2/L = 1$. From Figs. 15a–c and 16a–c, it can be realized that the effects of increasing lower plate length on $|R|$ and $|T|$ are almost similar to that of increasing upper plate length. Though the effect of lower plate length is comparatively less than that of the upper plate length.

The variations of the reflection coefficient for three different permeability of the lower plate for all three orientations of the dual porous plates are depicted in Fig. 17a–c with fixed parametric values d_1/L=0.3, d_2/L=0.1, $G_2 = 1, a_1/L = 1, a_2/L = 1$. It is evident from these figures that with increasing permeability of the lower plate, the reflection coefficient decreases. This decreasing nature is not uniform in Case 1 and Case 2, whereas in Case 3, the nature of the graphs are regular and the differences

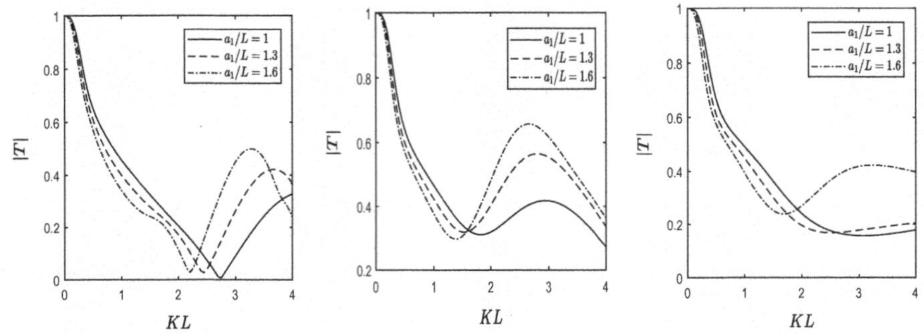

(a) Case 1 (b) Case 2 (c) Case 3

Fig. 16 $|T|$ versus dimensionless wavenumber KL for different values of a_1/L with $d_1/L = 0.3, d_2/L = 0.1, G_1 = 2, G_2 = 1, a_2/L = 1$

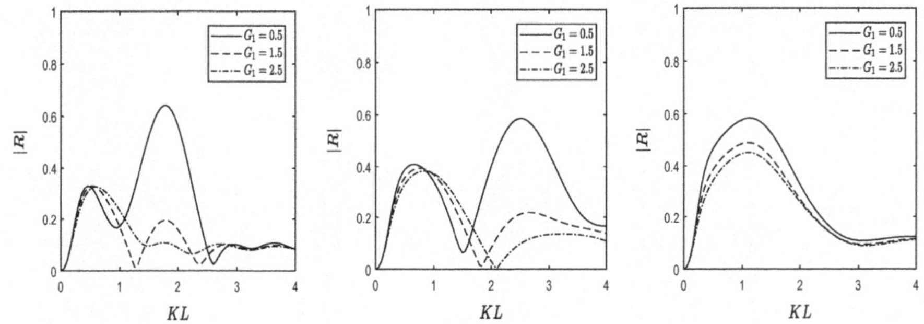

(a) Case 1 (b) Case 2 (c) Case 3

Fig. 17 $|R|$ versus dimensionless wavenumber KL for different values of G_1 with $d_2/L = 0.1, d_1/L = 0.3, G_2 = 1, a_1/L = 1, a_2/L = 1$

in $|R|$ values can be noticed clearly within a particular range ($0.5 < KL < 2.5$) of dimensionless wavenumber. But, from Fig. 17a–c, it can be concluded that the effect of the lower plate's permeability in reflecting incident wave is more prominent in the first two cases (Case 1 and Case 2).

In Fig. 18a–c, we demonstrate the effects of permeability of the upper plate on the absolute values of the reflection coefficient for each case. Here other parametric values are $d_1/L = 0.3, d_2/L = 0.1, G_1 = 0.5, a_1/L = 1, a_2/L = 1$. Figure 18a–c show that the system of dual horizontal porous plates reflect more amount of incoming waves with increasing permeability of the upper plate. Though this effect of permeability of the upper plate on the reflection coefficient is not uniform for all cases.

Comparisons on the values of hydrodynamic forces between dual horizontal porous plates of different orientations and single horizontal porous plate are given in Fig. 19a–c. In each of these figure, the solid line and the dashed line denote the values of hydrodynamic force acting on the upper and lower plates respectively for

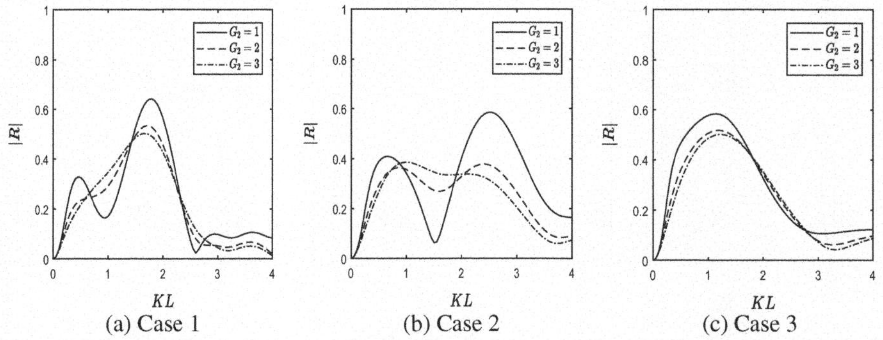

Fig. 18 $|R|$ versus dimensionless wavenumber KL for different values of G_2 with $d_1/L = 0.3, d_2/L = 0.1, G_1 = 0.5, a_1/L = 1, a_2/L = 1$

the dual porous plate system with $d_1/L = 0.3, d_2/L = 0.1, G_1 = 1 + 0.5i, G_2 = 1, a_1/L = 1, a_2/L = 1$, whereas the dash-dot line denotes the values of hydrodynamic force acting on the upper plate with $d_1/L = 0.3, d_2/L = 0.1, G_1 = 1 + 0.5i, G_2 = 1, a_1/L = 0.001, a_2/L = 1$ (i.e., there is no lower plate) and the dot-dot line denotes the values of hydrodynamic force acting on the lower plate with $d_1/L = 0.3, d_2/L = 0.1, G_1 = 1 + 0.5i, G_2 = 1, a_1/L = 1, a_2/L = 0.001$ (i.e., there is no upper plate). Similar to the case of finite depth water, here also we notice that for Case 1 and Case 2, the hydrodynamic forces exerted on the upper plate in presence of lower plate and in absence of it are nearly equal. On the other hand, for Case 3, the upper plate alone (in absence of the lower plate) experiences significantly more wave force than that when it lies above the lower plate. Also, it can be seen from Fig. 19a–c that almost in all three cases within a moderate range of dimensionless wavenumber ($KL < 2.5$), the addition of the upper plate causes reduction in wave force on the lower plate. Comparison of Fig. 19a–c reveals that the hydrodynamic force on the lower plate in a dual porous plate system is the least for Case 1 i.e. when the plates are arranged in a step-like manner.

In Fig. 20, the energy loss coefficient is plotted against the dimensionless wavenumber KL for three different cases of dual plates' orientations. Fixed parametric values are $d_1/L = 0.3, d_2/L = 0.1, G_1 = 1 + 0.5i, G_2 = 1, a_1/L = 1, a_2/L = 1$. Figure 20 reveals that in a lower frequency region ($KL \leq 0.5$), the amount of energy loss by the porous plates for each case are almost equal. In Case 3, as the dimensionless wavenumber KL increases, the energy loss coefficient monotonically increases and becomes almost stationary after a certain value of KL ($KL > 2.5$). On the other hand, in Case 1 and Case 2, an oscillating nature in the graphs of the energy loss coefficient can be observed.

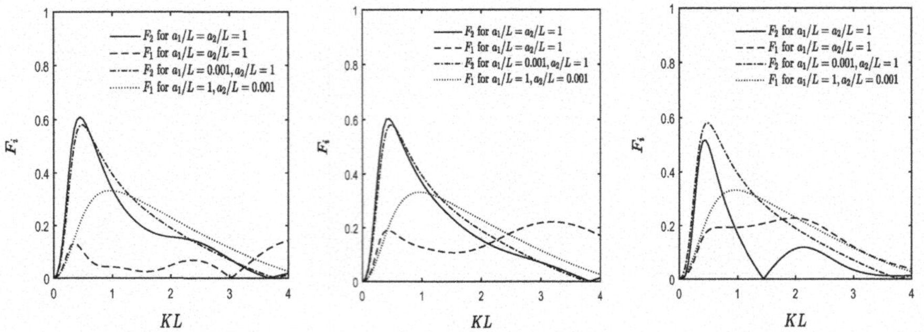

(a) Case 1 (b) Case 2 (c) Case 3

Fig. 19 Comparison of hydrodynamic forces acting on single and dual plate system in deep water for $d_1/L = 0.3$, $d_2/L = 0.1$, $G_1 = 1 + 0.5\text{i}$, $G_2 = 1$

Fig. 20 Energy loss coefficient J versus dimensionless wave-number KL in infinite depth water for $G_1 = 1 + 0.5\text{i}$, $G_2 = 1$, $d_1/L = 0.3$, $d_2/L = 0.1$, $a_1/L = 1$, $a_2/L = 1$

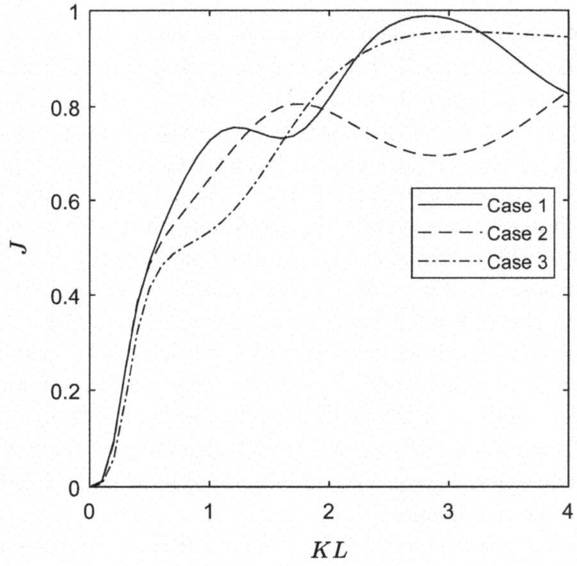

6 Conclusion

On the basis of the two-dimensional linear water wave theory, the problem of water wave scattering by dual horizontal porous plates submerged at different depths have been investigated. We have studied the hydrodynamic performances of a dual porous plates' system for the three different cases viz. **Case 1**. when the plates are kept in a step-like manner, **Case 2**. when there is a partial overlap between the plates and **Case 3**. when the plates are (almost) fully overlapped. Using the boundary condition for the fluid across the plates and with the appropriate use of Green's integral theorem in the water domain, we reduce the corresponding governing boundary value problem into two coupled hypersingular integral equations involving the unknown potential difference functions across the plates. The coupled integral equations are solved numerically by reducing them to a system of linear equations. These solutions are used to compute the numerical estimates for the reflection coefficient, transmission coefficient, energy loss coefficient and hydrodynamic forces exerted on the plates. The convergence tests, verification of the energy identity relations for the porous plates in finite as well as in infinite depth water and the comparison of present results with the published results for few cases have been presented which give several checks on the correctness of our analytical method and numerical results. The effects of the various parameters for each of the case are depicted through figures. The major conclusions drawn from the numerical results are:

- If the gap length between the plates is reduced, we get lower reflection when the pair of plates are placed in deep water. This is also true for finite depth water when the plates are either kept in steps or are partially overlapped. When they are completely overlapped in the water of uniform finite depth, reduction in gap length results in higher reflection.
- At a lower frequency level, more incoming waves are transmitted if the lengths of the plates are decreased. For higher frequencies, fluctuations in the values of transmission coefficient are noticed.
- With the increase in porosity, the reflection coefficient decreases for all the orientations of the plates and for deep as well as for finite depth water. However, when the plates are fully overlapped, the variation in the values of reflection coefficient is not much significant. But, if the porosity of the lower plate in the system of two partially overlapped plates is varied, a substantial lower reflection is observed. This is also true for step like plates placed in deep water.
- For all arrangements of the plates, it is observed that when the lower plate lies beneath the upper plate, it experiences less hydrodynamic force compared to the force that acts on it in absence of the upper plate. On the contrary, the absence/ presence of the lower plate does not much affect the hydrodynamic force acting on the upper plate for Case 1 and Case 2. However, when the upper plate lies completely above the lower plate (Case 3), the normal force on it is quite less.

Above observations might be helpful to the marine engineers in understanding the efficiency of submerged horizontal dual porous plates in scattering or dissipating incoming wave energy. These may be useful in giving some ideas about how to arrange a set of two porous plates in order to yield optimum reflection, transmission and to minimise hydrodynamic force. The present work also provides simple analytical and numerical methods by virtue of which the efficiency of a model comprising of two or more plates (porous or rigid or a combination of them) of arbitrary geometry and orientation can be analysed.

References

1. Mehlum, E., Stamnes, J.: On the focusing of ocean swells and its significance in power production. Cent. Inst. for Indust. Res., Blindern, Oslo, SI Rep. **77**, 01–38 (1978)
2. Newman, J.N.: Amplification of waves by submerged plates. In: 30th International Workshop on Water Waves and Floating Bodies (IWWWFB). Bristol, UK, pp. 153–156 (2015)
3. Heins, A.E.: Water waves over a channel of finite depth with a submerged plane barrier. Can. J. Math. **2**, 210–222 (1950)
4. Burke, J.E.: Scattering of surface waves on an infinitely deep fluid. J. Math. Phys. **5**, 805–819 (1964)
5. Patarapanich, M.: Maximum and zero reflection from submerged plate. J. Waterw. Port Coast. Ocean Eng. **110**, 171–181 (1984)
6. McIver, M.: Diffraction of water waves by a moored, horizontal, flat plate. J. Eng. Math. **19**, 297–319 (1985)
7. Patarapanich, M., Cheong, H.-F.: Reflection and transmission characteristics of regular and random waves from a submerged horizontal plate. Coast. Eng. **13**, 161–182 (1989)
8. Neelamani, S., Reddy, M.S.: Wave transmission and reflection characteristics of a rigid surface and submerged horizontal plate. Ocean Eng. **19**, 327–341 (1992)
9. Porter, R.: Linearised water wave problems involving submerged horizontal plates. Appl. Ocean Res. **50**, 91–109 (2015)
10. Yu, X., Chwang, A.T.: Water waves above submerged porous plate. J. Eng. Mech. **120**, 1270–1282 (1994)
11. Chwang, A.T., Wu, J.: Wave scattering by submerged porous disk. J. Eng. Mech. **120**, 2575–2587 (1994)
12. Cho, I.H., Kim, M.H.: Interactions of horizontal porous flexible membrane with waves. J. Waterw., Port, Coast., Ocean Eng. **126**, 245–253 (2000)
13. Liu, Y., Li, Y.C.: An alternative analytical solution for water-wave motion over a submerged horizontal porous plate. J. Eng. Math. **69**, 385–400 (2011)
14. Cho, I.H., Kim, M.H.: Transmission of oblique incident waves by a submerged horizontal porous plate. Ocean Eng. **61**, 56–65 (2013)
15. Liu, Y., Li, H.J., Li, Y.C.: A new analytical solution for wave scattering by a submerged horizontal porous plate with finite thickness. Ocean Eng. **42**, 83–92 (2012)
16. Wu, J., Wan, Z., Fang, Y.: Wave reflection by a vertical wall with a horizontal submerged porous plate. Ocean Eng. **25**, 767–779 (1998)
17. Yip, T.L., Chwang, A.T.: Perforated wall breakwater with internal horizontal plate. J. Eng. Mech. **126**, 533–538 (2000)
18. Neelamani, S., Gayathri, T.: Wave interaction with twin plate wave barrier. Ocean Eng. **33**, 495–516 (2006)
19. Liu, Y., Li, Y.C., Teng, B.: Wave interaction with a perforated wall breakwater with a submerged horizontal porous plate. Ocean Eng. **34**, 2364–2373 (2007)
20. Liu, Y., Li, Y.C., Teng, B., Dong, S.: Wave motion over a submerged breakwater with an upper horizontal porous plate and a lower horizontal solid plate. Ocean Eng. **35**, 1588–1596 (2008)
21. Cho, I.H., Koh, H.J., Kim, J.R., Kim, M.H.: Wave scattering by dual submerged horizontal porous plates. Ocean Eng. **73**, 149–158 (2013)
22. Liu, Y., Li, H.J.: Wave scattering by dual submerged horizontal porous plates: further results. Ocean Eng. **81**, 158–163 (2014)
23. Sasmal, A., De, S.: Oblique water wave diffraction by two vertical porous barriers with non-identical submerged gaps. Meccanica **54**, 1525–1544 (2019)
24. Sasmal, A., Paul, S., De, S.: Effect of porosity on oblique wave diffraction by two unequal vertical porous barriers. J. Mar. Sci. Appl. **18**, 417–432 (2019)
25. Chakrabarti, A., Martha, S.C.: Methods of solution of singular integral equations. Math. Sci. **6**, 1–29 (2012)
26. Abdulkawi, M.: Solution of Cauchy type singular integral equations of the first kind by using differential transform method. Appl. Math. Model. **39**, 2107–2118 (2015)

27. Banerjea, S., Chakraborty, R., Samanta, A.: Boundary element approach of solving Fredholm and Volterra integral equations. Int. J. Math. Model. Numer. Optim. **9**, 1–11 (2019)
28. Maurya, R.K., Devi, V., Srivastava, N., Singh, V.K.: An efficient and stable Lagrangian matrix approach to Abel integral and integro-differential equations. Appl. Math. Comput. **374**, 125005 (2020)
29. Singh, H., Baleanu, D., Srivastava, H., Dutta, H., Jha, N.K.: Solution of multi-dimensional Fredholm equations using Legendre scaling functions. Appl. Numer. Math. **150**, 313–324 (2020)
30. Ursell, F.: The effect of a fixed vertical barrier on surface waves in deep water. Math. Proc. Cambridge Philos. Soc. **43**, 1–12 (1947)
31. Morris, C.A.N.: A variational approach to an unsymmetric water-wave scattering problem. J. Eng. Math. **9**, 291–300 (1975)
32. Gayen, R., Gupta, S., Chakrabarti, A.: Approximate solution of the problem of scattering of surface water waves by a partially immersed rigid plane vertical barrier. Appl. Math. Lett. **58**, 19–25 (2016)
33. Chakrabarti, A.: On the solution of the problem of scattering of surface–water waves by the edge of an ice cover. Proc. R. Soc. London. Ser. A: Math., Phys, Eng. Sci. **456**, 1087–1099 (2000a)
34. Chakrabarti, A.: On the solution of the problem of scattering of surface water waves by a sharp discontinuity in the surface boundary conditions. ANZIAM J. **42**, 277–286 (2000b)
35. Gayen, R., Mondal, A.: A hypersingular integral equation approach to the porous plate problem. Appl. Ocean Res. **46**, 70–78 (2014)
36. Gayen, R., Mandal, B.N., Chakrabarti, A.: Water wave diffraction by a surface strip. J. Fluid Mech. **571**, 419–438 (2007)
37. Gayen, R., Mandal, B.N.: Scattering of surface water waves by a floating elastic plate in two dimensions. SIAM J. Appl. Math. **69**, 1520–1541 (2009)
38. De, S., Mandal, B.N., Chakrabarti, A.: Use of Abel integral equations in water wave scattering by two surface-piercing barriers. Wave Motion **47**, 279–288 (2010)
39. Gayen, R., Gupta, S.: Scattering of surface waves by a pair of asymmetric thin elliptic arc shaped plates with variable permeability. Eur. J. Mech.-B/Fluids **80**, 122–132 (2020)
40. Islam, N., Gayen, R.: Water wave scattering and energy dissipation by interface-piercing porous plates. J. Mar. Sci. Technol. (2020). https://doi.org/10.1007/s00773-020-00725-6
41. Parsons, N.F., Martin, P.A.: Scattering of water waves by submerged plates using hypersingular integral equations. Appl. Ocean Res. **14**, 313–321 (1992)
42. McIver, P.: Diffraction of water waves by a segmented permeable breakwater. J. Waterw., Port, Coast., Ocean Eng. **131**, 69–76 (2005)
43. Mondal, D., Banerjea, S.: Scattering of water waves by an inclined porous plate submerged in ocean with ice cover. Q. J. Mech. Appl. Mech. **69**, 195–213 (2016)
44. Kundu, S., Gayen, R., Datta, R.: Scattering of water waves by an inclined elastic plate in deep water. Ocean Eng. **167**, 221–228 (2018)
45. Paul, S., De, S.: Effects of vertical porous barrier on progressive waves in a two layered fluid. Ocean Eng. **156**, 153–166 (2018)
46. Yu, X.: Diffraction of water waves by porous breakwaters. J. Waterw., Port, Coast., Ocean Eng. **121**, 275–282 (1995)
47. Golberg, M.A.: The convergence of several algorithms for solving integral equations with finite part integrals. J. Integr. Equ. **5**, 329–340 (1983)
48. Golberg, M.A.: The convergence of several algorithms for solving integral equations with finite part integrals II. J. Integr. Equ. **9**, 267–275 (1985)
49. Mandal, B.N., Chakrabarti, A.: Water Wave Scattering by Barriers. WIT Press (2000)

The manufacturer's authorised representative in the EU is Springer Nature Customer Service Centre GmbH, Europaplatz 3, 69115 Heidelberg, Germany. If you have any concerns regarding our products, please contact ProductSafety@springernature.com

Printed and bound by CPI Group (UK) Ltd, Croydon, CR0 4YY
30/09/2026
02225699-0005